AF389344

Zika Virus and Host Interactions

Zika Virus and Host Interactions

Editors

Tom Hobman
Anil Kumar
Daniel Limonta

MDPI • Basel • Beijing • Wuhan • Barcelona • Belgrade • Manchester • Tokyo • Cluj • Tianjin

Editors

Tom Hobman
Department of Cell Biology,
University of Alberta
Canada

Anil Kumar
Department of Biochemistry,
Microbiology and Immunology,
University of Saskatchewan
Canada

Daniel Limonta
Department of Cell Biology,
University of Alberta
Canada

Editorial Office
MDPI
St. Alban-Anlage 66
4052 Basel, Switzerland

This is a reprint of articles from the Special Issue published online in the open access journal *Cells* (ISSN 2073-4409) (available at: https://www.mdpi.com/journal/cells/special_issues/zika_virus_host).

For citation purposes, cite each article independently as indicated on the article page online and as indicated below:

LastName, A.A.; LastName, B.B.; LastName, C.C. Article Title. *Journal Name* **Year**, *Volume Number*, Page Range.

ISBN 978-3-03943-949-2 (Hbk)
ISBN 978-3-03943-950-8 (PDF)

Cover image courtesy of Jonathan Turpin, Etienne Frumence, Philippe Desprès, Wildriss Viranaicken, and Pascale Krejbich-Trotot.

Contents

About the Editors

Tom Hobman is a professor in the department of Cell Biology at the University of Alberta in Edmonton Canada. His laboratory has studied cellular aspects of RNA virus host interactions for over 25 years. Recent pathogens of interest include emerging flaviviruses, HIV-1, alphaviruses and SARS-CoV-2.

Anil Kumar is an Assistant Professor at the Department of Biochemistry, Microbiology and Immunology at the University of Saskatchewan in Canada. He obtained his doctoral degree from the University of Heidelberg, Germany, training with Prof. Ralf Bartenschlager and completed his post-doctoral training in the lab of Prof. Tom Hobman at the University of Alberta, Canada. His research interests focus on the host–pathogen interactions of arboviruses; especially flaviviruses and alphaviruses. He is the author of several studies on the interactions of the host innate immune system with arboviruses.

Daniel Limonta is a Medical Microbiologist (MD., PhD.) with a strong background in virology. He completed a Cell and Molecular Biology Master's degree at the Oswaldo Cruz Institute in Brazil, and a Medical Microbiology Residency and Virology PhD Programs at the Pedro Kouri Institute of Tropical Medicine in Havana, Cuba. His main topic of research in Brazil and Cuba was the Dengue virus pathogenesis. He was awarded a Postdoctoral Fellowship in Canada to work at the Li Ka Shing Institute of Virology at the University of Alberta. His postdoctoral research has been focused on host and pandemic viruses' interactions, including the Zika virus and more recently SARS-CoV-2.

Editorial

Zika Virus and Host Interactions: From the Bench to the Bedside and Beyond

Daniel Limonta [1,2,*] **and Tom C. Hobman** [1,2,3,4]

[1] Department of Cell Biology, University of Alberta, Edmonton, AB T6G 2H7, Canada; tom.hobman@ualberta.ca
[2] Li Ka Shing Institute of Virology, University of Alberta, Edmonton, AB T6G 2E1, Canada
[3] Department of Medical Microbiology & Immunology, University of Alberta, Edmonton, AB T6G 2E1, Canada
[4] Women & Children's Health Research Institute, University of Alberta, Edmonton, AB T6G 1C9, Canada
* Correspondence: dlimonta@ualberta.ca; Tel.: +1-780-492-6595

Received: 6 November 2020; Accepted: 10 November 2020; Published: 12 November 2020

Keywords: zika; host; interferon; cell death; placenta; NS5; peroxisome; mosquito; guinea pig; tight junction

Before the emergence of SARS-CoV-2 (severe acute respiratory syndrome coronavirus 2), the causative agent of the current COVID-19 (coronavirus disease 2019) pandemic [1], another RNA virus, Zika virus (ZIKV) belonging to the flavivirus family, re-emerged and was declared a Public Health Emergency of International Concern by the WHO in February 2016 because of clusters of microcephaly cases and other neurological disorders in Brazil [2]. Until that time, ZIKV was known as a cause of a self-limited febrile disease similar to dengue fever and other arboviral febrile diseases that occur in tropical and subtropical regions [3]. The scientific community was surprised to find that a mosquito-borne flavivirus could also be transmitted to fetuses during pregnancy and to sexual partners. So far, no vaccines or antiviral drugs are licensed for ZIKV and the virus has spread and become endemic to many Latin-American countries. Furthermore, because of increased international travel and trade, this virus has also spread to some developed nations [4].

In this Special Issue in *Cells*, Zika virus and host interactions, thirteen articles that address diverse aspects of ZIKV–host interactions from basic science to clinical research are presented (Figure 1). Arboviral diseases are infectious diseases in which a third key element beside the pathogen and susceptible individuals is involved—arthropods. In this Special Issue, Martinez-Rojas et al. [5] showed how extracellular vesicles (EVs) released from ZIKV-infected mosquito cells (C6/36) may affect not only bystander mosquito cells, but also human cell types, that are important ZIKV targets. Isolated EVs can infect and promote a pro-inflammatory state in human monocytes and microvascular endothelial cells. These modifications of cellular behavior may contribute to ZIKV transmission and pathogenesis in vector and human host cells.

Animals models are critical for understanding viral pathogenesis and testing therapeutic interventions. To this end, Saver et al. [6] inoculated guinea pigs by subcutaneous and vaginal routes to mimic the infected mosquito bite and sexual transmission, respectively. By studying multiple tissues with respect to tropism and persistence, they identified sensory and autonomic ganglia in the nervous system as a potential novel niche of ZIKV persistence.

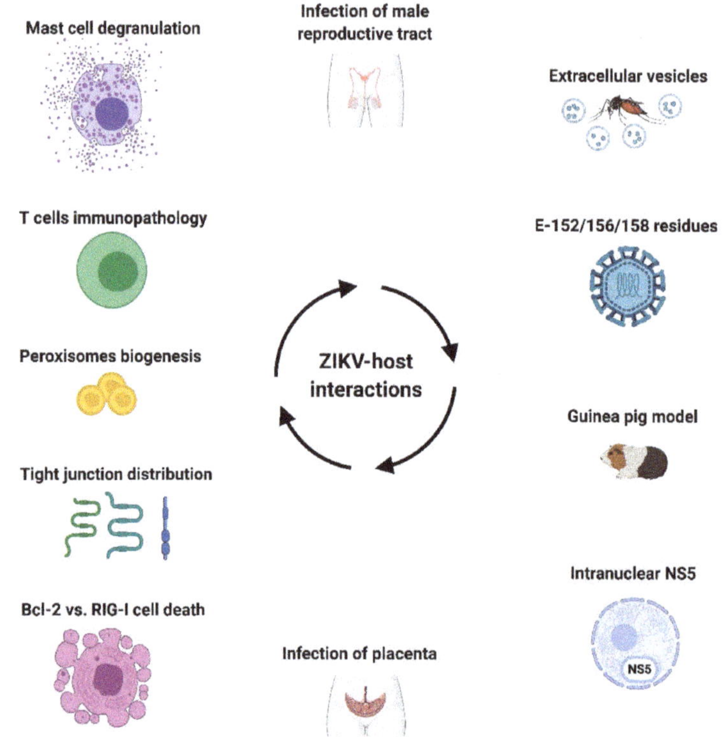

Figure 1. Zika virus (ZIKV)–host interactions. ZIKV and host interactions were addressed in this Special Issue by using in vitro, ex vivo, and animal models together with clinical samples. The complex interplay between host factors and viral components in different cell types and tissues dictates the outcomes of infection: aborted or productive infection, clinical presentation, and disease severity. Created with BioRender.com.

The unusual pathogenic link of ZIKV with the male reproductive system was reviewed by Almeida et al. [7]. The infection and persistence of ZIKV in the testes leading to sexual transmission may impact reproduction by a number of physiological and immunological mechanisms. During pregnancy, infection of the placenta after sex with the infected partner can lead to fetal infection and the development of congenital Zika syndrome [4].

Also included in this Special Issue are two reports focused on interactions of ZIKV with placental cells. Miranda et al. [8] showed that syncytiotrophoblast cells of placentae from ZIKV-infected women display increased paracellular permeability in comparison with placentae from non-infected women, a phenomenon that may be linked to distribution of tight junction proteins. Results from transepithelial electrical resistance assays using a human trophoblast cell line supported the in vivo findings. Rabelo et al. [9] analyzed another cell type from the placenta, the mast cells. These authors detected ZIKV NS1-positive mast cells in the placenta of infected women and then using a human mast cell line permissive for ZIKV infection, looked at virus-induced ultrastructural changes. Furthermore, degranulation kinetics and production of pro-inflammatory cytokines and vascular endothelial growth factor were triggered shortly after ZIKV infection of these cells. The authors suggest that mast cells may contribute to associated inflammation and viral spread in placenta.

A productive infection of host cells by ZIKV would not be possible without effective evasion of the innate immune response. Several articles in this Special Issue addressed different mechanisms used by ZIKV to overcome the antiviral response of the host. Apoptosis is effective in clearing viruses from tissues and any delay of this process may facilitate viral production and spread. Turpin et al. [10] demonstrated that ZIKV replication inhibits intrinsically and extrinsically induced apoptosis in A549 cells. They also showed that control of apoptosis during viral infection may be mediated by the anti-apoptotic bcl-2 family protein. Schilling et al. [11] used CRISPR/Cas9-mediated knockouts and transcriptomic analysis in A549 cells to show that RIG-I was the main sensor required to induce a protective type I IFN response during ZIKV infection. Surprisingly, RIG-I-mediated signaling also prevented A549 cells from viral-induced apoptosis. Specifically, loss of RIG-I led to apoptotic activation 4–6 days after infection, which may be related to reduced type I IFN signaling and increased virus multiplication in RIG-I-deficient cells. Furthermore, these authors showed that ZIKV non-structural protein NS5 not only suppresses IFNAR signaling, but also inhibits transcription of type I interferon genes.

The critical role of the flaviviral NS5 in evading the antiviral response against ZIKV was further explored by two other groups. Roby et al. [12] showed that NS5 of ZIKV as well as those from other flaviviruses, such as West Nile virus and Japanese encephalitis virus, antagonize host cell Janus kinase/signal transducer and activator of transcription (JAK/STAT) signaling downstream of interferons α/β. This is caused in part by NS5 binding to cellular heat shock protein 90 (HSP90) in replication complexes, resulting in suppression of JAK/STAT-dependent cytokine signaling. NS5 binding not only destabilized HSP90, but also disrupted the interaction of this chaperone with other client kinases. In a review article, Tan et al. [13] discussed how nuclear localization of ZIKV NS5 in placental and neural cells antagonizes type I interferon signaling and contributes to the transcriptional activation of pro-inflammatory genes. These findings support the possible role of ZIKV NS5 nuclear spherical-shell structures in the modulation of the host immune response.

The six other non-structural proteins encoded by ZIKV also interact with a number of human host proteins. Golubeva et al. [14] generated an interaction network built on previous interaction networks of ZIKV and a combination of tandem-affinity purification coupled to mass spectrometry with yeast two-hybrid screens to identify 150 human proteins interacting with ZIKV non-structural proteins. In validation experiments, they showed that the protein inhibitor of activated STAT1 (PIAS1) binds to NS5 and modulates its stability. Furthermore, this interactome study validated proteins known to be associated with microcephaly disorders.

Wong et al. [15] addressed how peroxisomes, which are metabolic organelles with key roles in antiviral defense, are affected by ZIKV infection. Peroxisomes function as signaling platforms for the interferon response, and dysregulation of peroxisome biogenesis can alter antiviral signaling. The authors demonstrated that ZIKV infection depletes peroxisomes in human fetal brain cells, while the peroxisome biogenesis factor PEX11B restricts ZIKV replication likely by increasing peroxisome numbers and enhancing downstream antiviral interferon signaling. ZIKV capsid protein was shown to interact with the peroxisome biogenesis factor PEX19, resulting in its degradation and causing loss of peroxisomes. As peroxisome function and numbers can be pharmacologically modulated, this study suggests that further investigation of peroxisome-based antiviral strategies for ZIKV and other viral infections is warranted.

Bos et al. [16] showed the importance of amino acid residues isoleucine 152 (Ile152), threonine 156 (Thr156), and histidine 158 (His158) (E-152/156/158) in the ZIKV envelope (E) protein for infection of host cells. Of note, these three amino acid residues surround an important N-glycosylation site (E-154) in E protein that is associated with ZIKV infectivity. To illustrate the importance of these E protein residues in ZIKV infection, the authors swapped them between the Brazilian epidemic strain BR15 and placed them with the corresponding amino acids from the pre-epidemic African strain MR766. The BR15 virus containing the E-152/156/158 residues from MR766 was less infectious in vitro than the parental virus. In contrast, the MR766 virus containing amino acid residues 152/156/158 from the Brazilian strain was more infectious. As changes to this region of E protein did not affect virus binding to host cells, virus-induced cell death, or the interferon response, the authors speculated that they may be important for infectivity by promoting viral membrane fusion inside infected cells. These findings may contribute to the design of innovative strategies to control ZIKV infection at early stages.

Finally, regarding T-cell immunity to ZIKV, the pre-existing immunity to the related flavivirus, dengue virus (DENV), has been discussed as a risk factor for immunopathological responses [4]. Pardy and Richer [17] reviewed current evidence from epidemiological studies and different mouse models about CD4 and CD8 T-cell responses to ZIKV. They concluded that the issue of whether prior DENV immunity provides cross-protection to ZIKV remains an open question that should be further investigated. Furthermore, a better understanding of T-cell responses to ZIKV and the mechanism by which this virus evades cell-mediated immunity will be important for designing effective anti-ZIKV vaccine candidates.

We hope that this Special Issue will be of wide interest, particularly to those researchers focused on different aspects of ZIKV–host interactions. It is expected that findings from these studies will contribute to a better understanding of the vector and human host cells interacting with ZIKV, and will serve as the basis for novel diagnostics, antiviral therapeutics, and vaccines.

Funding: This research received no external funding.

Conflicts of Interest: The authors declare no conflict of interest.

References

1. Hu, B.; Guo, H.; Zhou, P.; Shi, Z.L. Characteristics of SARS-CoV-2 and COVID-19. *Nat. Rev. Microbiol.* **2020**. [CrossRef] [PubMed]
2. WHO Director-General Summarizes the Outcome of the Emergency Committee Regarding Clusters of Microcephaly and Guillain-Barré Syndrome. 2016. Available online: https://www.who.int/news/item/01-02-2016-who-director-general-summarizes-the-outcome-of-the-emergency-committee-regarding-clusters-of-microcephaly-and-guillain-barr%C3%A9-syndrome (accessed on 4 November 2020).
3. Musso, D.; Gubler, D.J. Zika Virus. *Clin. Microbiol. Rev.* **2016**, *29*, 487–524. [CrossRef] [PubMed]
4. Pierson, T.C.; Diamond, M.S. The continued threat of emerging flaviviruses. *Nat. Microbiol.* **2020**, *5*, 796–812. [CrossRef] [PubMed]
5. Martínez-Rojas, P.P.; Quiroz-García, E.; Monroy-Martínez, V.; Agredano-Moreno, L.T.; Jiménez-García, L.F.; Ruiz-Ordaz, B.H. Participation of Extracellular Vesicles from Zika-Virus-Infected Mosquito Cells in the Modification of Naïve Cells' Behavior by Mediating Cell-to-Cell Transmission of Viral Elements. *Cells* **2020**, *9*, 123. [CrossRef]
6. Saver, A.E.; Crawford, S.A.; Joyce, J.D.; Bertke, A.S. Route of Infection Influences Zika Virus Shedding in a Guinea Pig Model. *Cells* **2019**, *8*, 1437. [CrossRef] [PubMed]
7. Almeida, R.D.N.; Braz-de-Melo, H.A.; Santos, I.O.; Corrêa, R.; Kobinger, G.P.; Magalhaes, K.G. The Cellular Impact of the ZIKA Virus on Male Reproductive Tract Immunology and Physiology. *Cells* **2020**, *9*, 1006. [CrossRef] [PubMed]
8. Miranda, J.; Martín-Tapia, D.; Valdespino-Vázquez, Y.; Alarcón, L.; Espejel-Nuñez, A.; Guzmán-Huerta, M.; Muñoz-Medina, J.E.; Shibayama, M.; Chávez-Munguía, B.; Estrada-Gutiérrez, G.; et al. Syncytiotrophoblast of Placentae from Women with Zika Virus Infection Has Altered Tight Junction Protein Expression and Increased Paracellular Permeability. *Cells* **2019**, *8*, 1174. [CrossRef] [PubMed]
9. Rabelo, K.; Gonçalves, A.J.D.S.; Souza, L.J.; Sales, A.P.; Lima, S.M.B.; Trindade, G.F.; Ciambarella, B.T.; Amorim Tasmo, N.R.; Diaz, B.L.; Carvalho, J.J.; et al. Zika Virus Infects Human Placental Mast Cells and the HMC-1 Cell Line, and Triggers Degranulation, Cytokine Release and Ultrastructural Changes. *Cells* **2020**, *9*, 975. [CrossRef] [PubMed]
10. Turpin, J.; Frumence, E.; Desprès, P.; Viranaicken, W.; Krejbich-Trotot, P. The ZIKA Virus Delays Cell Death Through the Anti-Apoptotic Bcl-2 Family Proteins. *Cells* **2019**, *8*, 1338. [CrossRef] [PubMed]
11. Schilling, M.; Bridgeman, A.; Gray, N.; Hertzog, J.; Hublitz, P.; Kohl, A.; Rehwinkel, J. RIG-I Plays a Dominant Role in the Induction of Transcriptional Changes in Zika Virus-Infected Cells, which Protect from Virus-Induced Cell Death. *Cells* **2020**, *9*, 1476. [CrossRef] [PubMed]

12. Roby, J.A.; Esser-Nobis, K.; Dewey-Verstelle, E.C.; Fairgrieve, M.R.; Schwerk, J.; Lu, A.Y.; Soveg, F.W.; Hemann, E.A.; Hatfield, L.D.; Keller, B.C.; et al. Flavivirus Nonstructural Protein NS5 Dysregulates HSP90 to Broadly Inhibit JAK/STAT Signaling. *Cells* **2020**, *9*, 899. [CrossRef] [PubMed]

13. Tan, M.J.A.; Chan, K.W.K.; Ng, I.H.W.; Kong, S.Y.Z.; Gwee, C.P.; Watanabe, S.; Vasudevan, S.G. The Potential Role of the ZIKV NS5 Nuclear Spherical-Shell Structures in Cell Type-Specific Host Immune Modulation during ZIKV Infection. *Cells* **2019**, *8*, 1519. [CrossRef] [PubMed]

14. Golubeva, V.A.; Nepomuceno, T.C.; Gregoriis, G.; Mesquita, R.D.; Li, X.; Dash, S.; Garcez, P.P.; Suarez-Kurtz, G.; Izumi, V.; Koomen, J.; et al. Network of Interactions between ZIKA Virus Non-Structural Proteins and Human Host Proteins. *Cells* **2020**, *9*, 153. [CrossRef] [PubMed]

15. Wong, C.P.; Xu, Z.; Hou, S.; Limonta, D.; Kumar, A.; Power, C.; Hobman, T.C. Interplay between Zika Virus and Peroxisomes during Infection. *Cells* **2019**, *8*, 725. [CrossRef] [PubMed]

16. Bos, S.; Viranaicken, W.; Frumence, E.; Li, G.; Desprès, P.; Zhao, R.Y.; Gadea, G. The Envelope Residues E152/156/158 of Zika Virus Influence the Early Stages of Virus Infection in Human Cells. *Cells* **2019**, *8*, 1444. [CrossRef] [PubMed]

17. Pardy, R.D.; Richer, M.J. Protective to a T: The Role of T Cells during Zika Virus Infection. *Cells* **2019**, *8*, 820. [CrossRef] [PubMed]

Publisher's Note: MDPI stays neutral with regard to jurisdictional claims in published maps and institutional affiliations.

Article

RIG-I Plays a Dominant Role in the Induction of Transcriptional Changes in Zika Virus-Infected Cells, which Protect from Virus-Induced Cell Death

Mirjam Schilling [1], Anne Bridgeman [1], Nicki Gray [2], Jonny Hertzog [1], Philip Hublitz [3], Alain Kohl [4] and Jan Rehwinkel [1,*]

[1] Medical Research Council Human Immunology Unit, Medical Research Council Weatherall Institute of Molecular Medicine, Radcliffe Department of Medicine, University of Oxford, Oxford OX3 9DS, UK; mirjam.schilling@ndm.ox.ac.uk (M.S.); anne.bridgeman@imm.ox.ac.uk (A.B.); jonny.hertzog@kellogg.ox.ac.uk (J.H.)

[2] MRC WIMM Centre for Computational Biology, MRC Weatherall Institute of Molecular Medicine, University of Oxford, Oxford OX3 9DS, UK; nicki.gray@imm.ox.ac.uk

[3] Genome Engineering Facility, Medical Research Council Weatherall Institute of Molecular Medicine, Radcliffe Department of Medicine, University of Oxford, John Radcliffe Hospital, Oxford OX3 9DS, UK; philip.hublitz@ndcls.ox.ac.uk

[4] MRC-Centre for Virus Research, University of Glasgow, Glasgow G61 1QH, UK; Alain.Kohl@glasgow.ac.uk

[*] Correspondence: jan.rehwinkel@rdm.ox.ac.uk

Received: 15 May 2020; Accepted: 13 June 2020; Published: 16 June 2020

Abstract: The Zika virus (ZIKV) has received much attention due to an alarming increase in cases of neurological disorders including congenital Zika syndrome associated with infection. To date, there is no effective treatment available. An immediate response by the innate immune system is crucial for effective control of the virus. Using CRISPR/Cas9-mediated knockouts in A549 cells, we investigated the individual contributions of the RIG-I-like receptors MDA5 and RIG-I to ZIKV sensing and control of this virus by using a Brazilian ZIKV strain. We show that RIG-I is the main sensor for ZIKV in A549 cells. Surprisingly, we observed that loss of RIG-I and consecutive type I interferon (IFN) production led to virus-induced apoptosis. ZIKV non-structural protein NS5 was reported to interfere with type I IFN receptor signaling. Additionally, we show that ZIKV NS5 inhibits type I IFN induction. Overall, our study highlights the importance of RIG-I-dependent ZIKV sensing for the prevention of virus-induced cell death and shows that NS5 inhibits the production of type I IFN.

Keywords: Zika virus; IFN; RIG-I; MDA5; apoptosis; NS5; IFNAR1

---

1. Introduction

The recent epidemic caused by Zika virus (ZIKV), a mosquito-borne flavivirus, revealed the potential of the virus to inflict severe harm on infected individuals. Whereas infection is often asymptomatic or causes a self-limiting acute febrile illness in adults, it has been linked to multiple neurodevelopmental defects, including microcephaly, in newborns [1,2]. ZIKV has furthermore been associated with cases of Guillain–Barré syndrome, an autoimmune disease resulting in rapid-onset muscle weakness [3,4].

Experiments in cell culture and mice showed that ZIKV infection is controlled by type I interferons (IFNs), which represent the first line of defense against viral infections [5–8]. Type I IFNs are induced by pattern recognition receptors (PRRs) upon sensing of pathogen associated molecular patterns. For example, viral RNAs are detected by RIG-I-like receptors (RLRs) including retinoic acid-inducible gene I (RIG-I) and melanoma differentiation-associated gene 5 (MDA5). These PRRs then activate mitochondrial antiviral-signaling protein (MAVS). Downstream of MAVS, transcription factors such as interferon regulatory factor 3 (IRF3) induce the expression and secretion of type I IFNs [9]. Binding of the secreted type I IFNs to their receptor in turn induces a signaling cascade involving signal transducer and activator of transcription (STAT) 1 and 2 hetero-dimerization that leads to the expression of interferon stimulated genes (ISGs). Some ISGs encode antiviral proteins that then restrict viral replication. Deficiency of PRRs increases ZIKV replication in human skin fibroblasts and mice lacking the type I IFN receptor (IFNAR), STAT2, MAVS or a combination of IRF transcription factors show higher viral replication and pathology [5,7,8,10–12]. ZIKV induces type I IFNs, and this is mainly dependent on MAVS, suggesting an important role of RIG-I and/or MDA5 in virus detection [7,13–18]. ZIKV antagonizes IFNAR signaling as viral replication is blocked only moderately if type I IFN is applied with or after ZIKV infection [19]. NS5, the RNA-dependent RNA polymerase that replicates the viral genome, is a potent viral antagonist of IFNAR signaling [6,13,20]. NS5 proteins from African and French Polynesian isolates were reported to interact with and target STAT2 for degradation [6,21]. Furthermore, NS5 prevents STAT1 and STAT2 phosphorylation [19,22,23].

Here, we explored the contribution of RIG-I and MDA5 to ZIKV sensing and found that RIG-I was the main sensor required to induce a protective type I IFN response upon virus infection. Loss of RIG-I-mediated type I IFN production in infected A549 cells led to the activation of apoptosis. We furthermore show that ZIKV NS5 not only interferes with IFNAR signaling, but additionally inhibited type I IFN induction upstream of type I IFN gene transcription.

2. Materials and Methods

2.1. Cell Lines

A549 cells (kind gift from G. Kochs, Freiburg), HEK293 and Vero cells (kind gifts from C. Reis e Sousa, London, UK) and A549 BVDV NPro cells (kind gift from R. Randall, St Andrews) were cultured at 37 °C in Dulbecco's modified Eagle's medium (DMEM), supplemented with 10% FCS and 2-mM L-Glutamine.

2.2. Generation of Knock-Out Cells

To stably knock out RIG-I and MDA5, sgRNAs cloned into pX458-Ruby (Addgene 110164, deposited by Dr. Philip Hublitz) and described earlier [13] were used. To stably knock out IFNAR1, a sgRNA targeting *IFNAR1* exon 3 was selected based on the MIT algorithm (crispr.mit.edu) and cloned into pX458 (Addgene 48138, deposited by Dr. Feng Zhang). A549 and HEK293 cells were single-cell FACS sorted according to the co-expressed fluorescent protein (Ruby$^+$ for cells transfected with the sgRNAs targeting RIG-I or MDA5, GFP$^+$ for IFNAR1) 48 h post transfection. After 4 weeks, cells that had grown out to confluency were subjected to cell line characterization. We extracted genomic DNA and analyzed the target locus with a PCR screening protocol using primers up- and downstream of the sgRNA target sites. Primer sequences were: RIG-I (fwd: ttacattgtctcagactaagaggc, rev: gtgaagaatgggcacagtcggcc), MDA5 (fwd: cgtcattgtcaggcacagag, rev: agctctgccactgtttttcc) and IFNAR (fwd: gtgtatgctaaaatgttaatagg, rev: cctttgcgaaatggtgtaaatgag). Full knock-out was verified by submission of sequencing reads to TIDE (https://tide.nki.nl), an algorithm that decomposes sequencing data and allows determination of the spectrum of indels and their respective frequencies. Additionally, whole cell lysates were analyzed by western blot after stimulation with recombinant type I IFN (IFN-A/D, Sigma, 100 U/mL).

2.3. ZIKV

The Brazilian ZIKV isolate ZIKV/*H. sapiens*/Brazil/PE243/2015 was originally described in [24] and was grown on Vero cells. Viral titers were determined by plaque assay on A549 BVDV NPro cells. These cells are optimized for virus growth as they stably express the NPro protein of bovine viral diarrhea virus (BVDV), which induces degradation of IRF3 [25].

2.4. IFNβ ELISA

Human IFNβ concentrations in cell culture supernatants were measured by enzyme-linked immunosorbent assay (ELISA) using LumiKine™ Xpress hIFN-β 2.0 (Invivogen, cat. nb.: luex-hifnbv2) according to the manufacturer's instructions.

2.5. Type I IFN Bioassay

To measure type I IFN bioactivity, HEK293 cells stably expressing the pGF1-ISRE reporter [26]—in which firefly luciferase expression is driven by interferon-stimulated response elements—were incubated with untreated supernatants of infected cells. Relative light units (RLUs) were measured with a luminometer after 24 h using ONE-Glo Luciferase Assay System (Promega) according to the manufacturer's instructions and OptiPlate™ 96-well plates (Perkin Elmer, cat. nb.: 6005299).

2.6. IFNβ Promoter Reporter Assay

HEK293 cells seeded in 96-well plates were transiently transfected with 50 ng of ZIKV-NS3 or -NS5 expression plasmid, 20 ng of a plasmid encoding firefly luciferase (F-Luc) under the control of the *IFNβ* promoter and 5 ng pRL-TK, a plasmid which constitutively expresses renilla luciferase (R-Luc). Twenty-four hours later, cells were transfected with 5 ng IVT–RNA or 50 ng Hela–EMCV–RNA per well [13]. F-Luc activity was determined 24 h after RNA transfection using Dual-Luciferase Reporter Assay System (Promega) and normalized to R-Luc activity.

2.7. Caspase Activity Assay

Caspase 3/7 Glo assay (Promega) was performed according to the manufacturer's instructions.

2.8. qRT-PCR

Cells were lysed and total RNA was extracted using the QIAshredder (Qiagen) and RNeasy Mini Kit (Qiagen) according to the manufacturer's instructions. RNA was reverse transcribed using SuperScript II Reverse Transcriptase (Invitrogen) into cDNA that was then used for qPCR with either TaqMan Universal PCR Master Mix (Applied Biosystems) or SYBR green PCR kit (Life Technologies). C_T values were normalized to GAPDH (ΔC_T). TaqMan primer probes used include GAPDH (Assay ID: Hs02758991 g1), IFIT1 (Assay ID: Hs03027069 s1) and MX1 (Assay ID: Hs00895608 m1). SYBR green primer probes used include GAPDH (fwd: CATGGCCTTCCGTGTTCCTA, rev: CCTGCTTCACCACCTTCTTGA) and ZIKV (fwd: CGAGGAACATCCAGACTC, rev: ATTGGAGATCCTGAAGTTCC).

2.9. 3′ mRNA Sequencing

A549 wt, RIG-I KO (clone B05) and MDA5 KO (clone c27) cells were lysed at 24 h post infection with ZIKV (multiplicity of infection (MOI): 5), and total RNA was extracted using the QIAshredder (Qiagen) and RNeasy Mini Kit (Qiagen) according to manufacturer's instructions. RNA concentration was measured by the Qubit RNA HS (high sensitivity) Assay Kit (Life Technologies) according to the manufacturer's instructions. Quality of the extracted RNA was controlled by the Agilent 2200 TapeStation System (Agilent Technologies).

cDNA was generated from total RNA using first oligo-dT and subsequently random priming. The prepared libraries were QC'ed and multiplexed before sequencing over one lane of the NextSeq flow cell (high output, 75 bp single reads).

Following QC analysis with the fastQC package, reads were aligned using STAR against the human genome assembly (GRCh38 (hg38) UCSC transcripts) [27]. Read counts were visualized using UCSC genome browser [28]. Gene expression levels were quantified as read counts using the featureCounts function from the Subread package with default parameters [29]. The read counts were used for the identification of global differential gene expression between specified populations using the edgeR package [30]. RPKM values were also generated using the edgeR package [30]. Genes were considered differentially expressed between populations if they had an adjusted *p*-value (false discovery rate, FDR) of less than 0.05. The edgeR package was also used to generate heatmaps and plots [30]. The Venn diagram was created using [31].

2.10. FACS

HEK293 cells were trypsinized, resuspended in FACS buffer (PBS, 1% FCS, 2 mM EDTA, 0.02% sodium azide) and fixed immediately by adding an equal volume of pre-warmed (to 37 °C) BD Cellfix (BD Biosciences). Cells were permeabilized by adding chilled BD Phosflow Perm Buffer III (BD Biosciences) drop-by-drop while vortexing. Intracellular staining was performed with anti-p-STAT1 antibody (mouse anti-human p-STAT1, BD clone 4a (RUO)). Cellular debris and doublets were gated out using forward scatter and side scatter channels and 50,000 live single cells were analyzed per sample using an Attune Flow Cytometer (ThermoFisher Scientific).

2.11. Western Blot Analysis

Cells were lysed in YG-lysis buffer (10 mM Trizma, 50 mM NaCl, 30 mM sodium pyrophosphate, 5 mM sodium fluoride, 5 μM $ZnCl_2$, 10% NP40, plus protease inhibitor) and the samples were incubated at 95 °C for 5 min. Protein lysates were separated on 10% SDS-PAGE gels and transferred onto polyvinylidene difluoride (PVDF) membranes. As primary antibodies we used anti-MDA5 (mouse, [13]), anti-Mx (mouse; M143 [32]), anti-PARP (rabbit, cell signaling), anti-RIG-I (mouse, AdipoGen), anti-IRF3 (D6I4C, rabbit, cell signaling), anti-p-IRF3-S396 (4D4G, rabbit, cell signaling), anti-ZIKV NS3 and anti-ZIKV NS5 sera (kind gift from A. Merits, Tartu) and anti-beta-actin-HRP (clone AC-15, Sigma-Aldrich). Detection of the primary antibodies was performed by the use of peroxidase-conjugated secondary antibodies (GE Healthcare).

3. Results

3.1. RIG-I Is the Main Sensor for ZIKV Infection in A549 Cells

Previously, we showed that total RNA extracted from cells infected with ZIKV contains immunostimulatory RNAs that activate the MAVS pathway when transfected into reporter cells [13]. To study the contribution of RIG-I and MDA5 to sensing of live ZIKV in infected cells, we screened a panel of cell lines for type I IFN induction after ZIKV infection. We found that the lung adenocarcinoma cell line A549 produced robust levels of type I IFNs upon infection (data not shown). Furthermore, this cell line was well established for studies of type I IFN responses and was therefore chosen for this work. We generated A549 cells lacking either RIG-I or MDA5 by CRISPR/Cas9-mediated knock-out (KO) and validated these by western blot analysis and sequencing (Figure 1A,B). We obtained one MDA5 KO clone designated c27. MDA5 protein expression was undetectable in these cells (Figure 1A). Sequencing of the region targeted by the sgRNA suggested a +1 insertion on one allele and a −5 deletion on the other, both of which disrupt the reading frame (Figure 1B). Three RIG-I KO clones were generated, and all had no detectable RIG-I protein (Figure 1A). RIG-I KO clone B05 was used in subsequent experiments; sequencing showed a +1 insertion as well as −1 and −2 deletions at the sgRNA target site, which all disrupt the reading frame (Figure 1B). The presence of three different alleles could be explained by the triploidy of much of the A549 genome [33,34].

In order to compare the amounts of type I IFN produced, wild-type (wt) and KO A549 cells were infected with ZIKV using a multiplicity of infection (MOI) of 0.1 or 1. After 24 h, we collected supernatants and measured IFNβ levels by ELISA. These virus doses and the timepoint were chosen to monitor type I IFN responses to incoming virus early after infection. Similar amounts of IFNβ were present in supernatants from wt and MDA5 KO cells (Figure 1C). In contrast, little or no IFNβ was detectable in samples from RIG-I KO cells. Next, we measured bioactive type I IFN levels in supernatants collected from cells infected (MOI 1) for 48 h by using a bioassay: supernatant samples were transferred onto HEK293 cells with a stably integrated pGF1-ISRE reporter [26]. These cells harbor an F-Luc gene under control of interferon-stimulated response elements (ISREs) that were bound and activated by STAT1/2 upon engagement of IFNAR. Cells stimulated with the supernatant of infected wt or MDA5 KO cells induced similar amounts of F-Luc, whereas the supernatant of infected RIG-I KO cells did not lead to significant F-Luc induction (Figure 1D). Furthermore, we tested the activation of IRF3 in infected cells by western blot using an antibody recognizing S396-phosphorylated IRF3 (p-IRF3). This analysis revealed IRF3 phosphorylation upon ZIKV infection in wt and MDA5 KO cells, but not in RIG-I KO cells (Figure 1E). At the selected MOIs and 24-h timepoint analyzed, infection levels were similar in cells of all genotypes as indicated by comparable levels of the viral NS3 protein (Figure 1E). In summary, these data demonstrated that loss of RIG-I abrogated the induction and secretion of type I IFN in A549 cells upon ZIKV infection. To examine the impact of reduced IRF3 activation and type I IFN secretion on ISG induction, A549 cells were infected with ZIKV (MOI 1 or 5) and *IFIT1* and *MX1* mRNA levels were quantified by RT-qPCR. *IFIT1* mRNA was robustly induced in A549 wt and MDA5 KO cells, whereas no induction was detectable in A549 RIG-I KO cells (Figure 1F). Similarly, induction of *MX1* transcripts was not detectable in RIG-I KO cells; however, in contrast to *IFIT1*, *MX1* induction was also reduced in MDA5 KO cells (Figure 1F). This suggested that a subset of ISGs was controlled by both RIG-I and MDA5. To determine the impact of individual RLRs on ZIKV replication in a setting where the infection spreads between cells, we infected cells with a low dose of ZIKV (MOI 0.1) and ZIKV RNA was quantified by RT-qPCR up to 5 days post infection. Virus replication was similar in A549 wt and MDA5 KO cells; however, the virus replicated more potently from day 3 onwards in RIG-I KO cells (Figure 1G). Taken together, these data suggest that RIG-I was the main sensor that detects ZIKV infection in A549 cells leading to the induction of type I IFNs and ISGs. In turn, absence of RIG-I facilitated virus replication.

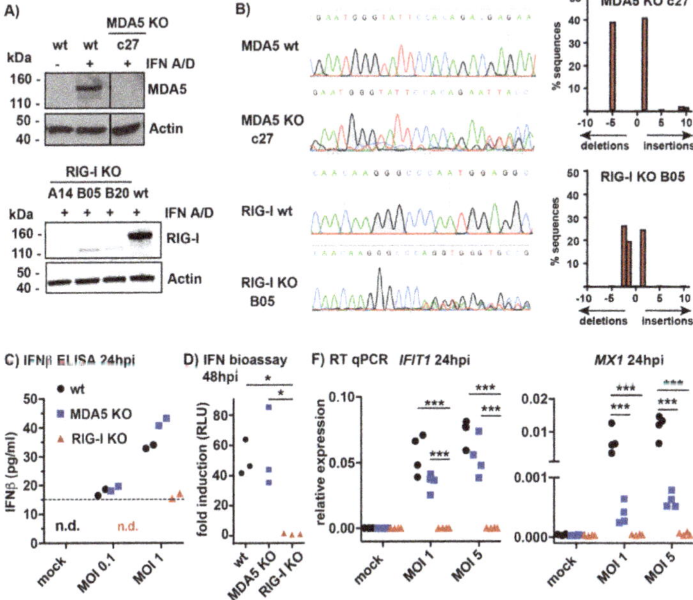

Figure 1. *Cont.*

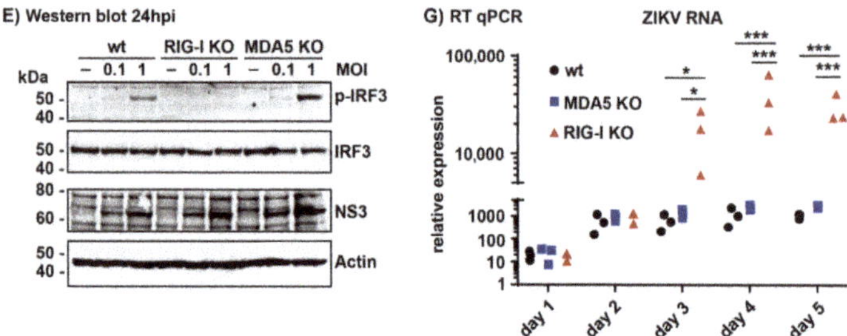

Figure 1. Retinoic acid-inducible gene I (RIG-I) is the main sensor for Zika virus (ZIKV) infection in A549 cells. (**A**) A549 cells were knocked-out for melanoma differentiation-associated gene 5 (MDA5) or RIG-I as described in Materials and Methods. Cells were stimulated with recombinant type I IFN (IFN-A/D) to induce MDA5 and RIG-I to detectable levels and lysates were analyzed by western blot using the indicated antibodies. Actin served as a loading control. The vertical line indicates a cut combining two parts of the same blot. c27, A14, B05 and B20 are individual clones. wt, wild type; (**B**) Genomic DNA was extracted from MDA5 KO clone c27 or RIG-I KO clone B05 cells. A fragment of DNA surrounding the targeted area was amplified by PCR and sequenced (left). Sequences were analyzed using TIDE (right). The number of nucleotides inserted or deleted, and the percentage of sequences affected are shown; (**C**) A549 MDA5 KO or RIG-I KO cells (clone B05) were infected with ZIKV (MOI 0.1 or 1), supernatant was collected 24 h later and IFNβ levels were analyzed by ELISA. The horizontal dashed line indicates the detection limit; n.d., not detectable; (**D**) A549 cells were infected with ZIKV (MOI 1) and supernatant was collected 48 h later. HEK293 cells stably expressing the pGF1-ISRE reporter were incubated with the supernatant and F-Luc activity was measured after 24 h. Shown is the fold induction relative to supernatant from mock infected cells. (**E**) A549 cells were infected with ZIKV (MOI 0.1 or 1), protein samples were collected 24 h later and analyzed by western blot using the indicated antibodies. Actin served as a loading control; (**F**) A549 cells were infected with ZIKV (MOI 1 or 5) and RNA was isolated 24 h later. Levels of *IFIT1* and *MX1* mRNAs were determined with RT-qPCR and C_T values normalized to *GAPDH*; (**G**) A549 cells were infected with ZIKV (MOI 0.1) and RNA was isolated at the indicated time points. RT-qPCR was performed and ZIKV RNA levels are presented relative to *GAPDH*. Data in **A**, **C** and **E** are representative of two independent experiments. Data in **D**, **F** and **G** are pooled from three (**D,G**) and four (**F**) independent experiments. Each data point is the mean value of two technical replicates. Statistical analysis: One-way (**D**) and two-way (**F,G**) ANOVA with Tukey's multiple comparison (* $p < 0.05$, *** $p < 0.001$).

3.2. Transcriptomic Analysis of ZIKV-Infected Cells Indicates that RIG-I Plays a Dominant Role in ISG Induction

In light of our observation that the induction of *MX1* transcripts after ZIKV infection was not only RIG-I, but also partially MDA5-dependent, we wanted to further investigate how the individual receptors influence transcriptomic changes after virus infection. We therefore performed 3′ mRNA sequencing of total RNA that was extracted from ZIKV-infected cells 24 h after infection using an MOI of 5 to robustly induce ISGs. Our analysis included four biologic replicates each for uninfected and infected wt, RIG-I KO or MDA5 KO cells. Across cells of all genotypes, a total of 236 genes were differentially regulated upon ZIKV infection (Figure 2A). Most genes were differentially expressed in A549 wt and MDA5 KO cells with a substantial overlap between the two. Only 24 genes were differentially regulated in RIG-I KO cells, indicating that transcriptomic changes upon ZIKV infection were largely driven by RIG-I. This was also evident from the heatmap in Figure 2B where ZIKV-infected RIG-I KO cells rather clustered with the uninfected instead of the ZIKV-infected cells of other genotypes. Most the 236 genes differentially expressed upon ZIKV infection were upregulated while only few were downregulated (Figure 2B). Next, we analyzed ISGs using the gene set defined in [35]. A total of 98 genes differentially expressed in ZIKV-infected cells were ISGs. Most ISGs—such as *IFIT1* or *RSAD2* (also known as viperin)—were upregulated in a RIG-I-dependent and MDA5-independent manner (Figure 2C,D). Induction of a small number of ISGs not only required RIG-I but was also partially MDA5-dependent (Figure 2C). As predicted from our RT-qPCR results shown in Figure 1D, this included *MX1* (Figure 2D). Other ISGs were induced to a similar extent upon infection in all three cell lines, including *CCL5* (Figure 2C,D). RIG-I and MDA5 may be redundant for activation of these genes—or their induction could require other signaling pathways. Taken together, these data show that most transcriptional changes in A549 cells upon ZIKV infection occurred in a RIG-I-dependent manner and that RIG-I was particularly important for the induction of ISGs.

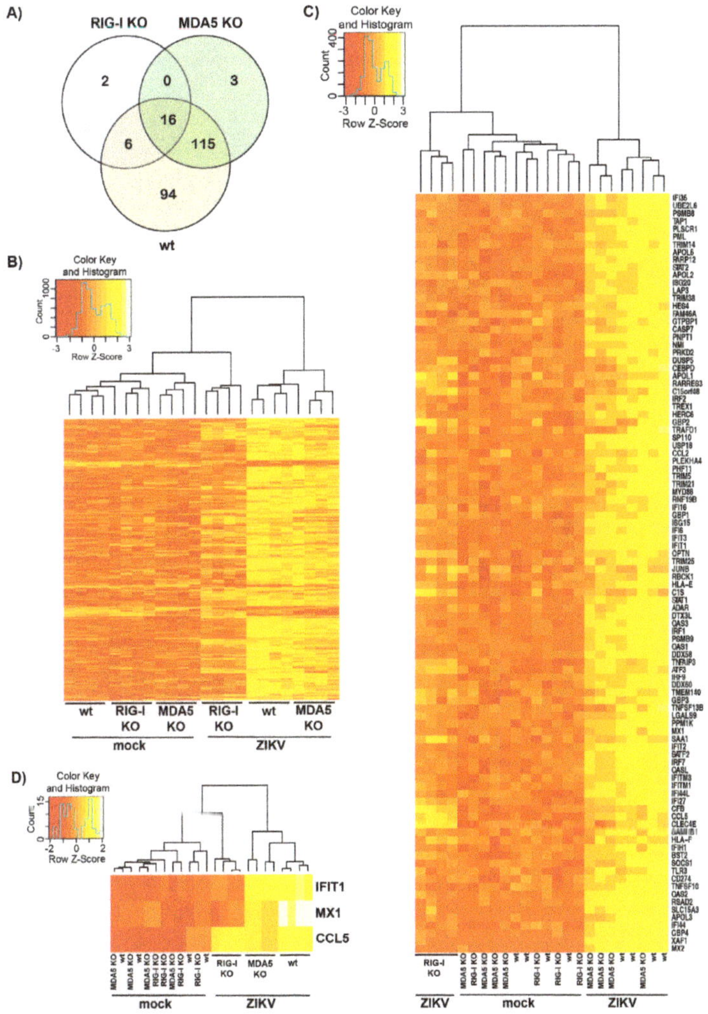

Figure 2. RIG-I sensing drives most transcriptional changes upon ZIKV infection. (**A**) Venn diagram showing differentially expressed genes in wt, RIG-I KO and MDA5 KO cells 24 h after infection; (**B–D**) Heat maps depicting all 236 differentially expressed genes (**B**), 98 differentially expressed ISGs (**C**) and three exemplary ISGs with different expression profiles in A549 wt, RIG-I KO and MDA5 KO cells upon ZIKV infection (**D**). Colors represent z-scores that indicate a value's relationship to the mean, measured as standard deviations from the mean. z-scores calculated for each row (i.e., each gene) and were plotted instead of the normalized expression values to ensure that expression patterns are not overwhelmed by absolute expression values. Data in (**A–D**) pooled from four independent biologic samples.

3.3. RIG-I-Mediated Signaling Protects A549 Cells from Apoptosis

ZIKV infection causes apoptosis [36–43]. We therefore asked whether reduced innate immune recognition of ZIKV in RIG-I KO cells impacts virus-induced cell death. A549 wt, RIG-I KO and MDA5 KO cells were infected with a low dose of ZIKV (MOI 0.1) to analyze a spreading infection and the confluency of the cells was measured for 6 days using an in-incubator imaging system (Incucyte). Interestingly, we found that after 6 days of infection the confluency of RIG-I KO cells was decreased by about 50%, whereas the confluency of wt and MDA5 KO cells was not affected by ZIKV infection (Figure 3A). Furthermore, crystal violet staining revealed virus-induced cell death in two different RIG-I KO clones, but not in wt or MDA5 KO cells six days after infection with two different doses of virus (MOI 0.1 and 0.01) (Figure 3B). To determine whether apoptosis was induced during ZIKV infection in RIG-I KO cells, we performed western blot analysis of PARP cleavage, a molecular signature of apoptosis. Indeed, ZIKV infection resulted in increased levels of cleaved PARP in RIG-I KO cells, but not in wt and MDA5 KO cells (Figure 3C). In addition, we monitored activity of the apoptotic caspases-3 and -7. Four days after ZIKV infection, RIG-I KO cells showed a 4-fold induction of caspase-3/7 activity, while only a 2-fold induction was observed in wt and MDA5 KO cells (Figure 3D). Furthermore, ZIKV-infected cells displayed shrinkage and membrane blebbing, morphologic changes typical for apoptotic cells (Figure 3E). Taken together, these data showed that a lack of RIG-I signaling in A549 cells led to a loss of protection from ZIKV-induced apoptosis, which may be due to reduced type I IFN production and increased virus replication in RIG-I-deficient cells.

3.4. ZIKV NS5 Inhibits Type I IFN Induction

ZIKV NS5 inhibits type I IFN signaling by inducing degradation of STAT2 and by blocking phosphorylation of STAT1 [6,13,19,21,44]. Results of overexpression studies suggested that NS5 also blocks the innate immune response by inhibiting the induction of type I IFN [6,13,44,45]. However, it is possible that the latter effect is indirect as RLRs and many proteins involved in their downstream signaling are encoded by ISGs [46]. As such, lower levels of type I IFN induction in cells expressing NS5 could be explained by reduced levels of RLRs or other proteins involved in type I IFN induction. To distinguish between such indirect effects of NS5 and direct inhibition of type I IFN induction, we generated HEK293 cells lacking IFNAR1 and obtained one clone designated c27. As expected, IFNAR1 KO cells were incapable to phosphorylate STAT1 in response to IFNα2a (Figure 4A). We further validated our IFNAR1 KO cells by sequencing and found −1, −3 and −5 deletions at the sgRNA target site (Figure 4B; HEK293 are largely triploid [47,48]). While the −1 and −5 deletions result in a frameshift, the −3 deletion removes one amino acid keeping the reading frame intact. Our functional data in Figure 4A showing the absence of response to IFNα2a suggest that the protein encoded by the −3 mutant allele is either non-functional or rapidly degraded.

We then transfected wt and IFNAR1 KO HEK293 cells with an expression plasmid for ZIKV NS5. We used empty vector and ZIKV NS3, which we previously found not to block RLR signaling [13] or EMCV L, which blocks IRF3 [49], as negative and positive controls, respectively. Alongside these expression plasmids, cells were co-transfected with an *IFNβ* promoter F-Luc reporter construct and R-Luc as a transfection control. Next, we stimulated RIG-I or MDA5 by transfecting 5′-triphosphate containing in vitro transcribed RNA (IVT–RNA) or RNA extracted from EMCV-infected Hela cells (Hela–EMCV–RNA), respectively (Figure 4C, [13]). Luciferase activities were measured 24 h after RNA transfection. As expected, both immunostimulatory RNAs induced the *IFNβ* promoter in wt cells that had been transfected with empty vector or ZIKV NS3, while the response was strongly reduced by ZIKV NS5 and EMCV L (Figure 4D). Importantly, NS5 inhibited induction of the *IFNβ* promoter to a similar extent in IFNAR1 KO cells stimulated with IVT–RNA (Figure 4D). In addition, the response to Hela–EMCV–RNA also appeared to be reduced by NS5 in IFNAR1 KO cells, although this trend did not reach statistical significance. We conclude that ZIKV NS5 blocked RIG-I-mediated IFN induction in the absence of IFNAR signaling. These observations suggest that ZIKV NS5 not only inhibits antiviral responses downstream of IFNAR signaling, but also has a direct effect on the induction of type I IFNs by RLRs.

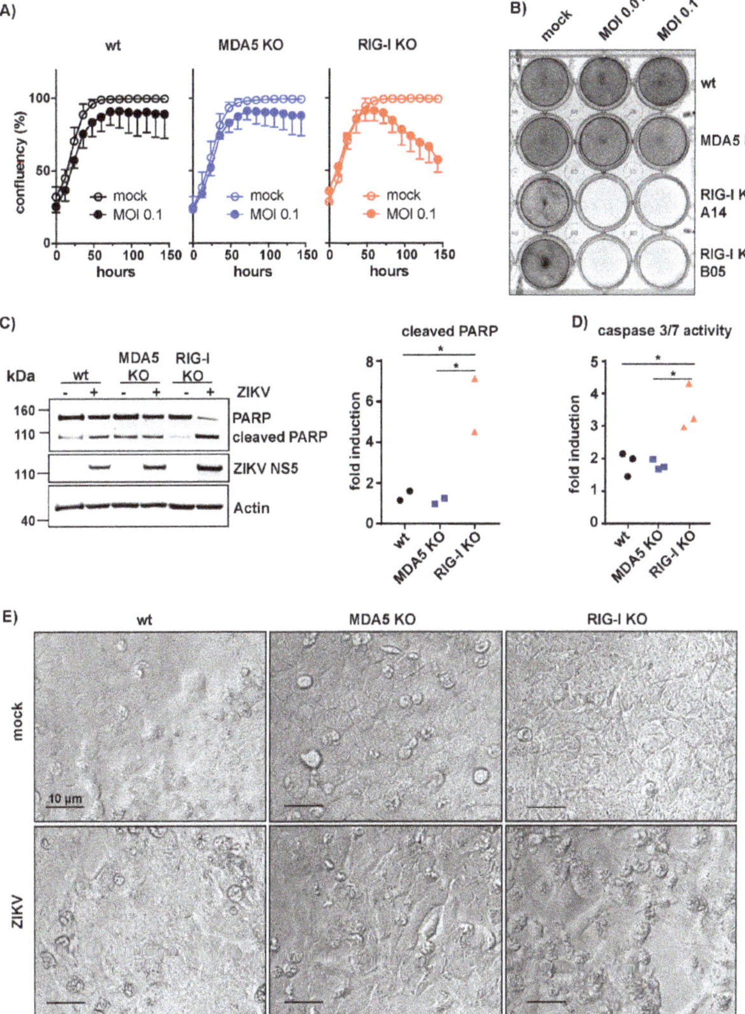

Figure 3. RIG-I signaling protects A549 cells from ZIKV-induced cell death. (**A**) A549 cells were infected with ZIKV (MOI 0.1) and cell confluency was measured for 6 days in the Incucyte; (**B**) A549 cells were infected with ZIKV, fixed 6 days after infection and stained with crystal violet; (**C**) A549 cells were infected with ZIKV (MOI 0.1) and lysed 4 days after infection. Cell lysates were analyzed by western blot using the indicated antibodies (left). Actin served as a loading control. Signal intensity of full length and cleaved PARP was quantified relative to background and the fold induction of cleaved PARP was calculated (right); (**D**) A549 cells were infected with ZIKV (MOI 0.1) and the activity of caspase 3 and 7 was determined 4 days after infection using the Promega Caspase-3/7 Glo assay; (**E**) A549 cells were infected with ZIKV (MOI 0.1) and images were acquired 4 days after infection. Scale bar: 10 μm. Data in **A**, **C** (right) and **D** are pooled form two (**C**) or three (**A,D**) independent experiments. In **A**, mean and SD are plotted; in **C** and **D**, data points correspond to individual experiments and statistical analysis was with one-way ANOVA with Tukey's multiple comparison (* $p < 0.05$). Data in **B**, **C** (left) and **E** are representative of two (**C**) or three independent experiments (**B,E**).

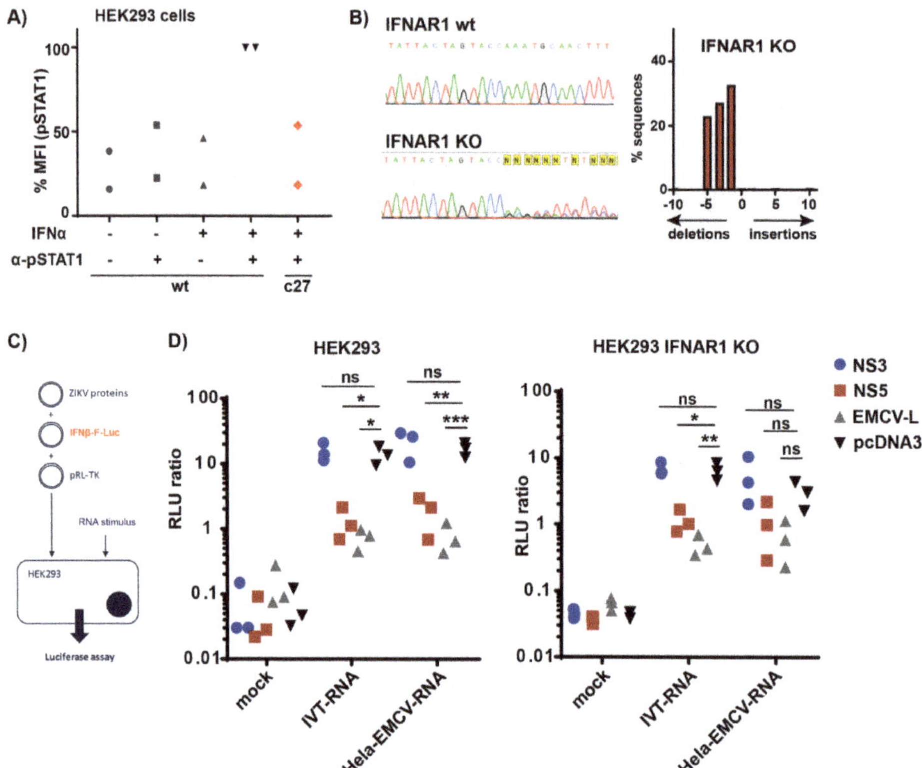

Figure 4. ZIKV NS5 inhibits IFN induction. (**A**) wt HEK293 and IFNAR1 KO clone c27 cells were stimulated with 5000 U/mL IFNα2a for 30 min before fixation and intracellular staining with α-pSTAT1 antibody. As controls, some wt cells were not stimulated or the α-pSTAT1 antibody was omitted. pSTAT1 levels were determined by flow cytometry. Mean fluorescence intensity (MFI) of the pSTAT1 signal was calculated and set to 100 in wt cells; (**B**) genomic DNA was extracted from IFNAR1 KO clone c27 cells. A fragment of DNA surrounding the targeted area was amplified by PCR and sequenced (left). Sequences were analyzed using TIDE (right). Number of nucleotides inserted or deleted, and the percentage of sequences affected are shown; (**C**) schematic of the experiment in **D**; (**D**) wt and IFNAR1 KO cells were transfected with the indicated expression plasmids, a plasmid encoding F-Luc under the control of the *IFNβ* promoter and a plasmid, which expresses R-Luc. Twenty-four hours later, cells were transfected with 5 ng IVT–RNA or 50 ng Hela–EMCV–RNA per well. F-Luc activity was determined 24 h after RNA transfection and normalized to R-Luc. Data in **A** and **D** are pooled from two and three independent experiments, respectively. Data points correspond to individual experiments and in **D** are mean values of technical triplicates. Statistical analysis: Two-way ANOVA with Tukey's multiple comparison (ns $p \geq 0.05$, * $p < 0.05$, ** $p < 0.01$, *** $p < 0.001$).

4. Discussion

Our data demonstrate that RIG-I is the main sensor for ZIKV infection in A549 cells. Genetic ablation of RIG-I in A549 cells led to a loss of type I IFN production and ISG induction as well as to an increase in virus titer. Transcriptomic analysis corroborated that knockdown of RIG-I strongly reduced differential gene expression upon ZIKV infection. Importantly, most ISGs induced after ZIKV infection were RIG-I-dependent. This is likely due to a combination of reduced type I IFN secretion by RIG-I-deficient cells and reduced activation of IRF3, which directly regulates some ISGs [46]. This RIG-I-dependency of ZIKV sensing is in line with a recent publication by Esser-Nobis and colleagues [16]. RIG-I is thought to recognize the conserved 5′-triphosphate group found on nascent RNAs of flaviviruses as shown by Chazal et al. [50]. This is interesting as flaviviruses replicate in complexes formed in invaginations of ER membranes, raising the question as to how RIG-I gains access to viral RNAs [51]. Replication factories are likely to be dynamic structures and viral RNAs can potentially leak into the cytoplasm. Studies on dengue virus, West Nile virus and tick-borne encephalitis virus furthermore revealed 10-nm-wide openings of these invaginations to the cytoplasm using electron tomography [51]. ZIKV RNA is thought to exit to the cytoplasm through these pores to be packaged into virions and for protein translation [52]. RIG-I can be activated by less than 20 RNA molecules per cell [53]. A few nascent positive-stranded RNA molecules that escape replication factories would therefore be sufficient to induce an IFN response by RIG-I.

Our study furthermore revealed that RIG-I prevents ZIKV-induced apoptosis, likely due to RIG-I-induced innate immunity that curtails virus replication. ZIKV-infected A549 RIG-I KO cells succumbed to cell death 4 days post infection with a low MOI and showed increased cleavage of PARP as well as activation of caspases-3 and -7. Several studies suggested that ZIKV-induced cell death in neuronal cells is responsible for neurodevelopmental defects such as microcephaly [36,43,54]. An increase in cell death upon ZIKV infection was modeled in brain-specific organoids derived from human induced pluripotent stem cells (iPSCs) [55,56]. Studies in different cell lines suggest that the induction of apoptosis may be cell-type specific. A549 cells were shown to succumb to apoptosis 48 h after infection when infected with a high MOI [57]. In contrast, human monocyte-derived dendritic cells, Vero cells or mosquito C6/36 cells infected with several different African or Asian ZIKV strains did not induce apoptosis 24 and 48 h after infection [58]. Our work now shows that a functional immune response to ZIKV infection protects cells from apoptosis. The observed cell-type specific differences could therefore correlate with how efficiently a cell senses the virus and how potently the virus is restricted by the initiated type I IFN response. It is thus important to study the levels and functionality of RIG-I in cell types infected by ZIKV, including neuronal cells. Furthermore, ZIKV delays apoptosis by modulating the activities of anti-apoptotic Bcl-2 family proteins [59]. It will be interesting for future studies to determine if and how viral targeting of RIG-I and Bcl-2 proteins is functionally linked.

The importance of RIG-I and type I IFN to ZIKV infection is also evident from the presence of viral antagonists. Here, we confirmed that NS5—one of the most potent viral antagonists—not only blocks type I IFN signaling, but also efficiently and directly inhibited type I IFN production triggered by RIG-I, as suggested previously [6,13,44,45,60]. It is now important to identify the precise mechanisms by which the RLR signaling cascade is targeted by NS5. Recently, Li et al. described that NS5 directly represses K63-linked polyubiquitination of RIG-I [61]. In addition, an inhibitory effect of NS5 on IRF3 activation has been suggested [44,62,63]. Interestingly, Lin and colleagues reported an interaction of NS5 with TBK1 that results in reduced phosphorylation of IRF3 [45]. The latter findings were made by NS5 overexpression in HEK293 cells. It would be very interesting in future studies to confirm these findings at endogenous protein level during infection with live ZIKV. In ZIKV-infected cells, RIG-I-dependent responses are induced despite the presence of NS5 (Figures 1–3). It is therefore likely that NS5's ability to block RIG-I signaling is not absolute. Whether this relates to relative protein levels in infected cells or to cell-to-cell variability remains to be determined. A virus expressing an NS5 mutant that fails to interfere with RIG-I signaling but maintains other functions of NS5 would be useful for such studies. It is noteworthy that a recent report using immortalized human fetal astrocytes and an siRNA approach found that both RIG-I and MDA5 were required for induction of *IFNβ* and *ISG* transcripts [17]. This work further described an inhibitory effect of ZIKV NS3 on RIG-I and MDA5 signaling, in apparent contradiction to our data shown here in Figure 4D and in [13]. It is possible that these differences relate to cell-type specific expression of co-factors of RIG-I and MDA5, an interesting hypothesis for future studies.

Taken together, our study emphasizes the importance of RIG-I-mediated-ZIKV-sensing in controlling virus replication and virus-induced cell death. Targeting viral antagonists to support ZIKV sensing by RIG-I may open up novel treatment options and limit the severity of ZIKV associated neurological symptoms.

Author Contributions: Conceptualization, M.S. and J.R.; methodology, all authors; software, n/a; validation, M.S. and J.R.; formal analysis, M.S. and J.R.; investigation, M.S., A.B., N.G., J.H.; resources, P.H. and A.K.; data curation, M.S. and N.G.; writing—original draft preparation, M.S. and J.R.; writing—review and editing, all authors; visualization, M.S. and N.G.; supervision, J.R.; project administration, J.R.; funding acquisition, J.R. and A.K. All authors have read and agreed to the published version of this manuscript.

Funding: This work was funded by the UK Medical Research Council [MRC core funding of the MRC Human Immunology Unit; J.R.], [MC_UU_12014/8, MR/N017552/1; A.K.] and by the Wellcome Trust [Grant Number 100954; J.R.]. J.H. is supported by the European Commission under the Horizon2020 program H2020 MSCA-ITN GA 675278 EDGE. The funders had no role in study design, data collection and analysis, decision to publish or preparation of the manuscript.

Acknowledgments: We thank members of the Rehwinkel lab for discussion. We thank the Oxford Genomics Center at the Wellcome Center for Human Genetics (funded by Wellcome Trust grant reference 203141/Z/16/Z) for the generation and initial processing of the sequencing data. We further acknowledge G. Kochs, C. Reis e Sousa, R. Randall and A. Merits for their kind gifts of cell lines and antibodies.

Conflicts of Interest: The authors declare no conflicts of interest.

References

1. Faria, N.R.; Azevedo, R.; Kraemer, M.U.G.; Souza, R.; Cunha, M.S.; Hill, S.C.; Theze, J.; Bonsall, M.B.; Bowden, T.A.; Rissanen, I.; et al. Zika virus in the Americas: Early epidemiological and genetic findings. *Science* **2016**, *352*, 345–349. [CrossRef]

2. Liu, Z.Y.; Shi, W.F.; Qin, C.F. The evolution of Zika virus from Asia to the Americas. *Nat. Rev. Microbiol.* **2019**, *17*, 131–139. [CrossRef]

3. Li, H.; Saucedo-Cuevas, L.; Shresta, S.; Gleeson, J.G. The Neurobiology of Zika Virus. *Neuron* **2016**, *92*, 949–958. [CrossRef]

4. Cao-Lormeau, V.M.; Blake, A.; Mons, S.; Lastere, S.; Roche, C.; Vanhomwegen, J.; Dub, T.; Baudouin, L.; Teissier, A.; Larre, P.; et al. Guillain-Barre Syndrome outbreak associated with Zika virus infection in French Polynesia: A case-control study. *Lancet* **2016**, *387*, 1531–1539. [CrossRef]

5. Hamel, R.; Dejarnac, O.; Wichit, S.; Ekchariyawat, P.; Neyret, A.; Luplertlop, N.; Perera-Lecoin, M.; Surasombatpattana, P.; Talignani, L.; Thomas, F.; et al. Biology of Zika Virus Infection in Human Skin Cells. *J. Virol.* **2015**, *89*, 8880–8896. [CrossRef] [PubMed]

6. Kumar, A.; Hou, S.; Airo, A.M.; Limonta, D.; Mancinelli, V.; Branton, W.; Power, C.; Hobman, T.C. Zika virus inhibits type-I interferon production and downstream signaling. *EMBO Rep.* **2016**, *17*, 1766–1775. [CrossRef] [PubMed]

7. Lazear, H.M.; Govero, J.; Smith, A.M.; Platt, D.J.; Fernandez, E.; Miner, J.J.; Diamond, M.S. A Mouse Model of Zika Virus Pathogenesis. *Cell Host Microbe* **2016**, *19*, 720–730. [CrossRef] [PubMed]

8. Miner, J.J.; Cao, B.; Govero, J.; Smith, A.M.; Fernandez, E.; Cabrera, O.H.; Garber, C.; Noll, M.; Klein, R.S.; Noguchi, K.K.; et al. Zika Virus Infection during Pregnancy in Mice Causes Placental Damage and Fetal Demise. *Cell* **2016**, *165*, 1081–1091. [CrossRef]

9. Hartmann, G. Nucleic Acid Immunity. *Adv. Immunol.* **2017**, *133*, 121–169. [CrossRef]

10. Yockey, L.J.; Varela, L.; Rakib, T.; Khoury-Hanold, W.; Fink, S.L.; Stutz, B.; Szigeti-Buck, K.; Van den Pol, A.; Lindenbach, B.D.; Horvath, T.L.; et al. Vaginal Exposure to Zika Virus during Pregnancy Leads to Fetal Brain Infection. *Cell* **2016**, *166*, 1247–1256.e1244. [CrossRef]

11. Ma, W.; Li, S.; Ma, S.; Jia, L.; Zhang, F.; Zhang, Y.; Zhang, J.; Wong, G.; Zhang, S.; Lu, X.; et al. Zika Virus Causes Testis Damage and Leads to Male Infertility in Mice. *Cell* **2016**, *167*, 1511–1524.e10. [CrossRef]

12. Tripathi, S.; Balasubramaniam, V.R.; Brown, J.A.; Mena, I.; Grant, A.; Bardina, S.V.; Maringer, K.; Schwarz, M.C.; Maestre, A.M.; Sourisseau, M.; et al. A novel Zika virus mouse model reveals strain specific differences in virus pathogenesis and host inflammatory immune responses. *PLoS Pathog.* **2017**, *13*, e1006258. [CrossRef]

13. Hertzog, J.; Dias Junior, A.G.; Rigby, R.E.; Donald, C.L.; Mayer, A.; Sezgin, E.; Song, C.; Jin, B.; Hublitz, P.; Eggeling, C.; et al. Infection with a Brazilian isolate of Zika virus generates RIG-I stimulatory RNA and the viral NS5 protein blocks type I IFN induction and signaling. *Eur. J. Immunol.* **2018**. [CrossRef]

14. Ma, J.; Ketkar, H.; Geng, T.; Lo, E.; Wang, L.; Xi, J.; Sun, Q.; Zhu, Z.; Cui, Y.; Yang, L.; et al. Zika Virus Non-structural Protein 4A Blocks the RLR-MAVS Signaling. *Front. Microbiol.* **2018**, *9*, 1350. [CrossRef]

15. Hu, Y.; Dong, X.; He, Z.; Wu, Y.; Zhang, S.; Lin, J.; Yang, Y.; Chen, J.; An, S.; Yin, Y.; et al. Zika virus antagonizes interferon response in patients and disrupts RIG-I-MAVS interaction through its CARD-TM domains. *Cell Biosci.* **2019**, *9*, 46. [CrossRef]

16. Esser-Nobis, K.; Aarreberg, L.D.; Roby, J.A.; Fairgrieve, M.R.; Green, R.; Gale, M., Jr. Comparative analysis of African and Asian lineage-derived Zika Virus strains reveals differences in activation of and sensitivity to antiviral innate immunity. *J. Virol.* **2019**. [CrossRef]

17. Riedl, W.; Acharya, D.; Lee, J.H.; Liu, G.; Serman, T.; Chiang, C.; Chan, Y.K.; Diamond, M.S.; Gack, M.U. Zika Virus NS3 Mimics a Cellular 14-3-3-Binding Motif to Antagonize RIG-I- and MDA5-Mediated Innate Immunity. *Cell Host Microbe* **2019**, *26*, 493–503.e6. [CrossRef]

18. Liao, X.; Xie, H.; Li, S.; Ye, H.; Li, S.; Ren, K.; Li, Y.; Xu, M.; Lin, W.; Duan, X.; et al. 2′, 5′-Oligoadenylate Synthetase 2 (OAS2) Inhibits Zika Virus Replication through Activation of Type Iota IFN Signaling Pathway. *Viruses* **2020**, *12*, 418. [CrossRef]

19. Bowen, J.R.; Quicke, K.M.; Maddur, M.S.; O'Neal, J.T.; McDonald, C.E.; Fedorova, N.B.; Puri, V.; Shabman, R.S.; Pulendran, B.; Suthar, M.S. Zika Virus Antagonizes Type I Interferon Responses during Infection of Human Dendritic Cells. *PLoS Pathog.* **2017**, *13*, e1006164. [CrossRef] [PubMed]

20. Roby, J.A.; Esser-Nobis, K.; Dewey-Verstelle, E.C.; Fairgrieve, M.R.; Schwerk, J.; Lu, A.Y.; Soveg, F.W.; Hemann, E.A.; Hatfield, L.D.; Keller, B.C.; et al. Flavivirus Nonstructural Protein NS5 Dysregulates HSP90 to Broadly Inhibit JAK/STAT Signaling. *Cells* **2020**, *9*, 899. [CrossRef]

21. Grant, A.; Ponia, S.S.; Tripathi, S.; Balasubramaniam, V.; Miorin, L.; Sourisseau, M.; Schwarz, M.C.; Sanchez-Seco, M.P.; Evans, M.J.; Best, S.M.; et al. Zika Virus Targets Human STAT2 to Inhibit Type I Interferon Signaling. *Cell Host Microbe* **2016**, *19*, 882–890. [CrossRef]

22. Cumberworth, S.L.; Clark, J.J.; Kohl, A.; Donald, C.L. Inhibition of type I interferon induction and signalling by mosquito-borne flaviviruses. *Cell Microbiol.* **2017**, *19*. [CrossRef]

23. Best, S.M. The Many Faces of the Flavivirus NS5 Protein in Antagonism of Type I Interferon Signaling. *J. Virol.* **2017**, *91*. [CrossRef]

24. Donald, C.L.; Brennan, B.; Cumberworth, S.L.; Rezelj, V.V.; Clark, J.J.; Cordeiro, M.T.; Freitas de Oliveira Franca, R.; Pena, L.J.; Wilkie, G.S.; Da Silva Filipe, A.; et al. Full Genome Sequence and sfRNA Interferon Antagonist Activity of Zika Virus from Recife, Brazil. *PLoS Negl. Trop. Dis.* **2016**, *10*, e0005048. [CrossRef]

25. Hilton, L.; Moganeradj, K.; Zhang, G.; Chen, Y.H.; Randall, R.E.; McCauley, J.W.; Goodbourn, S. The NPro product of bovine viral diarrhea virus inhibits DNA binding by interferon regulatory factor 3 and targets it for proteasomal degradation. *J. Virol.* **2006**, *80*, 11723–11732. [CrossRef] [PubMed]

26. Bridgeman, A.; Maelfait, J.; Davenne, T.; Partridge, T.; Peng, Y.; Mayer, A.; Dong, T.; Kaever, V.; Borrow, P.; Rehwinkel, J. Viruses transfer the antiviral second messenger cGAMP between cells. *Science* **2015**, *349*, 1228–1232. [CrossRef]

27. Dobin, A.; Gingeras, T.R. Mapping RNA-seq Reads with STAR. *Curr. Protoc. Bioinform.* **2015**, *51*, 11–19. [CrossRef]

28. Kuhn, R.M.; Haussler, D.; Kent, W.J. The UCSC genome browser and associated tools. *Brief Bioinform.* **2013**, *14*, 144–161. [CrossRef]

29. Liao, Y.; Smyth, G.K.; Shi, W. featureCounts: An efficient general purpose program for assigning sequence reads to genomic features. *Bioinformatics* **2014**, *30*, 923–930. [CrossRef] [PubMed]

30. Robinson, M.D.; McCarthy, D.J.; Smyth, G.K. edgeR: A Bioconductor package for differential expression analysis of digital gene expression data. *Bioinformatics* **2010**, *26*, 139–140. [CrossRef]

31. Oliveros, J.C. Venny. An Interactive Tool for Comparing Lists with Venn's Diagrams. Available online: https://bioinfogp.cnb.csic.es/tools/venny/index.html (accessed on 21 October 2019).

32. Flohr, F.; Schneider-Schaulies, S.; Haller, O.; Kochs, G. The central interactive region of human MxA GTPase is involved in GTPase activation and interaction with viral target structures. *FEBS Lett.* **1999**, *463*, 24–28. [CrossRef]

33. Park, S.Y.; Choi, H.C.; Chun, Y.H.; Kim, H.; Park, S.H. Characterization of chromosomal aberrations in lung cancer cell lines by cross-species color banding. *Cancer Genet. Cytogenet.* **2001**, *124*, 62–70. [CrossRef]

34. Isaka, T.; Nestor, A.L.; Takada, T.; Allison, D.C. Chromosomal variations within aneuploid cancer lines. *J. Histochem. Cytochem.* **2003**, *51*, 1343–1353. [CrossRef] [PubMed]

35. Schoggins, J.W.; Wilson, S.J.; Panis, M.; Murphy, M.Y.; Jones, C.T.; Bieniasz, P.; Rice, C.M. A diverse range of gene products are effectors of the type I interferon antiviral response. *Nature* **2011**, *472*, 481–485. [CrossRef] [PubMed]

36. Ho, C.Y.; Ames, H.M.; Tipton, A.; Vezina, G.; Liu, J.S.; Scafidi, J.; Torii, M.; Rodriguez, F.J.; du Plessis, A.; DeBiasi, R.L. Differential neuronal susceptibility and apoptosis in congenital Zika virus infection. *Ann. Neurol.* **2017**, *82*, 121–127. [CrossRef] [PubMed]

37. Ribeiro, M.R.; Moreli, J.B.; Marques, R.E.; Papa, M.P.; Meuren, L.M.; Rahal, P.; de Arruda, L.B.; Oliani, A.H.; Oliani, D.; Oliani, S.M.; et al. Zika-virus-infected human full-term placental explants display pro-inflammatory responses and undergo apoptosis. *Arch. Virol.* **2018**, *163*, 2687–2699. [CrossRef]

38. Liu, H.; Liao, H.M.; Li, B.; Tsai, S.; Hung, G.C.; Lo, S.C. Comparative Genomics, Infectivity and Cytopathogenicity of American Isolates of Zika Virus that Developed Persistent Infections in Human Embryonic Kidney (HEK293) Cells. *Int. J. Mol. Sci.* **2019**, *20*, 3035. [CrossRef]

39. Sherer, M.L.; Khanal, P.; Talham, G.; Brannick, E.M.; Parcells, M.S.; Schwarz, J.M. Zika virus infection of pregnant rats and associated neurological consequences in the offspring. *PLoS ONE* **2019**, *14*, e0218539. [CrossRef]

40. Anfasa, F.; Goeijenbier, M.; Widagdo, W.; Siegers, J.Y.; Mumtaz, N.; Okba, N.; van Riel, D.; Rockx, B.; Koopmans, M.P.G.; Meijers, J.C.M.; et al. Zika Virus Infection Induces Elevation of Tissue Factor Production and Apoptosis on Human Umbilical Vein Endothelial Cells. *Front. Microbiol.* **2019**, *10*, 817. [CrossRef]

41. Li, G.; Bos, S.; Tsetsarkin, K.A.; Pletnev, A.G.; Despres, P.; Gadea, G.; Zhao, R.Y. The Roles of prM-E Proteins in Historical and Epidemic Zika Virus-mediated Infection and Neurocytotoxicity. *Viruses* **2019**, *11*, 157. [CrossRef]

42. Liu, J.; Li, Q.; Li, X.; Qiu, Z.; Li, A.; Liang, W.; Chen, H.; Cai, X.; Chen, X.; Duan, X.; et al. Zika Virus Envelope Protein induces G2/M Cell Cycle Arrest and Apoptosis via an Intrinsic Cell Death Signaling Pathway in Neuroendocrine PC12 Cells. *Int. J. Biol. Sci.* **2018**, *14*, 1099–1108. [CrossRef] [PubMed]

43. De Sousa, J.R.; Azevedo, R.S.S.; Martins Filho, A.J.; Araujo, M.T.F.; Moutinho, E.R.C.; Baldez Vasconcelos, B.C.; Cruz, A.C.R.; Oliveira, C.S.; Martins, L.C.; Baldez Vasconcelos, B.H.; et al. Correlation between Apoptosis and in Situ Immune Response in Fatal Cases of Microcephaly Caused by Zika Virus. *Am. J. Pathol.* **2018**, *188*, 2644–2652. [CrossRef] [PubMed]

44. Xia, H.; Luo, H.; Shan, C.; Muruato, A.E.; Nunes, B.T.D.; Medeiros, D.B.A.; Zou, J.; Xie, X.; Giraldo, M.I.; Vasconcelos, P.F.C.; et al. An evolutionary NS1 mutation enhances Zika virus evasion of host interferon induction. *Nature Commun.* **2018**, *9*, 414. [CrossRef] [PubMed]

45. Lin, S.; Yang, S.; He, J.; Guest, J.D.; Ma, Z.; Yang, L.; Pierce, B.G.; Tang, Q.; Zhang, Y.J. Zika virus NS5 protein antagonizes type I interferon production via blocking TBK1 activation. *Virology* **2018**, *527*, 180–187. [CrossRef]

46. Schoggins, J.W. Interferon-Stimulated Genes: What Do They All Do? *Annu. Rev. Virol.* **2019**, *6*, 567–584. [CrossRef]

47. Lin, Y.C.; Boone, M.; Meuris, L.; Lemmens, I.; Van Roy, N.; Soete, A.; Reumers, J.; Moisse, M.; Plaisance, S.; Drmanac, R.; et al. Genome dynamics of the human embryonic kidney 293 lineage in response to cell biology manipulations. *Nat. Commun.* **2014**, *5*, 4767. [CrossRef]

48. Bylund, L.; Kytola, S.; Lui, W.O.; Larsson, C.; Weber, G. Analysis of the cytogenetic stability of the human embryonal kidney cell line 293 by cytogenetic and STR profiling approaches. *Cytogenet. Genome Res.* **2004**, *106*, 28–32. [CrossRef]

49. Hato, S.V.; Ricour, C.; Schulte, B.M.; Lanke, K.H.; de Bruijni, M.; Zoll, J.; Melchers, W.J.; Michiels, T.; van Kuppeveld, F.J. The mengovirus leader protein blocks interferon-alpha/beta gene transcription and inhibits activation of interferon regulatory factor 3. *Cell. Microbiol.* **2007**, *9*, 2921–2930. [CrossRef]

50. Chazal, M.; Beauclair, G.; Gracias, S.; Najburg, V.; Simon-Lorière, E.; Tangy, F.; Komarova, A.V.; Jouvenet, N. RIG-I Recognizes the 5′ Region of Dengue and Zika Virus Genomes. *Cell Rep.* **2018**, *24*, 320–328. [CrossRef]

51. Paul, D.; Bartenschlager, R. Architecture and biogenesis of plus-strand RNA virus replication factories. *World J. Virol.* **2013**, *2*, 32–48. [CrossRef]

52. Cortese, M.; Goellner, S.; Acosta, E.G.; Neufeldt, C.J.; Oleksiuk, O.; Lampe, M.; Haselmann, U.; Funaya, C.; Schieber, N.; Ronchi, P.; et al. Ultrastructural Characterization of Zika Virus Replication Factories. *Cell Rep.* **2017**, *18*, 2113–2123. [CrossRef] [PubMed]

53. Zeng, W.; Sun, L.; Jiang, X.; Chen, X.; Hou, F.; Adhikari, A.; Xu, M.; Chen, Z.J. Reconstitution of the RIG-I pathway reveals a signaling role of unanchored polyubiquitin chains in innate immunity. *Cell* **2010**, *141*, 315–330. [CrossRef] [PubMed]

54. Li, H.; Saucedo-Cuevas, L.; Regla-Nava, J.A.; Chai, G.; Sheets, N.; Tang, W.; Terskikh, A.V.; Shresta, S.; Gleeson, J.G. Zika Virus Infects Neural Progenitors in the Adult Mouse Brain and Alters Proliferation. *Cell Stem Cell* **2016**, *19*, 593–598. [CrossRef] [PubMed]

55. Qian, X.; Nguyen, H.N.; Song, M.M.; Hadiono, C.; Ogden, S.C.; Hammack, C.; Yao, B.; Hamersky, G.R.; Jacob, F.; Zhong, C.; et al. Brain-Region-Specific Organoids Using Mini-bioreactors for Modeling ZIKV Exposure. *Cell* **2016**, *165*, 1238–1254. [CrossRef] [PubMed]

56. Gabriel, E.; Ramani, A.; Karow, U.; Gottardo, M.; Natarajan, K.; Gooi, L.M.; Goranci-Buzhala, G.; Krut, O.; Peters, F.; Nikolic, M.; et al. Recent Zika Virus Isolates Induce Premature Differentiation of Neural Progenitors in Human Brain Organoids. *Cell Stem Cell* **2017**, *20*, 397–406.e5. [CrossRef] [PubMed]

57. Frumence, E.; Roche, M.; Krejbich-Trotot, P.; El-Kalamouni, C.; Nativel, B.; Rondeau, P.; Misse, D.; Gadea, G.; Viranaicken, W.; Despres, P. The South Pacific epidemic strain of Zika virus replicates efficiently in human epithelial A549 cells leading to IFN-beta production and apoptosis induction. *Virology* **2016**, *493*, 217–226. [CrossRef]

58. Vielle, N.J.; Zumkehr, B.; Garcia-Nicolas, O.; Blank, F.; Stojanov, M.; Musso, D.; Baud, D.; Summerfield, A.; Alves, M.P. Silent infection of human dendritic cells by African and Asian strains of Zika virus. *Sci. Rep.* **2018**, *8*, 5440. [CrossRef]

59. Turpin, J.; Frumence, E.; Despres, P.; Viranaicken, W.; Krejbich-Trotot, P. The ZIKA Virus Delays Cell Death Through the Anti-Apoptotic Bcl-2 Family Proteins. *Cells* **2019**, *8*, 1338. [CrossRef]

60. Lundberg, R.; Melen, K.; Westenius, V.; Jiang, M.; Osterlund, P.; Khan, H.; Vapalahti, O.; Julkunen, I.; Kakkola, L. Zika Virus Non-Structural Protein NS5 Inhibits the RIG-I Pathway and Interferon Lambda 1 Promoter Activation by Targeting IKK Epsilon. *Viruses* **2019**, *11*, 1024. [CrossRef]

61. Li, A.; Wang, W.; Wang, Y.; Chen, K.; Xiao, F.; Hu, D.; Hui, L.; Liu, W.; Feng, Y.; Li, G.; et al. NS5 Conservative Site Is Required for Zika Virus to Restrict the RIG-I Signaling. *Front. Immunol.* **2020**, *11*, 51. [CrossRef]

62. Wu, Y.; Liu, Q.; Zhou, J.; Xie, W.; Chen, C.; Wang, Z.; Yang, H.; Cui, J. Zika virus evades interferon-mediated antiviral response through the co-operation of multiple nonstructural proteins in vitro. *Cell Discov.* **2017**, *3*, 17006. [CrossRef] [PubMed]

63. Zhao, Z.; Tao, M.; Han, W.; Fan, Z.; Imran, M.; Cao, S.; Ye, J. Nuclear localization of Zika virus NS5 contributes to suppression of type I interferon production and response. *J. Gen. Virol.* **2019**. [CrossRef] [PubMed]

Article

Zika Virus Infects Human Placental Mast Cells and the HMC-1 Cell Line, and Triggers Degranulation, Cytokine Release and Ultrastructural Changes

Kíssila Rabelo [1,*], Antônio José da Silva Gonçalves [2], Luiz José de Souza [3], Anna Paula Sales [3], Sheila Maria Barbosa de Lima [4], Gisela Freitas Trindade [4], Bianca Torres Ciambarella [1], Natália Recardo Amorim Tasmo [5], Bruno Lourenço Diaz [5], Jorge José de Carvalho [1], Márcia Pereira de Oliveira Duarte [2,*] and Marciano Viana Paes [2,*]

[1] Laboratório de Ultraestrutura e Biologia Tecidual, Universidade do Estado do Rio de Janeiro, Rio de Janeiro 20551-030, Brazil; bia.btc2@gmail.com (B.T.C.); jjcarv@gmail.com (J.J.d.C.)
[2] Laboratório Interdisciplinar de Pesquisas Médicas, Instituto Oswaldo Cruz, Rio de Janeiro 21040-900, Brazil; ajsgoncalves@gmail.com
[3] Faculdade de Medicina de Campos, Campos dos Goytacazes, Rio de Janeiro 28035-581, Brazil; luizjosedes@gmail.com (L.J.d.S.); ap.sss@hotmail.com (A.P.S.)
[4] Laboratório de Tecnologia Virológica, Biomanguinhos, Rio de Janeiro 21040-900, Brazil; smaria@bio.fiocruz.br (S.M.B.d.L.); gisela.freitas@bio.fiocruz.br (G.F.T.)
[5] Laboratório de Inflamação, Instituto de Biofísica Carlos Chagas Filho, Universidade Federal do Rio de Janeiro, Rio de Janeiro 21941-170, Brazil; nataliaamorim1@hotmail.com (N.R.A.T.); bldiaz@biof.ufrj.br (B.L.D.)
*   Correspondence: kissilarabelo91@gmail.com (K.R.); marciapo@gmail.com (M.P.d.O.D.); marciano@ioc.fiocruz.br (M.V.P); Tel.: +55-21-25621038 (M.V.P.)

Received: 6 March 2020; Accepted: 5 April 2020; Published: 16 April 2020

Abstract: Zika virus (ZIKV) is an emergent arthropod-borne virus whose outbreak in Brazil has brought major public health problems. Infected individuals have different symptoms, including rash and pruritus, which can be relieved by the administration of antiallergics. In the case of pregnant women, ZIKV can cross the placenta and infect the fetus leading to congenital defects. We have identified that mast cells in the placentae of patients who had Zika during pregnancy can be infected. This led to our investigation on the possible role of mast cells during a ZIKV infection, using the HMC-1 cell line. We analyzed their permissiveness to infection, release of mediators and ultrastructural changes. Flow cytometry detection of ZIKV-NS1 expression 24 h post infection in 45.3% of cells showed that HMC-1 cells are permissive to ZIKV infection. Following infection, β-hexosaminidase was measured in the supernatant of the cells with a notable release at 30 min. In addition, an increase in TNF-α, IL-6, IL-10 and VEGF levels were measured at 6 h and 24 h post infection. Lastly, different intracellular changes were observed in an ultrastructural analysis of infected cells. Our findings suggest that mast cells may represent an important source of mediators that can activate other immune cell types during a ZIKV infection, which has the potential to be a major contributor in the spread of the virus in cases of vertical transmission.

Keywords: flavivirus; immune response; inflammatory mediator

---

1. Introduction

Zika fever is an important *Arbovirus*-caused disease that has surfaced in numerous countries in Asia, Africa and America [1]. The etiological agent of this disease, Zika virus (ZIKV), was initially isolated in 1947 from the blood of sentinel *Rhesus* monkeys during a study on yellow fever transmission in the Zika forest of Uganda, which gave rise to its name [2,3]. Transmission of the ZIKV is primarily through bites of infected *Aedes* mosquitos, with the most common vectors being *Aedes aegypti* and *Aedes albopictus,* but it can also happen by vertical transmission [4,5]. As a result of vertical transmission, there were alarming cases of Congenital Zika Syndrome, as the virus could cause damage to the placenta, infect placental cells and reach the fetus [6]. A ZIKV particle has a diameter of 25–30 nm and is a member of the *Flaviviridae* family that shares many similarities with other more widely known related viruses such as dengue, West Nile, Japanese encephalitis and yellow fever [4,7]. It has a single-stranded RNA genome with a positive polarity of 11 Kb and encodes a polyprotein precursor that is processed into the structural proteins such as capsid (C), pre-membrane (prM) and envelope (E) along with seven non-structural proteins (NS1, NS2A, NS2B, NS3, NS4A, NS4B and NS5) [8,9].

Mast cells are resident immunological cells found abundantly in tissues such as skin, endometrium and placenta that have prominent roles in immunologic reactions [10–13]. Their presence and prevalence in these tissues, along with their proximity to blood vessels, predispose these cells to be among the first immune cells that can be infected by ZIKV after a mosquito bite penetrates the skin. As some of the most frequent symptoms of zika are rash and pruritus, which are relieved by the administration of antiallergic drugs (anti-histamines), this has led us to believe that mast cells can play a role, although not yet elucidated, in the pathogenesis of the disease [14–16]. We hypothesize that it may be one of the cells involved in placental infections, which can directly contribute to vertical transmission.

Although there are no studies in the literature that have investigated the involvement of mast cells in a ZIKV infection to date, mast cells have a proven role in infections by dengue, another *Arbovirus*. Several products originating from mast cells are found at high levels in patients infected by dengue, especially those with plasma leakage [17,18]. While mast cells are permissive to dengue infection, it is most probable that they display a low level of the specific receptors required since the quantity of virus necessary to successfully infect this cell type is always higher than is needed for macrophages and dendritic cells [17,19,20].

HMC-1 cells are a lineage of human mast cells that characteristically express the cytokine receptor c-Kit abundantly and release different cytokines after degranulation stimuli. This cell line possesses the features necessary to serve as an in vitro model for the development of studies on mast cells [21]. HMC-1 has been widely used in studies on degranulation studies, endothelial activation and its interaction with other arboviruses [22–24].

Here, we present our observations on the presence of mast cells in ZIKV-infected human placentae and observed viral replication in these cells. Additionally, we investigated the potential for ZIKV to infect HMC-1 cells as a model system for mast cells and quantified the percentage of infected cells in different MOIs. We further studied the degranulation of these cells after contact/infection with ZIKV by measuring β-hexosaminidase release as well as the expression profiles of TNF-α (tumor necrosis factor-α), IL-6 (interleukin-6), IL-10 (interleukin-10) and VEGF (vascular endothelial growth factor). As a final point, we evaluated the effects of ZIKV infection on the ultrastructure of HMC-1 cells. Together, the findings validate a critical and, to our knowledge, previously unrecognized role for mast cells in the infection and propagation of ZIKV in humans.

2. Materials and Methods

2.1. Placentae Collection, Patient Clinical History and Ethical Approval

At delivery, samples from the placentae were collected and fixed in 10% formaldehyde. Samples were collected at the Hospital Plantadores de Cana, Campos dos Goytacazes, RJ, Brazil. As a control, a sample of a full-term placenta from a healthy donor was included.

Case 1: A 23-year-old patient. Symptoms: fever, arthralgia, exanthema and pruritus in the third trimester of gestation. At 38 weeks of gestation, her baby girl was born by cesarean delivery, with 37 cm of cephalic circumference. The mother's IgM serology was positive for Zika. The test for dengue NS1 was negative.

Case 2: A 34-year-old patient. Symptoms: exanthema and pruritus in the third trimester of gestation. Her baby girl was born at term, by cesarean delivery, with 38 weeks of gestation. She presented with a normal 34 cm of cephalic circumference. The mother's IgM serology was positive for Zika.

Patient recruitment and the procedures performed were pre-approved by the Ethics Committee of the Oswaldo Cruz Foundation/FIOCRUZ (CAEE: 65924217.4.0000.5248) and by the Ethics Committee of Faculty of Campos Medicine/Benedito Pereira Nunes Foundation (CAEE: 65924217.4.3001.5244). The patients were fully informed of the research plans and provided written consent to participate, which included permission to publish all data without identifying information.

2.2. Histopathology and Histological Detection of Mast Cells in ZIKV Infected Placentae

All histological processing of the sample was performed as described previously by our group [25]. The histopathological analysis was performed on the images observed and captured by hematoxylin and eosin (H&E) staining. The staining used to highlight the mast cells was Toluidine Blue 1%. Stained specimens were visualized by light microscopy (Olympus, Tokyo, Japan), and digital images were obtained using Image-Pro Plus software version 7.0.

2.3. Immunofluorescence Assay

Immunofluorescence was performed as described in Rabelo et al., 2017 [25]. Antibodies were used at a dilution of 1:200 for a mouse monoclonal anti-Zika NS1 IgG (Arigo Biolaboratories, Taiwan, Republic of China), and a rabbit polyclonal antihuman c-Kit IgG (Santa Cruz, Texas, USA). After staining with primary antibodies, sections were incubated with an Alexa 488-conjugated rabbit anti-mouse IgG, Alexa 555-conjugated goat anti-rabbit IgG, or Alexa 555-conjugated goat anti-mouse IgG (ThermoFisher, Waltham, MA, USA). Slides were visualized by fluorescence microscopy (Olympus, Tokyo, Japan), and digital images were obtained using Image-Pro Plus software version 7.0.

2.4. Immunohistochemistry

The protocol for immunohistochemistry was described previously by our group [25]. Briefly, the slides were incubated overnight at 4 °C with a 1:200 dilution of the mouse monoclonal antibody IgG antibody against Zika NS1 (Arigo Biolaboratories, Taiwan, Republic of China). Then, sections were maintained with a rabbit anti-mouse IgG conjugated to horseradish peroxidase (Spring Bioscience Corporation, CA, USA) for 40 min at room temperature. We visualized the sections by light microscopy (Olympus, Tokyo, Japan), and digital images were obtained using Image-Pro Plus software version 7.0.

2.5. Cell Line

The HMC-1 cell line was kindly provided by Dr. Joseph H. Butterfield (Mayo Clinic, Rochester, NY, USA) and cultured in Iscove's Modified Dulbecco's Medium (IMDM- Thermo Fisher, Waltham, MA, USA) supplemented with 10% fetal bovine serum (FBS, Cultilab, Campinas, SP, Brazil), 40 U/mL penicillin/streptomycin (Sigma, St. Louis, MS, USA) and 1.2 mM α-thioglycerol (Sigma, St. Louis, MS, USA). Cells were maintained at 37 °C in a humidified incubator at 5% CO_2. Culture media was exchanged every 3–4 days with splitting of cultures at a confluency of 80–90%.

2.6. ZIKV Viral Stock Productuion

A primary clinical virus specimen was isolated from a serum sample of a patient from Paraiba. The virus was propagated in a culture of C6/36 *Ae. albopictus* mosquito cells and harvested virus was tittered by the infection of Vero cells (CCL-81) followed by RT-PCRq, which determined a titer of 5.8×10^6 PFU/mL. Copy numbers were assessed by using a standard curve in the RT-PCRq reaction containing 1×10^8 copies/reaction. The oligonucleotide set utilized targeted the intergenic region of the Membrane/Envelope as described by Lanciotti, 2008 [26] (Table 1).

Table 1. Oligonucleotide sets to amplify ZIKV genome.

Genome Position	Region		Sequence
835–857	M/E	sense	TTGGTCATGATACTGCTGATTGC
911–890	M/E	reverse	CCTTCCACAAAGTCCCTATTGC
860–886	M/E	probe	FAM-CGGCATACAGCATCAGGTGCATAGGAG-NFQ

2.7. ZIKV Infections

Infections were performed by varying the multiplicity of infection (MOI) at 0.1, 0.2 and 1.0. ZIKV viral particles per host cell. Virus was added to cell culture and incubated for 1 h at 37 °C prior to removal of unattached viral particles and a further incubation of 6 h or 24 h. For a 30 min time point, virus was incubated with cells for 30 min before rinsing and preparation of flow cytometry. As a negative control, cells were incubated in the same conditions with a mock viral stock consisting of a supernatant of non-infected Vero cells.

2.8. Flow Cytometry Analysis

The expression of NS1 protein in infected HMC-1 cells was analyzed by flow cytometry. Cells were collected by centrifugation, and suspended in PBS for 30 min, 6 h or 24 h after infection with different MOIs. Approximately 10^6 cells/well were fixed in 4% formaldehyde for 25 min and permeabilized with 0.05% saponin for 30 min. Next, cells were incubated with a 1:1000 dilution of the mouse monoclonal IgG antibody against ZIKV non-structural protein NS1 (Arigo Biolaboratories, Taiwan, Republic of China) for 1 h at 37 °C before being washed with PBS. This was followed by an incubation with a 1:200 dilution of an Alexa 488-conjugated anti-mouse (Thermo Fisher, Waltham, MA, USA) for 30 min. After washing with PBS, cells were suspended in PBS and applied to a flow cytometer (Facs Calibur; BD Biosciences, San Jose, CA, USA) to measure fluorescence, which was analyzed offline with Summit 6.1 software.

2.9. Measurement of Mast Cell Degranulation

Mast cell degranulation was evaluated by measuring the activity of the granule-stored enzyme-β-hexosaminidase that was secreted into the extracellular medium. Cells were infected with MOI 0.1, 0.2 or 1 in 6-well plates (1×10^6/well) for 30 min. Aliquots of the supernatant (15 μl) were transferred to 96-well plates and incubated with 60 μL of substrate (1 mM p-nitrophenyl-N-acetyl-b-D-glucosaminide) in 0.05 M sodium citrate (pH 4.5) for 60 min at 37 °C. In addition, we used 60 μL of substrate solution (1 mM p-nitrophenyl-N-acetyl-β-D-glucosaminide (Sigma, St. Louis, MS, USA) in 100 mM sodium citrate, pH 4.5) and incubated for 60 min at 37 °C. Reactions were stopped by adding 150 μL of 0.1 M Na_2CO_3-$NaHCO_3$ buffer (pH 10). Enzyme activity was measured as the absorbance at 405 nm. Total β-hexosaminidase activity was determined by releasing all enzyme through lysis with 0.1% Triton X-100 and measuring activity from a 15 μl aliquot. As a positive control for degranulation, we used 20 μg/mL of 48/80 compound (Sigma, St. Louis, MS, USA). The results are presented as the percentage of total β-hexosaminidase content of the cells.

2.10. ELISA Assays

The quantity of cytokines and factors released from mast cells by infection with ZIKV was measured by ELISA. Supernatants from HMC-1 cells infected at a MOI of 1 for 30 min, 6 h or 24 h were evaluated for IL-6 (900-T16), IL-10 (900-K21), TNF-α (900-T25) and VEGF (900-K10) with commercial ELISA assay kits (Peprotech Inc. Rocky Hill, NJ, USA), according to the manufacturer's instructions.

2.11. Transmission Electron Microscopy Procedure

HMC-1 cells were infected with ZIKV at a MOI of 1 for 30 min or 24 h and then fixed with 2.5% glutaraldehyde in 0.1 M sodium cacodylate buffer (pH 7.2). Cells were post-fixed with 1% buffered osmium tetroxide, dehydrated in an acetone series (30, 50, 70, 90, and 100%) and then embedded in EPON (Electron Microscopy Sciences, Hatfield, PA, USA) through polymerization at 60 °C for 3 days. Ultrathin sections (60–90 nm) were contrasted with uranyl acetate and lead citrate before visualization using a JEOL 1001 transmission electron microscope (Jeol Ltd., Tokyo, Japan).

2.12. Statistical Analysis

Data were analyzed in GraphPad Prism software v 6.0 (GraphPad Software, San Diego, CA, USA) using non-parametric statistical tests. Significant differences between the analyzed groups were determined using the Mann–Whitney test with a threshold of $p < 0.05$.

3. Results

3.1. Detection of Mast Cells, Histopathology and ZIKV Replication in Placental Infected Tissues

First, we evaluated the presence of mast cells in the placentae of ZIKV infected women during pregnancy in comparison to a non-infected control sample. To detect mast cells, we performed immunohistochemistry with a Toluidine Blue stain and identified these cells in placental sections of these patients by the prominent purple coloration (Figure 1A–C, arrows). Next, fluorescence microscopy images (Figure 1D–F) were used to identify cells that displayed both the mast cell marker c-Kit (red) and ZIKV NS1 protein (green). As expected, no evidence of ZIKV NS1 protein was observed in control placenta (Figure 1D). In constrast, dually labeled cells were readily observed in placenta from both ZIKV seropostive patients (Figure 1E,F), which suggested that these cells were infected and supported virus replication (Figure 1E,F). To examine the histopathological aspects, H&E stainging was used to identify maternal portions (basal decidua) and fetal portions (chorionic villi), which were normal in the control placenta (Figure 1G). Within the placentae from the ZIKV infected patients, case 1 presented areas with immature chorionic villi, chronic villositis and chronic deciduitis with lymphocytes in chorionic villi and decidua (Figure 1H). The placenta from case 2 showed intervillitis with lymphocytes in the intevillous space and immature chorionic villi (Figure 1I). To extentd the search for cells supporting ZIKV replication, immunohistochemistry was used to provide broad staining of NS1 protein both in the maternal and fetal portions of the placentae. Again, the control, non-infected samples showed no reactivity against NS1. Within placentae from infected mothers, extensive reactivity was seen in not only immune cells, but also trophoblasts and decidual cells suggesting that they are also permissive to infection (Figure 1K,L).

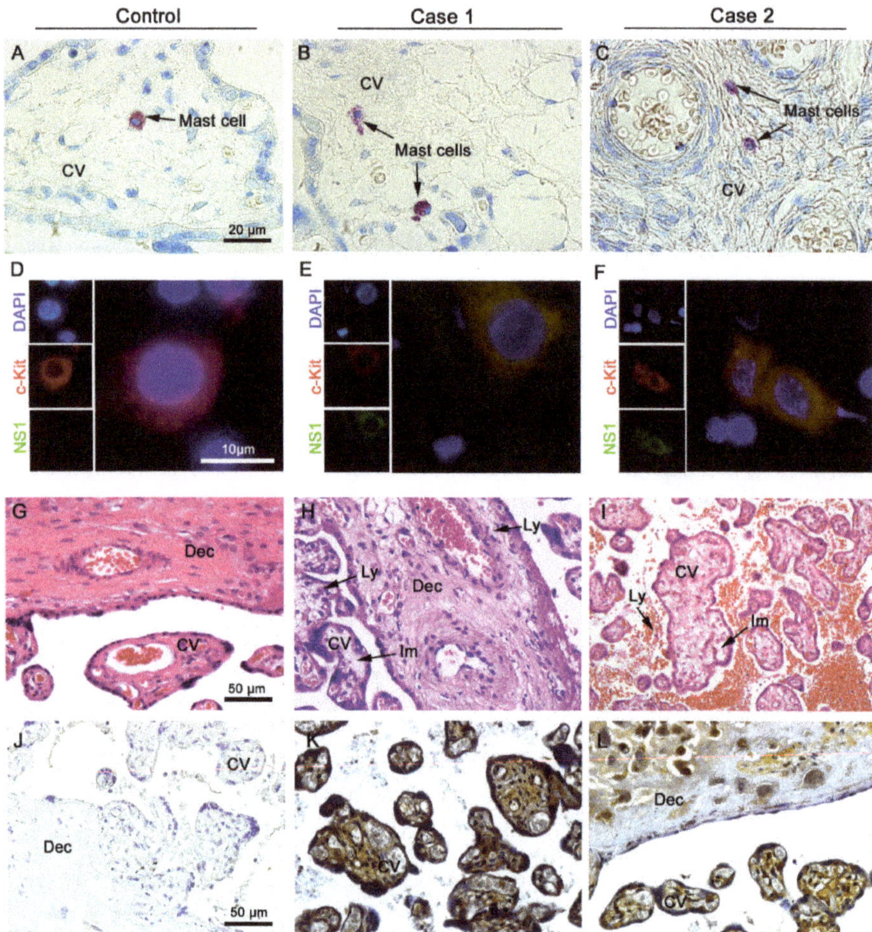

Figure 1. Detection of ZIKV infected mast cells in placental tissue from seropositve mothers. Placentae were collected from mothers infected or not with ZIKV immediately after childbirth and preserved in formadehyde. (**A–C**) Brightfield images of sections stained with Toluidine Blue showing metacromatic granules (purple, arrows) in mast calls. (**D–F**) Immunofluorescent images of DAPI (nuclei; blue), c-Kit (mast cell marker; red) and NS1 (ZIKV marker; green) showing ZIKV infected mast cells with both red and green fluorscence. No NS1 antigen was observed in any sections from the control placenta. The histopathological analysis of the H&E stained placentae showed normal aspects in decidua and chorionic villi within the control placenta (**G**), whereas infected placentae showed areas with lymphocytic infiltrates and immature chorionic villi (**H,I**). Detection of ZIKV NS1 protein by immunohistochemisty did not identify any positive cells in control placentae (**J**). Numerous cells positive for NS1 were detected in placentae from infected mothers, in both maternal and fetal portions (**K–L**). CV, chorionic villi; Dec, decidua; Im, immature chorionic villi; Ly, lymphocytes.

3.2. Infection Rate of ZIKV at Different MOIs

After observing that placental mast cells were infected with ZIKV during a natural infection, the susceptibility to ZIKV entry and permissiveness to its replication was evaluated using the HMC-1 cell line under controlled conditions. Cells were exposed to three different MOIs (0.1, 0.2 and 1) of virus or an equal volume of mock as a control to determine conditions of infections. A mock viral stock was generated from supernatants of Vero cells that were not exposed to ZIKV as a control. The percentage of cells infected by ZIKV was determined by counting the number of cells displaying the fluorescent detection of NS1, a protein that is present only after viral replication, by flow cytometry. Cells were either incubated with virus or mock for 30 min and processed for analysis, or for 1 h with a subsequent incubation for 6 h or 24 h. NS1 was detected under all conditions (Figure 2A), even after 30 min, which suggests that ZIKV can rapidly enter cells and begin replication. Considering that the percentage of cells was nearly equivalent across the three MOIs at 30 min, the results further suggest that only a subset of cells were susceptible to rapid infection. By increasing the virus binding and entry time to 1 h, followed by a 6 h incubation, the percent of cells infected increased with a maximum percent observed with a MOI of 1. A slight increase in the percentage of cells was measured when the post-infection incubation increased to 24 h. Averaged histograms of the three conditions (Figure 2B–D) show a nearly equivalent low background from the mock and the highest levels of infection with a MOI of 1 in 6 h and 24 h, with a mean of 40.10 ± 4.81 and 45.30 ± 3.44% of infected cells in three independent experiments, respectively.

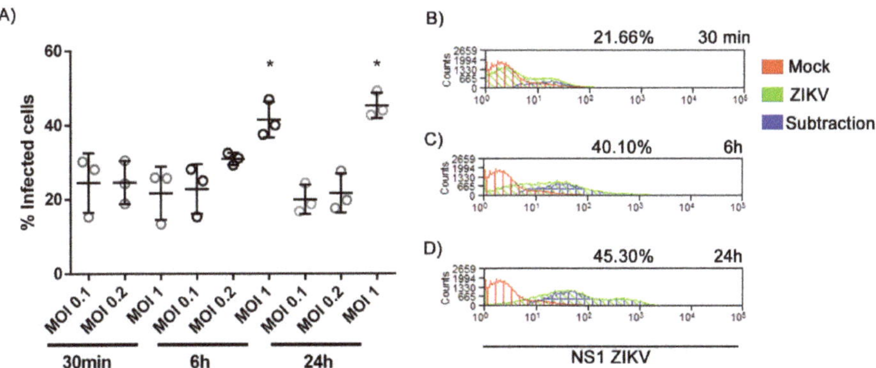

Figure 2. Percentage of HMC-1 cells infected with different MOIs of ZIKV. HMC-1 cells were incubated with ZIKV at MOIs of 0.1, 0.2 or 1 for 30 min and prepared for flow cytometry, or 1 h followed by 6 h or 24 h incubation before analysis. Cells were permeabilized, fixed and stained with the mouse monoclonal IgG antibody against ZIKV non-structural protein NS1 followed by incubation with the Alexa 488-conjugated anti-mouse. Panel (**A**) presents the individual percentages of HMC-1 cells expressing the NS1 protein under the different conditions from three independent experiments. Averaged histograms from the experiments with an MOI of 1 are shown in (**B**) 30 min, (**C**) 1 h with 6 h and (**D**) 1 h with 24 h infection. For negative control, cells were incubated with mock viral stocks. * Statistically significant differences between groups (same time of infection) assessed by a Mann–Whitney test ($p < 0.05$).

3.3. ZIKV Interaction Induces Degranulation

The results from the infection of mast cells by ZIKV suggested that the response of HMC–1 could be contributing to the observations. We chose to explore the activation and degranulation of mast cells by β-hexosaminidase, a resident enzyme released in response to degranulation. Initially, flow cytometry was used to analyze the percentage of cells that display degranulation following incubations of HMC-1 cells with ZIKV at different MOIs for different times. After a 30 min incubation, all three MOIs showed similar percentages (Figure 3A). The percentage of cells decreased following a 6 h incubation and returned to the levels of mock infections after a 24 h incubation suggesting that, at the later time points, the granulosome recuperated or the released enzyme lost activity.

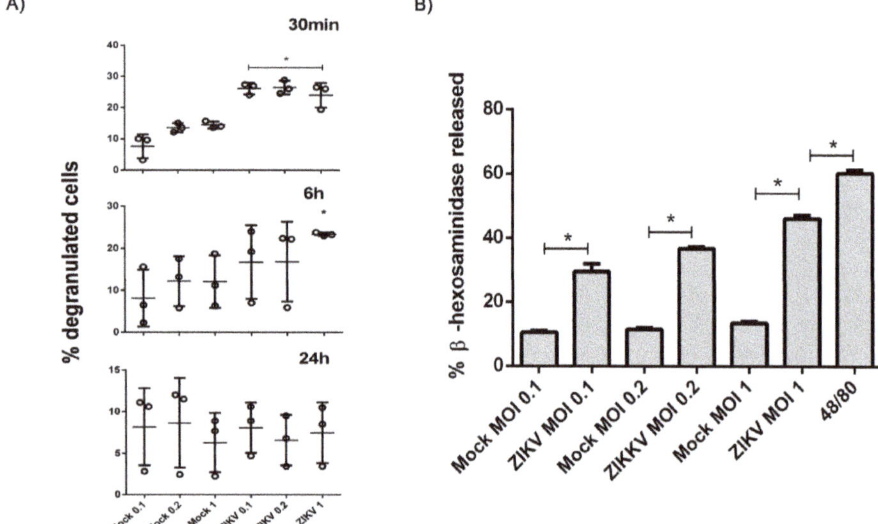

Figure 3. Kinetics of mast cell degranulation after interaction with ZIKV. (**A**) Percentage of degranulated cells after incubation with different MOIs of ZIKV in 30 min, 6 h and 24 h by flow cytometry. (**B**) Percentage of β-hexosaminidase release with different MOIs of ZIKV after 30 min. The synthetic compound 48/80 was used to elicite mast cell degranulation. * Statistically significant differences between groups (same time of infection) assessed by a Mann–Whitney test ($p < 0.05$). Data represent the mean of duplicate values for each sample, in three independent experiments.

To evaluate the early kinetics of mast cell activation, the amount of β-hexosaminidase, normalized to the total cellular β-hexosaminidase, was measured at 30 min for each of the MOIs (Figure 3B). Despite the β-hexosaminidase levels not reaching the percentage of the cells stimulated with the synthetic compound 48/80, the release was gradually increased according to the amount of viral particles, suggesting that the activation of these cells actually occurs due to adsorption of the virus to cell receptors.

3.4. ZIKV Led to Release of Cytokines and VEGF

To analyze the release of the cytokines TNF- α, IL-6, and IL-10, along with VEGF, during infection with ZIKV, we performed ELISAs on the supernatant of mast cells activated with 30 min of contact with the virus as well as the extracellular levels produced by 1 h of virus presence and an incubation of 6 or 24 h. After the shortest interaction time, the levels of TNF-α, IL-6 and IL-10 increased greater in response to exposure to the control than with ZIKV (Figure 4). The levels of VEGF were nearly equal. The levels of the cytokines and VEGF in the supernatant were significantly greater after a hour incubation with ZIKV stocks with an additional 6 h incubation than the mock stocks. This difference grew with the increase in the secondary incubation time to 24 h although the absolute levels of these cytokines and VEGF were lower compared to 6 h. The observed release of TNF-α, IL-6, IL-10 and VEGF at 30 min suggested that they responded to a range of external stimuli. Meanwhile, the elevated levels of these mediators 6 h or 24 h after the infection with ZIKV infection suggests a stimulation in expression and release of these mediators after infection.

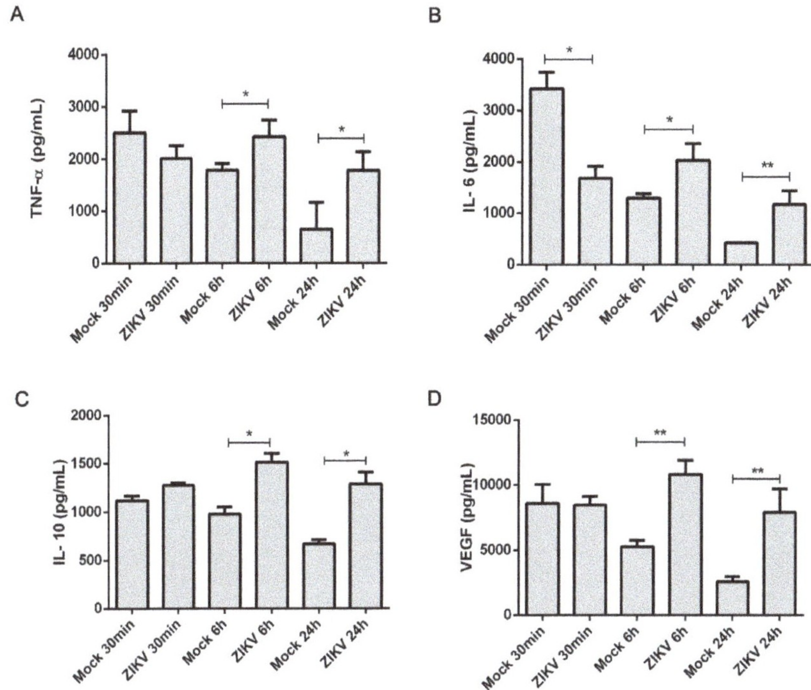

Figure 4. Cytokine and VEGF release by HMC-1 cells in response to ZIKV interactions. The supernatants of HCM-1 cells were collected after incubation with ZIKV or mock viral stocks for 30 min or following a 1 h incubation with an additional 6 or 24 h incubation. Commercial ELISAs were used to measure the level of released (**A**) TNF-α, (**B**) IL-6, (**C**) IL-10 levels and (**D**) VEGF. Data represent the mean of triplicate values for each sample obtained from three independet experiments. * Statistically significant differences assessed by a Mann–Whitney test ($p < 0.05$).

3.5. Ultrastructural Changes Caused by ZIKV Infection

To explore changes to aspects of the ultrastrutucture of mast cells in response to ZIKV infections, an infection with an MOI of 1 was used for the best conditions of infection as well as activation and degranulation of HMC-1. As a control for the analysis, the ultrastructure of cells incubated with the mock viral stock for 30 min was evaluted. Representative cells presented normal aspects for a mast cell in terms of the formation of the nucleus, and the volume of the mitochondria and normal endoplasmic reticulum with a high density of granules (Figure 5A–C). While cells incubated with the ZIKV for 30 min have a lower rate of infection, our previous data show they are at the optimal moment of adsorption and trigger degranulation. The ultrastructure of representative cells shows a decrease in cellular granules (Figure 5D–F), with no other major alterations. After 24 h incubation with the mock viral stock, we observed that the mastocytes continued to have a high density of granules, endoplasmic reticulum with closed cisterns, and mitochondria with some structural alterations, such as swollen and ruptured (Figure 5G–I). After the same period of incubation, the infected mast cells presented various organelle alterations observed as the formation of numerous vesicles, dilated endoplasmic reticulum cisterns, swollen mitochondria, ruptures in cellular membranes and, in some cells, the absence of a nucleus suggesting that a subset of cells may no longer be viable (Figure 5J–K). In several instances, the presence of viral-like particles were detected that match with the size of a ZIKV particle (Figure 5L).

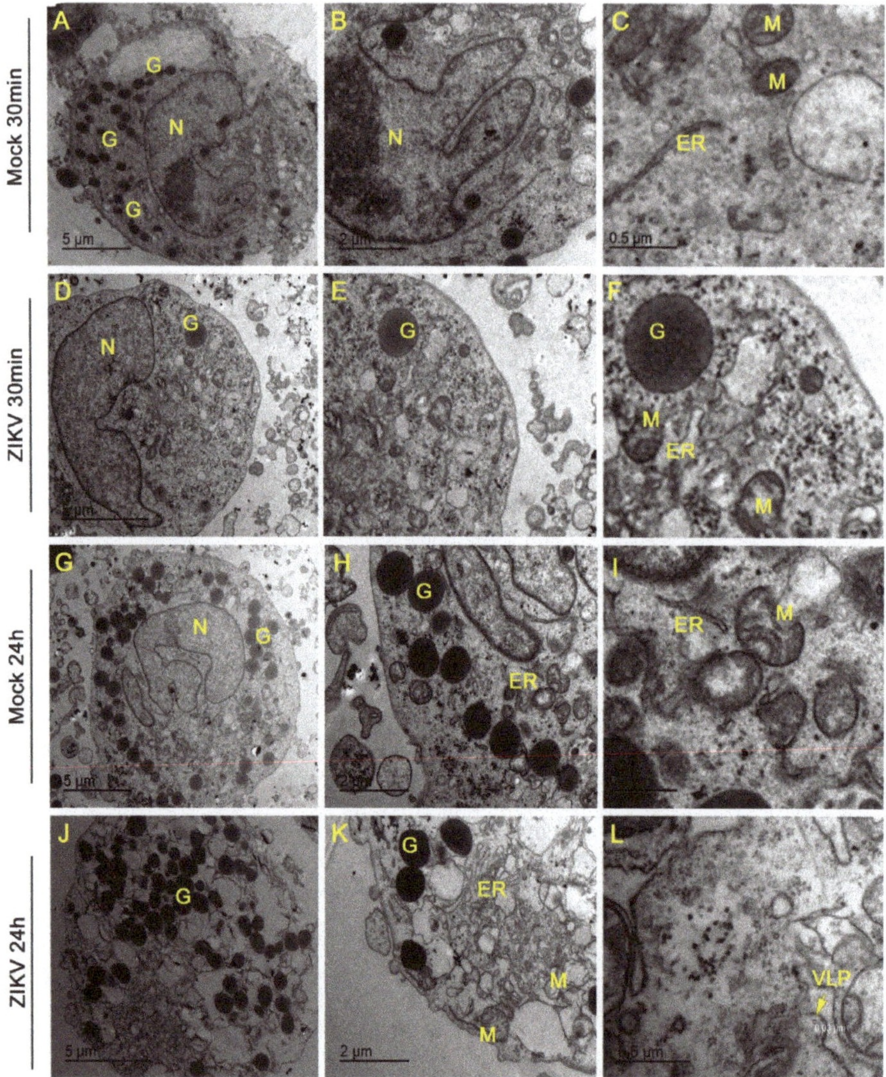

Figure 5. Ultrastructural changes in HMC-1 mast cells infected with ZIKV. HCM-1 cells were exposed to ZIKV or mock for 30 min or 1 h with a post 24 h incubation before processing and imaging of ultrathin sections by electron microscopy. (**A–C**) Control HMC-1s incubated with mock for 30 min. (**D–F**) An HMC-1 cell incubated with ZIKV for 30 min with decreased granules. (**G–I**) An HMC-1 cell incubated with mock for 24 h with a high density of granules. (**J–L**) HMC-1 cell infected with ZIKV for 24 h. Panel L shows a virus-like particle (VLP) with a diameter of approximately 30 nm, consistent with ZIKV. Granules (**G**), nucleus (**N**), mitochondria (**M**) and endoplasmic reticulum (ER).

4. Discussion

Mast cells have an important function in developing an inflammatory process and are present in a variety of tissues such as skin and mucous membranes that include the placenta. However, there have been no studies that have investigated the role of mast cells in ZIKV infection and its pathogenesis. Many studies have described a permissiveness and replication of ZIKV in different placental and immune cells [25,27,28]. These descriptions are extensive in relation to Hofbauer cells and deciduous macrophages [29–31]. Here, we report for the first time the detection of virus in vivo in mast cells present in placental tissue from two women seropositive for a ZIKV infection through the NS1 protein of Zika. Mast cells are resident cells in the endometrium and placenta, and it is believed that they can play multiple roles from implantation to placental immune response during pregnancy, including trophoblastic migration and angiogenesis [32,33]. The implications of ZIKV infections in placental mast cells could have some importance in understanding the inflammatory process and vertical transmission.

Based on the in situ results, we performed a series of experiments in vitro using the HMC-1 cell line as a model system for mast cells to unveil aspects of their interactions and reaction to infections by ZIKV. First, we observed by a flow cytometer analysis that the HMC-1 cells are able to support the entry of the virus as well as its rapid replication within 30 min. Replication was inferred by the detection of the NS1 protein of Zika, which is a non-structural protein that is not a constituent of the virus particle and is only present after its synthesis at the time of replication [9]. It is known that mast cells have the requisite receptors, such as FcγR, HSP70 and others, that could mediate the entry of ZIKV, and also mediate the entry of other arboviruses like dengue, as well as being involved in the transduction signals for the degranulation cascade [21,34].

One of the most abundant proteases present in mast cell granules widely used to assess degranulation is β-hexosaminidase, a glycolytic enzyme that is released into the tissues and triggers typical reactions in allergy and inflammatory responses [35]. We used the quantification of β-hexosaminidase in cell supernatants as a measurement of mast cell degranulation as a result of incubations with ZIKV. We used the synthetic compound 48/80, which is a standard degranulator, to elicit β-hexosaminidase release by HMC-1 cells [36]. There was a significant increase in the release of β-hexosaminidase by the HMC-1 cell line after contact with ZIKV, which was only detected at 30 min, which leads us to believe that viral adsorption is a stimulus for degranulation. The time frame of 30 min is consistent with that of the adsorption and internalization of flavivirus particles, which occurs rapidly in 13 to 15 min as observed for the intracellular localization of DENV particles [37]. In MOI 1, β-hexosaminidase levels were near that of the positive control with 48/80. Degranulation, detected by the release of β-hexosaminidase, has been associated with the injection of DENV in other studies [19]. The cleavage of some substrates of this enzyme has been associated with NKT cell differentiation, and the high activity of β-hexosaminidase has already been observed in placental dysfunction [35,38].

In addition to the enzymes released during degranulation, mast cells are responsible for the production and release of different pro-inflammatory cytokines. We evaluated the production of TNF-α, IL-6 and IL-10 at different times from viral adsorption to 6 h and 24 h post infection. At the moment of initial contact of the mast cells with the mock or the virus, there was a release of these cytokines and VEGF, which is consistent with mast cells having internal stores that are primed for release in response to a stimulus. As the supernatant of Vero cells (mock) has a rich secretion of proteins, this stimulus appears to have been sufficient for the release within 30 min. However, at the end of other incubation times, the mock viral stock controls were associated with low secretion levels of these mediators, which contrasted with the ZIKV infected cells. There was a significant increase in levels of both cytokines at 6 h, which would be expected to generate an environment conducive to the recruitment and differentiation of other immune cells. TNF-α is produced for optimal defense against pathogens in inflammation resolution and orchestrates the tissue recruitment of immune cells and promotes tissue remodeling and destruction [39]. IL-6 is a cytokine with a crucial role in inflammation. It also leads to recruitment and differentiation of mast cells, as well as monocytes, CD4+ and CD8+ T cells, and B lymphocytes, and it stimulates the production of VEGF by fibroblasts. IL-6 expression affects the homeostatic processes that is related to tissue injury and activation of stress-related responses [40–43]. Despite being an anti-inflammatory cytokine, the expression of IL-10 was increased in ZIKV-infected HMC-1 cells, which corroborates what was observed in another study, in the serum of Zika positive patients [44]. Moreover, it is a cytokine normally produced by triggered mast cells, which leads to activation of other mast cells and is present in allergic responses [45,46]. In support of our findings, an increase in cytokines related to the inflammatory environment in placental ZIKV infection has already been observed in another study performed by our group, with an increase in TNF-α and the VEGFR-2 receptor [28]. TNF-α combined with VEGF were similarly related to vascular placental dysfunction, leading to plasma overflow and preeclampsia [47]. The stabilization of mast cells can decrease their response and minimize the severity in dengue, which is related to the release of VEGF and vascular permeability [48].

The ultrastructural changes that occur in the infection can be quite enlightening in relation to the processes that the cell undergoes against the pathogen. We observed degranulation of HMC-1 after 30 min of contact with the virus, but the alterations in organelles were only evident 24 h after infection. The changes caused by ZIKV were already observed in placental cells, and are consistent with those that occur in DENV, even in other cell types [25,49,50]. These changes suggest damage, mainly to mitochondria and the endoplasmic reticulum, which could impinge on the energy and protein production machinery that are necessary for viral replication. In addition, we detected the presence of virus-like particles, with the size expected for ZIKV particles, ~ 30 nm, which confirms the permittivity and ability of mast cells to replicate the virus. These observations, together with the characteristics of mast cells as an immune system component, would suggest that they would be capable of circulating throughout an affected organism, or being resident in the tissue could be responsible for cell-to-cell infection that could underlie vertical transmission.

5. Conclusions

Our data serves as evidence that mast cells are permissive to ZIKV infection, since a non-structural protein, NS1, was detected 24 h post infection. ZIKV can induce degranulation on its first contact and can produce cytokines and VEGF both short term and over a few hours of infection. This response of mast cells can facilitate the installation of a pro-inflammatory environment in the sites where these cells are found, such as in mucous membranes like the placenta. In addition, the fact that they can support the replication of the virus in the human placenta suggests that this type of cell may contribute to vertical transmission. Further studies are needed to fully elucidate the role of mast cells in ZIKV infection.

Author Contributions: Conceptualization, K.R., M.P.O.D., A.J.D.S.G. and M.V.P.; material, L.J.D.S., S.M.B.d.L., G.F.T., M.P.d.O.D., A.J.D.S.G., A.P.S., B.L.D. and M.V.P, methodology, K.R., M.P.d.O.D., B.T.C., N.R.A.T., B.L.D. and A.J.D.S.G.; formal analysis, K.R., M.P.O.D. and A.J.D.S.G.; writing—original draft preparation, K.R.; writing—review and editing, all authors; supervision, M.P.O.D., J.J.C., M.V.P.; project administration, M.P.O.D., J.J.d.C., M.V.P.; funding acquisition, M.P.d.O.D., J.J.d.C., M.V.P. All authors have read and agreed to the published version of the manuscript.

Funding: This research was funded by CNPq and the FAPERJ, grant number (E-26/010.001.498/2016 and E-26/110.511/2014).

Acknowledgments: We thank the Platform of Confocal and Electron Microscopy at the State University of Rio de Janeiro and the Platform of Electron Microscopy in Fiocruz. We are grateful to D. Willian Provance for the manuscript review. This work was supported by the *CNPq* (308780/2015-9) and the *FAPERJ* (E-26/110.511/2014, E-26/010.001.498/2016 and E26/202.003/2016).

Conflicts of Interest: The authors declare no conflict of interest.

References

1. Musso, D.; Ko, A.I.; Baud, D. Zika Virus Infection—After the Pandemic. *N. Engl. J. Med.* **2019**, *381*, 1444–1457. [CrossRef]
2. Musso, D.; Gubler, D.J. Zika Virus. *Clin. Microbiol. Rev.* **2016**, *29*, 487–524. [CrossRef]
3. McCrae, A.W.R.; Kirya, B.G. Yellow fever and Zika virus epizootics and enzootics in Uganda. *Trans. R. Soc. Trop. Hyg.* **1982**, *76*, 552–562. [CrossRef]

4. Gubler, D.; Kuno, D.; Markoff, L. *Fields Virology*, 5th ed.; Knipe, D.M., Howley, P.M., Griffin, D.E., Lamb, R.A., Eds.; Lippincott Williams & Wilkins Publishers: Philadelphia, PA, USA, 2007.

5. Lazear, H.M.; Diamond, M.S. Zika Virus: New Clinical Syndromes and Its Emergence in the Western Hemisphere. *J. Virol.* **2016**, *90*, 4864–4875. [CrossRef] [PubMed]

6. Teixeira, F.M.E.; Pietrobon, A.J.; de Mendonça Oliveira, L.; da Silva Oliveira, L.M.; Sato, M.N. Maternal-Fetal Interplay in Zika Virus Infection and Adverse Perinatal Outcomes. *Front. Immunol.* **2020**, *11*, 1–15. [CrossRef] [PubMed]

7. Faye, O.; Freire, C.C.M.; Iamarino, A.; Faye, O.; De Oliveira, J.V.C.; Diallo, M.; Zanotto, P.M.A.; Sall, A.A. Molecular Evolution of Zika Virus during Its Emergence in the 20 th Century. *PLoS Negleted Trop. Dis.* **2014**, *8*, 1–10.

8. Abushouk, A.I.; Negida, A.; Ahmed, H. An updated review of Zika virus. *J. Clin. Virol.* **2016**, *84*, 53–58. [CrossRef]

9. Chambers, T.J.; Hahn, C.S.; Galler, R.; Rice, C.M. Flavivirus genome organization, expression and replication. *Rev. Microbiol.* **1990**, *44*, 649–688. [CrossRef]

10. Eady, R.A.J.; Cowen, T.; Marshall, T.F.; Plummer, V.; Greaves, M.W. Mast cell population density, blood vessel density and histamine content in normal human skin. *Br. J. Dermatol.* **1979**, *100*, 623–633. [CrossRef]

11. Babina, M.; Guhl, S.; Artuc, M.; Trivedi, N.N.; Zuberbier, T. Phenotypic variability in human skin mast cells. *Exp. Dermatol.* **2016**, 434–439. [CrossRef]

12. Derbala, Y.; Elazzamy, H.; Bilal, M.; Reed, R.; Dambaeva, S.; Dinorah, M.; Garcia, S.; Skariah, A.; Kim, J.K.; Fernandez, E.; et al. Mast cell—Induced immunopathology in recurrent pregnancy losses. *Am. J. Reprod. Immunol.* **2019**, *82*, e13128. [CrossRef] [PubMed]

13. Matsuno, T.; Toyoshima, S.; Sakamoto-sasaki, T. Allergology International Characterization of human decidual mast cells and establishment of a culture system. *Allergol. Int.* **2018**, *67*, S18–S24. [CrossRef] [PubMed]

14. Guanche, H.; Gutiérrez, F.; Ramirez, M.; Ruiz, A.; Pérez, C.R.; González, A. Clinical relevance of Zika symptoms in the context of a Zika Dengue epidemic. *J. Infect. Public Health* **2020**, *13*, 173–176. [CrossRef] [PubMed]

15. Paixão, E.S.; Barreto, F.; Teixeira, G.; Costa, C.N.; Rodrigues, L.C. History, Epidemiology, and Clinical Manifestations of Zika: A Systematic Review. *Am. J. Public Health* **2016**, *106*, 606–612. [CrossRef] [PubMed]

16. Munjal, A.; Khandia, R.; Dhama, K.; Sachan, S. Advances in Developing Therapies to Combat Zika Virus: Current Knowledge and Future Perspectives. *Front. Microbiol.* **2017**, *8*, 1–19. [CrossRef]

17. Syenina, A.; Jagaraj, C.J.; Aman, S.A.; Sridharan, A.; St John, A.L. Dengue vascular leakage is augmented by mast cell degranulation mediated by immunoglobulin Fcγ receptors. *Elife* **2015**, *4*, 1–16. [CrossRef] [PubMed]

18. Furuta, T.; Murao, L.A.; Thi, N.; Lan, P.; Huy, N.T.; Thi, V.; Huong, Q.; Thuy, T.T.; Tham, V.D.; Thi, C.; et al. Association of Mast Cell-Derived VEGF and Proteases in Dengue Shock Syndrome. *PLoS Neglected Trop. Dis.* **2012**, *6*. [CrossRef]

19. Troupin, A.; Shirley, D.; Londono-renteria, B.; Watson, A.M.; Mchale, C.; Hall, A.; Klimstra, W.B.; Gomez, G.; Colpitts, T.M.; Troupin, A.; et al. A Role for Human Skin Mast Cells in Dengue Virus Infection and Systemic Spread. *J. Immunol.* **2016**, *197*, 4382–4391. [CrossRef]

20. Londono-renteria, B.; Marinez-angarita, J.C.; Troupin, A.; Colpitts, T.M. Role of Mast Cells in Dengue Virus Pathogenesis 1 2 3. *DNA Cell Boil.* **2017**, *36*, 423–427. [CrossRef]

21. Nilsson, G.; Blom, T.; Kjellenf, M.K.L.; Butterfieldj, J.H.; Sundstrom, C.; Nilsson, K.; Hellman, L. Phenotypic Characterization of the Human Mast-Cell Line HMC-1. *Scand. J. Immunol.* **1994**, 489–498. [CrossRef]

22. Cochrane, D.E.; Carraway, R.E.; Harrington, K.; Laudano, M.; Rawlings, S.; Feldberg, R.S. HMC-1 human mast cells synthesize neurotensin (NT) precursor, secrete bioactive NT-like peptide (s) and express NT receptor NTS1. *Inflamm. Res.* **2011**, *60*, 1139–1151. [CrossRef] [PubMed]

23. Gilchrist, M.; Befus, A.D. Interferon- c regulates chemokine expression and release in the human mast cell line HMC 1: Role of nitric oxide. *Immunology* **2007**, *123*, 209–217. [PubMed]

24. Brown, M.G.; King, C.A.; Sherren, C.; Marshall, J.S.; Anderson, R. A dominant role for FcγRII in antibody-enhanced dengue virus infection of human mast cells and associated CCL5 release Abstract: Dengue virus is a major mosquito- borne human pathogen with four known serotypes. The presence of antidengue virus antibod. *J. Leukoc. Biol.* **2006**, *80*, 1242–1250. [CrossRef] [PubMed]

25. Rabelo, K.; de Souza, C.F.; de Souza, L.J.; de Souza, T.L.; dos Santos, F.B.; Nunes, P.C.G.; de Azeredo, E.L.; Salomão, N.G.; Trindade, G.F.; Basílio-de-Oliveira, C.A.; et al. Placental Histopathology and clinical presentation of severe congenital Zika syndrome in a human immunodeficiency virus-exposed uninfected infant. *Front. Immunol.* **2017**, *8*, 1–8. [CrossRef]
26. Lanciotti, R.S.; Kosoy, O.L.; Laven, J.J.; Velez, J.O.; Lambert, A.J.; Johnson, A.J.; Stan, S.M.; Duffy, M.R. Genetic and Serologic Properties of Zika Virus Associated with an Epidemic, Yap State. *Emerg. Infect. Dis.* **2008**, *14*, 1232–1239. [CrossRef]
27. Tabata, T.; Petitt, M.; Puerta-Guardo, H.; Michlmayr, D.; Wang, C.; Fang-Hoover, J.; Harris, E.; Pereira, L. Zika Virus Targets Different Primary Human Placental Cells, Suggesting Two Routes for Vertical Transmission. *Cell Host Microbe* **2016**, *20*, 155–166. [CrossRef]
28. Rabelo, K.; Souza, L.J.; Salomão, N.G.; Oliveira, E.R.A.; de Sentinelli, L.P.; Lacerda, M.S.; Saraquino, P.B.; Rosman, F.C.; Basílio-de-Oliveira, R.; Carvalho, J.J.; et al. Placental inflammation and fetal injury in a rare Zika case associated with Guillain-Barré Syndrome and abortion. *Front. Microbiol.* **2018**, *9*, 1–10. [CrossRef]
29. Quicke, K.M.; Bowen, J.R.; Johnson, E.L.; Schinazi, R.F.; Chakraborty, R.; Suthar, M.S.; Quicke, K.M.; Bowen, J.R.; Johnson, E.L.; Mcdonald, C.E.; et al. Zika Virus Infects Human Placental Macrophages. *Cell Host Microbe* **2016**, 1–8. [CrossRef]
30. Rosenberg, A.Z.; Yu, W.; Hill, D.A.; Reyes, C.A.; Schwartz, D.A. Placental Pathology of Zika Virus: Viral Infection of the Placenta Induces Villous Stromal Macrophage (Hofbauer Cell) Proliferation and Hyperplasia. *Arch. Pathol. Lab. Med.* **2017**, *141*, 43–48. [CrossRef]
31. Zulu, M.Z.; Martinez, O.; Gordon, S.; Gray, M. The Elusive Role of Placental Macrophages: The Hofbauer Cell. *J. Innate Immun.* **2019**, *11*, 447–456. [CrossRef]
32. Woidacki, K.; Jensen, F.; Zenclussen, A.C. Mast cells as novel mediators of reproductive processes. *Front. Immunol.* **2013**, *4*, 1–6. [CrossRef] [PubMed]
33. Faas, M.M.; Vos, P. De Innate immune cells in the placental bed in healthy pregnancy and preeclampsia. *Placenta* **2018**, *69*, 125–133. [CrossRef] [PubMed]
34. Chang, H.W.; Kanegasaki, S.; Jin, F.; Deng, Y.; You, Z.; Chang, J.-H.; Kim, D.Y.; Timilshina, M.; Kim, J.-R.; Lee, Y.J.; et al. A common signaling pathway leading to degranulation in mast cells and its regulation by CCR1-ligand. *Basic Transl. Allergy Immunol.* **2019**, *1*, 705–717. [CrossRef] [PubMed]
35. Fukuishi, N.; Murakami, S.; Ohno, A.; Matsui, N.; Fukutsuji, K.; Itoh, K.; Akagi, M.; Alerts, E. Does β -Hexosaminidase Function Only as a Degranulation Indicator in Mast Cells? The Primary Role of β -Hexosaminidase in Mast Cell Granules. *J. Immunol. Does* **2014**, *193*, 1886–1894. [CrossRef]
36. Schemann, M.; Kugler, E.M.; Buhner, S.; Eastwood, C.; Donovan, J.; Jiang, W.; Grundy, D. The Mast Cell Degranulator Compound 48/80 Directly Activates Neurons. *PLoS ONE* **2012**, *7*, e52104. [CrossRef]
37. Van Der Schaar, H.M.; Rust, M.J.; Chen, C.; Van Der Ende-Metselaar, H.; Wilschut, J.; Zhuang, X.; Smit, J.M. Dissecting the cell entry pathway of dengue virus by single-particle tracking in living cells. *PLoS Pathog.* **2008**, *4*. [CrossRef]
38. Arciuch, L.; Bielecki, D.; Borzym, M.; Południewski, G.; Arciszewki, K.; Rózański, A.; Zwierz, K. Isoenzymes of N-acetyl-beta-hexosaminidase in complicated pregnancy. *Acta Biochim. Pol.* **1999**, *46*, 977–983. [CrossRef]
39. Kalliolias, G.D.; Ivashkiv, L.B. TNF biology, pathogenic mechanisms and emerging therapeutic strategies. *Nat. Rev. Rheumatol.* **2015**, *12*, 49–62. [CrossRef]
40. Cop, N.; Ebo, D.G.; Bridts, C.H.; Elst, J.; Hagendorens, M.M.; Mertens, C.; Faber, M.A.; De Clerck, L.S.; Sabato, V. Influence of IL-6, IL-33, and TNF- a on Human Mast Cell Activation: Lessons from Single Cell Analysis by Flow Cytometry. *Int. Clin. Cytom. Soc.* **2018**, *411*, 405–411. [CrossRef]
41. Tanaka, T.; Narazaki, M.; Kishimoto, T. IL-6 in Inflammation, Immunity, and Disease. *Cold Spring Harb. Perspect. Boil.* **2014**, *6*. [CrossRef]
42. Mchale, C.; Mohammed, Z.; Deppen, J.; Gomez, G. BBA—General Subjects Interleukin-6 potentiates Fc ε RI-induced PGD 2 biosynthesis and induces VEGF from human in situ-matured skin mast cells. *Biochim. Biophys. Acta Gen. Subj.* **2018**, *1862*, 1069–1078. [CrossRef]
43. Hunter, C.A.; Jones, S.A. IL-6 as a keystone cytokine in health and disease. *Nat. Immunol.* **2015**, *16*, 448–457. [CrossRef] [PubMed]

44. Kam, Y.W.; Leite, J.A.; Lum, F.M.; Tan, J.J.L.; Lee, B.; Judice, C.C.; De Toledo Teixeira, D.A.; Andreata-Santos, R.; Vinolo, M.A.; Angerami, R.; et al. Specific biomarkers associated with neurological complications and congenital central nervous system abnormalities from Zika virus-infected patients in Brazil. *J. Infect. Dis.* **2017**, *216*, 172–181. [CrossRef]
45. Polukort, S.H.; Rovatti, J.; Carlson, L.; Thompson, C.; Ser-Dolansky, J.; Kinney, S.R.M.; Schneider, S.S.; Mathias, C. IL-10 enhances IgE-mediated mast cell responses and is essential for the development of experimental food allergy in IL-10-deficient mice. *J Immunol.* **2017**, *196*, 4865–4876. [CrossRef] [PubMed]
46. Pagliari, C.; Fernandes, E.R.; Guedes, F.; Alves, C.; Sotto, M.N. Role of mast cells as IL10 producing cells in paracoccidioidomycosis skin lesions. *Mycopathologia* **2006**, *162*, 331–335. [CrossRef] [PubMed]
47. Haase, K.; Gillrie, M.R.; Hajal, C.; Kamm, R.D. Pericytes Contribute to Dysfunction in a Human 3D Model of Placental Microvasculature through VEGF-Ang-Tie2 Signaling. *Adv. Sci.* **2019**, *6*. [CrossRef]
48. Sherif, N.A.; Ghozy, S.; Zayan, A.H.; Elkady, A.H. Mast cell mediators in relation to dengue severity: A systematic review and meta-analysis. *Rev. Med. Virol.* **2019**, *2084*, 1–12. [CrossRef]
49. Fish-Low, C.-Y.; Abubakar, S.; Othman, F.; Chee, H. Ultrastructural aspects of sylvatic dengue virus infection in Vero cell. *Malays. J. Pathol.* **2019**, *41*, 41–46.
50. Tadkalkar, N.; Prasad, S.; Gangodkar, S.; Ghosh, K. Dengue Virus NS1 Exposure Affects von Willebrand Factor Profile and Platelet Adhesion Properties of Cultured Vascular Endothelial Cells. *Indian J. Hematol. Blood Transfus.* **2018**, *35*, 502–505. [CrossRef]

Article

Flavivirus Nonstructural Protein NS5 Dysregulates HSP90 to Broadly Inhibit JAK/STAT Signaling

Justin A. Roby [1,†,*], Katharina Esser-Nobis [1], Elyse C. Dewey-Verstelle [1,‡],
Marian R. Fairgrieve [1,‡], Johannes Schwerk [1], Amy Y. Lu [1,2], Frank W. Soveg [1],
Emily A. Hemann [1], Lauren D. Hatfield [1], Brian C. Keller [3], Alexander Shapiro [4], Adriana Forero [1],
Jennifer E. Stencel-Baerenwald [1], Ram Savan [1] and Michael Gale Jr.[1,2,*]

[1] Center for Innate Immunity and Immune Disease, Department of Immunology, University of Washington
    School of Medicine, Seattle, WA 98109, USA; ken10@uw.edu (K.E.-N.); edewey@uw.edu (E.C.D.-V.);
    marianfairgrieve@gmail.com (M.R.F.); jschwerk@uw.edu (J.S.); amyyhlu@uw.edu (A.Y.L.);
    fsoveg@uw.edu (F.W.S.); eahemann@uw.edu (E.A.H.); ldaarreberg@gmail.com (L.D.H.);
    forera@uw.edu (A.F.); Jen.Baerenwald@gmail.com (J.E.S.-B.); savanram@uw.edu (R.S.)
[2] Departments of Global Health and Microbiology, University of Washington, Seattle, WA 98195, USA
[3] Division of Pulmonary, Critical Care & Sleep Medicine, The Ohio State University College of Medicine,
    Columbus, OH 43210, USA; Brian.Keller2@osumc.edu
[4] Shorewood High School, Shoreline, WA 98133, USA; ags245@cornell.edu
*   Correspondence: jroby@csu.edu.au (J.A.R.); mgale@uw.edu (M.G.J); Tel.: +61-269-332-026 (J.A.R.);
    +1-206-543-8514 (M.G.J)
†   Current address: School of Biomedical Sciences, Charles Sturt University, Wagga Wagga 2678, Australia.
‡   These authors contributed equally.

Received: 29 February 2020; Accepted: 1 April 2020; Published: 7 April 2020

Abstract: Pathogenic flaviviruses antagonize host cell Janus kinase/signal transducer and activator of transcription (JAK/STAT) signaling downstream of interferons α/β. Here, we show that flaviviruses inhibit JAK/STAT signaling induced by a wide range of cytokines beyond interferon, including interleukins. This broad inhibition was mapped to viral nonstructural protein 5 (NS5) binding to cellular heat shock protein 90 (HSP90), resulting in reduced Janus kinase–HSP90 interaction and thus destabilization of unchaperoned JAKs (and other kinase clients) of HSP90 during infection by *Zika virus*, West Nile virus, and Japanese encephalitis virus. Our studies implicate viral dysregulation of HSP90 and the JAK/STAT pathway as a critical determinant of cytokine signaling control during flavivirus infection.

Keywords: flavivirus; JAK/STAT; cytokine; West Nile virus; *Zika virus*; HSP90; NS5; virus–host interactions; anti-viral signaling; immune response

---

1. Introduction

The genus *Flavivirus* comprises several important human pathogens including West Nile virus (WNV), dengue virus (DENV), Japanese encephalitis virus (JEV), and *Zika virus* (ZIKV). These (+)-sense, single-stranded RNA viruses are transmitted to humans by mosquito bite and represent (re)emerging viruses. Roughly half of the world's population is at risk of DENV infection, causing ≈390 million infections and 21,000 deaths per year [1]. Since emergence in the USA, WNV has spread to all contiguous states causing ≈2340 deaths (The US Centers for Disease Control and Prevention (CDC) records as of August 2019), and is a model example of emerging infectious disease [2]. Recently, ZIKV emerged with outbreaks in Oceania and Latin America [3,4]. ZIKV infection causes symptoms ranging from flu-like illness to Gullain–Barré syndrome [4]. ZIKV can undergo maternal–fetal transmission, causing congenital Zika syndrome (CZS) marked by microencephaly and post-natal cognitive disorders [5–9].

ZIKV is of global concern but the virus–host interactions of ZIKV infection linked with CZS are not fully understood. Defining the processes in which flaviviruses employ to evade and antagonize host immune responses is paramount to informing vaccine and therapeutic strategies to combat infection. No approved vaccines or specific antivirals are currently available to prevent or treat human infections with DENV, WNV, or ZIKV.

The immune response to flavivirus infection initiates upon viral recognition by host cell pattern recognition receptors (PRRs) [10,11]. PRR signaling in the infected cell leads to activation of interferon regulatory factor (IRF) 3 and induction of interferons (IFNs) α, β, and λ. IFN secretion and binding to the cognate receptor then triggers the Janus kinase/signal transducer and activator of transcription (JAK/STAT) signaling cascade for rapid induction of IFN-stimulated genes (ISGs). Janus kinases JAK1, JAK2, and Tyk2 phosphorylate STAT1 and STAT2 on specific tyrosine residues (pY) in response to IFN/receptor interaction. pY-STATs heterodimerize and move to the nucleus to bind target promoters inducing ISG expression [12]. ISG products establish an antiviral state in responding cells to restrict virus replication and spread. The innate immune response is essential for control of flavivirus infection [13]. Flavivirus antagonism of IFN-responsive JAK/STAT signaling is a feature associated with virulence [14–19]. Flavivirus nonstructural protein 5 (NS5) has been identified as the primary mediator of IFN signaling inhibition [14]. Roles in JAK/STAT antagonism have also been attributed to other NS proteins and subgenomic flavivirus RNA [16,20–23].

In addition to IFN signaling, the JAK/STAT pathway is utilized by several distinct cytokines that activate various Janus kinases (JAKs; including JAK1, JAK2, JAK3, and Tyk2) upon receptor interaction [24,25]. However, flaviviral inhibition of JAK/STAT signaling by cytokines other than IFN remains unexplored. We examined flavivirus regulation of JAK/STAT signaling induced by IFNs β, γ, and $\lambda3$ as well as inflammatory (IL6), and also immune-regulatory (IL4, IL10) cytokines that utilize this pathway. We revealed the fact that flaviviruses broadly inhibit JAK/STAT signaling across STAT1-6 to disrupt responses to IFNs, as well as pro-inflammatory and immune-regulatory cytokines. Mechanistically, this broad antagonism of JAK/STAT signaling was mediated by flavivirus NS5 binding to host heat shock protein 90 (HSP90), leading to dysregulation of chaperone stabilization (and eventual subsequent degradation) of JAKs, as well as abrogation of cytokine-induced pY-STAT. Unlike previous reports of flavivirus interaction with other heat shock proteins [26–28], we showed that flaviviruses including ZIKV do not usurp chaperone activity to promote viral protein function and genome replication. Rather, flaviviruses appear to specifically target and disrupt HSP90 chaperone activity to suppress host JAK/STAT signaling, thus controlling the actions of a broad range of immunoregulatory and antiviral cytokines. Importantly, this broad-acting innate immune evasion strategy has never before been identified.

Thus, HSP90 is a key virus-targeted host factor, wherein interaction with viral NS5 prevents its interaction with client kinases, resulting in suppression of JAK/STAT-dependent cytokine signaling. NS5 targeting of HSP90 is thus the mechanism that underlies the disruption of broad cytokine signaling by flaviviruses.

2. Materials and Methods

2.1. Contact for Reagent and Resource Sharing

Further information and requests for resources and reagents should be directed to and will be fulfilled by the Lead Contact, Prof. Michael Gale Jr. (mgale@uw.edu).

2.2. Cell Lines

A549, Vero, HEK-293T, Huh-7 replicon-cured, PH5CH8 cells with interferon alpha receptor 1 knockout (IFNAR1$^{-/-}$), and murine embryonic fibroblast (MEF) cells were routinely cultured in complete Dulbecco's modified Eagle's medium (cDMEM; Corning, USA) supplemented with 10% (v/v) heat-inactivated fetal bovine serum (FBS; Hyclone, Logan, UT, USA), 2 mM L-glutamine,

1 mM sodium pyruvate, 10 mM 4-(2-hydroxyethyl)-1-piperazineethanesulfonic acid (HEPES), 1X antibiotic/antimycotic (Corning), and 1X non-essential amino acids (Corning, Corning, NY, USA). U251-MG cells were cultured in DMEM supplemented with 10% (v/v) FBS (Hyclone, Logan, UT, USA), 2 mM L-glutamine, 1.14 mM sodium pyruvate, 1X antibiotic/antimycotic (Corning), and 1X non-essential amino acids (Corning). THP-1 cells were cultured in complete Roswell Park Memorial Institute medium (cRPMI) supplemented with 10% (v/v) heat-inactivated fetal bovine serum (FBS; Hyclone, Logan, UT, USA), 2 mM L-glutamine, 1 mM sodium pyruvate, 10 mM HEPES, 1X antibiotic/antimycotic (Corning), and 1X non-essential amino acids (Corning). Prior to use in experiments, THP-1 cells were differentiated into macrophage-like cells via overnight stimulation with 40 nM phorbol myristate acetate (PMA) in cRPMI. Huh-7 WNrep cells were cultured in cDMEM supplemented with 1 mg/mL G418 to select for maintainence of the replicon. During cytokine stimulation, replicon cells were switched to cDMEM without G418 to standardize conditions between these and replicon-cured cells. All cell lines were confirmed as being free of mycoplasma contamination.

2.3. Virus Strains

Flaviviruses WNV-TX, WNV-MAD, ZIKV MR766, ZIKV FSS13025, ZIKV Fortaleza, JEV Nakayama, DENV-2 New Guinea C, and DENV-4 H241 were grown and titred on Vero cell monolayers in cDMEM. All infections were initiated using a low volume inoculum of virus in complete media (media dependent on cell type) incubated at 37 °C for 2 h with rocking. Inoculum was subsequently removed and replaced with fresh complete media. The encephalomyocarditis virus (EMCV) strain Mengo infectious clone was a kind gift from Ann C. Palmenberg (UW-Madison). Virus stocks were grown and titred on Vero cell monolayers in cDMEM. Sendai virus (SeV) Cantell strain was sourced from Charles River Laboratories (USA). All virus stocks were confirmed to be free of mycoplasma contamination.

2.4. Transcript Analysis by qRT-PCR

Cells were lysed and homogenized using QIAshredders (QIAGEN, Hilden, Germany), and total cellular RNA was isolated using a RNeasy Kit (QIAGEN) with DNase I (QIAGEN) digestion on column. Purified RNA was converted to complimentary DNA (cDNA) using the iScript cDNA Synthesis Kit (Bio-Rad, Hercules, CA, USA). The resulting cDNA was analyzed by qRT-PCR using SYBR Green Master Mix (ThermoFisher, Waltham, MA, USA) and gene-specific primers on the ABI 7500 Real-Time PCR System.

2.5. Protein Analysis by Western Blot

Cells were washed 1 x with phosphate buffered saline (PBS) and then lysed in radioimmunoprecipitation assay (RIPA) buffer (50 mM Tris pH 7.4, 150 mM NaCl, 1% (v/v) Triton X-100, 0.5% (w/v) sodium deoxycholate, 1:100 phosphatase inhibitor cocktail (VWR, Radnor, PA, USA), 1:100 protease inhibitor cocktail (Sigma, St. Louis, MO, USA), 250 nM okadaic acid) on ice. Lysates were scraped into microfuge tubes and immediately frozen at −80 °C. Lysates were subsequently thawed on ice for >30 min, then sonicated in a water bath ice slurry for 3 × 30 s bursts on a high setting with 2 × 20 s pauses between. Nuclei and other cellular debris were cleared from lysate via centrifugation at 14,000 rpm for 15 min at 4 °C. Protein content was quantified using a Bio-Rad Protein Assay (Bio-Rad), and samples were stored at −80 °C until required.

Prior to loading, lysates were incubated with 4X Laemmli buffer +/− β-mercaptoethanol at 95 °C for 3 min. Samples were loaded (≈7–14 µg total protein per lane) onto 4–20% Criterion TGX gels (Bio-Rad) and electophoresed at 97 V in Western running buffer (25 mM Tris, 192 mM glycine, 0.1% (w/v) SDS). Subsequently, proteins were transferred to nitrocellulose membranes at 90 V for 1 h under submerged conditions in Western transfer buffer (25 mM Tris, 192 mM glycine, 0.01% (w/v) SDS, 20% (v/v) methanol). Membranes were blocked for 1 h at RT in Tris-buffered saline (TBS)-based Odyssey blocking buffer (LI-COR, Lincoln, NE, USA), and subsequently stained overnight with primary antibody (see key resources table) in TBS-based Licor blocking buffer at 4 °C. The following day, membranes were washed 3x with TBS plus Tween 20 (TBST), and probed with secondary antibody (either horseradish peroxidase (HRP)-, Alexa680-, or Alexa790-conjugated antibody; key resources table) in TBS-based Odyssey blocking buffer for 1 h at RT. Finally, membranes were washed 3x with TBST and 2x with TBS prior to either direct imaging on an Odyssey CLx imager (LI-COR), or band development using enhanced chemiluminescence (ECL) prime Western blotting reagent (Fisher Scientific) and detection on a ChemiDoc XRS+ system (Bio-Rad).

2.6. Immunofluorescence Assays

Cells seeded on coverslips or chamber slides and used for experiments were washed 1x with PBS and were fixed with 4% paraformaldehyde in TBS for 30 min at room temperature. Following this, cells were washed with filtered 1X TBS and permeabilized (for pY-STAT staining, fixation was for 10 min at −20 °C with 100% ice-cold methanol and drying cells completely, followed by a further 3 washes with TBS; for all other samples, fixed cells were permeabilized via addition of 0.1% Triton X-100 to the blocking buffer, with permeabilization occurring during the blocking step). Blocking of cells was for 30 min at RT with filtered 5% normal goat serum (NGS) in TBS. Cells were then stained for 1 h at RT or overnight at 4 °C with primary antibody in 5% NGS/TBS. Subsequently, cells were washed 3x with TBS and probed for 1 h at RT with secondary antibody plus 1:10,000 DAPI (4′,6-diamidino-2-phenylindole dihydrochloride) in 5% NGS/TBS. Finally, cells were washed 3x with TBS; briefly washed 1x with deionized water (dH$_2$O) to remove excess salt; and carefully dried on the back of the coverslip, removing excess water. Coverslips were mounted onto slides using ProLong Gold mounting media and dried at least overnight at RT. Confocal immunofluorescence images were acquired on a Nikon Eclipse Ti microscope and analyzed using the NIS-Elements imaging system software (version 4.51) (Nikon Instruments, Tokyo, Japan). For some images, the red channel was modified equally via a saved look-up table (LUT) setting across all samples per experiment to aid visual clarity post-acquisition.

2.7. Conditioned Media Experiments

A549 cells were mock infected or infected with WNV-TX at multiplicity of infection (MOI) = 5 for 24 h. Cell-free culture media was harvested, divided into two equal volumes per condition, and half subjected to UV inactivation for 30 min in a Spectrolinker XL-1000 UV crosslinker (Spectronics Corporation, Westbury, NY, USA) at full power on ice. Normal and UV-inactivated conditioned media was then aliquoted and stored at −80 °C. Thawed conditioned media was subsequently used neat in the treatment of seeded A549 cells for 24 h prior to acute, 30 min cytokine treatment. Treated cells were lysed post-stimulation and analyzed via Western blot.

2.8. Generation of Recombinant DNA Constructs

Plasmids pCAGGS-HA and pCAGGS-WNV-NS4B were kindly provided by Adolfo Garcia-Sastre (Munoz-Jordan et al., 2005). Each gene from WNV strain TX02 was amplified from cDNA and cloned into the mammalian expression vector pCAGGS-HA in-frame with the 3′-HA tag. The 5′ end of amplified WNV genes included either EcoRI, NsiI, or SacI restriction sites followed by an AUG start codon while 3′ end contained KpnI or NsiI restriction sites for insertion into the multiple cloning site of pCAGGs-HA. Because the amino acid sequence of WNV strain NY99 NS4B from pCAGGS-WNV-NS4B (Munoz-Jordan et al., 2005) was identical to that of TX02 NS4B, the NY99-based construct pCAGGS-WNV-NS4B was used in these studies.

Plasmid pTwist-CMV-HSP90-HA was designed in-house and custom synthesized by Twist Bioscience (USA). The vector plasmid pTwist-CMV was created via excision of the HSP90-HA coding sequence with a Not-I/Nhe-I digestion, followed by blunting of the terminal ends via digestion with Mung Bean Nuclease (New England Biolabs, Ipswich, MA, USA), and finally blunt-end ligation. pTwist-CMV-FLAG-JAK1 was created via amplification of human JAK1 from a cDNA open reading frame (ORF) cloning vector (Sino Biological, Beijing, China) with gene-specific primers incorporating an N-terminal FLAG tag. This PCR product was then digested with Not-I/Nhe-I and ligated into the pTwist–CMV vector.

The vector pcDNA3.1(+) was purchased from Thermo Fisher Scientific (USA) and the plasmids pcDNA3.1-ZIKV(C)-FLAG and pcDNA3.1-ZIKV(NS5)-FLAG were the kind gift of Tom C. Hobman [29].

2.9. Transfection of Nucleic Acids

Plasmid DNA was transfected into cells using Lipofectamine 3000 (ThermoFisher) according to the manufacturer's instructions. A549 cells seeded into 24-well plates on coverslips for immunofluorescence assays were transfected with 1 µg/well DNA, and HEK-293T cells seeded into 6-well plates for co-immunoprecipitation (co-IP) assays were transfected with 5 µg/well total DNA (when co-transfected, this amount was split equally between plasmids; i.e., 2.5 µg/well each). The ratio DNA (µg)/lipofectamine 3000 (µL)/P3000 reagent (µL)/Opti-MEM (µL) was 1:3:2:50. Cell media was freshly changed prior to transfection. Plasmid DNA and P3000 were added to half the volume Opti-MEM (ThermoFisher) and mixed by inversion; lipofectamine 3000 was added to the other half of Opti-MEM and mixed. The two solutions were combined, mixed by inversion, and incubated 15 min at RT prior to drop-wise addition to cells.

2.10. Co-IP of Recombinant FLAG- and HA-tagged Proteins

HEK-293T cells seeded in 6-well plates and used for experiments were washed 1x with PBS and then lysed in 250 µL/well co-IP lysis buffer + inhibitors (50 mM Tris HCl pH 7.5, 250 mM NaCl, 5 mM ethylenediaminetetraacetic acid (EDTA), 0.02% (w/v) sodium azide, 1% (v/v) NP40, 1:100 phosphatase inhibitor cocktail (VWR), 1:100 protease inhibitor cocktail (Sigma), 250 nM okadaic acid) on ice. Lysates were scraped into new microcentrifuge tubes and incubated on ice for at least 30 min. Lysates were cleared of cellular debris via centrifugation at 14,000 rpm, 15 min, 4 °C, with supernatant transferred to fresh tube on ice and quantified as above. For each co-IP, 250 µg of each sample (in a 200 µL volume—excess lysis buffer was used to make volume) was incubated overnight at 4 °C in tubes on rotator with the following conditions. *For anti-HA*: 10 µL of thoroughly vortexed anti-HA magnetic beads (Cell Signaling Technology, Danvers, MA, USA) were added to each 200 µL sample, and were rotated overnight at 4°C. *For anti-FLAG*: Pre-binding of 1 µL anti-FLAG antibody (Sigma) to each 200 µL sample was performed on rotator at 4 °C for 15 min. Subsequently, 10 µL of vortexed protein G magnetic beads (ThermoFisher) were added to each 201 µL sample, and were rotated overnight at 4 °C.

The following day samples were immunoprecipitated using a magnet. Supernatant was discarded and 500 µL co-IP lysis buffer + inhibitors was added to each tube. These were briefly inverted to mix and rotated at 4 °C for 5 min. Magnetic precipitation and washing was repeated two more times (three washes total) using 300 µL for repeat washes.

Co-IP lysis buffer + inhibitors were used to dilute 4X Laemmli sample buffer (Bio-Rad) + β-mercaptoethanol to form a 1X sample buffer. A total of 50 µL of this buffer was added per sample as a final resuspension/elution. Samples were stored at −80 °C.

When loading gels, samples were incubated 3 min at 95 °C and then placed immediately on ice. Samples were spot centrifuged >10 s to pellet beads. Samples were then analyzed by Western blot as appropriate, with 10–15 µL co-immunoprecipitate loaded per lane.

2.11. ReCLIP Analysis of Endogenous Proteins

The protocol was modified from [30]. A549 cells infected with flaviviruses at MOI = 5 for 24 h were washed 2x with PBS prior to crosslinking proteins for 30 min at RT with 0.5 mM dithiobis(succinimidyl propionate) (DSP) in PBS. Crosslinked cells were subsequently quenched with incubation for 10 min at RT with TBS. Cells were then washed 1x with TBS and lysed on ice for 30 min with RIPA buffer followed by scraping into microcentrifuge tubes. Lysates were briefly sonicated in an ice-slurry bath (2 × 20 s pulses on the high setting with a 30 s pause). Lysate was cleared via centrifugation at 14,000 rpm for 15 min at 4 °C.

Lysate was pre-cleared by incubating 600 µg of total protein with 15 µL Protein G Dynabeads (ThermoFisher) in 1300 µL final volume for 30 min at 4 °C with rotation. For each IP, 150 µg of this cleared lysate was incubated with appropriate antibodies (2 µg mouse immunoglobulin G 2a (IgG2a) isotype control, 2 µg mouse monoclonal anti-HSP90α/β clone F-8, or 10 µL hybridoma supernatant mouse monoclonal anti-WNV NS5 clone 5D4) in 500 µL total volume overnight at 4 °C with rotation. The following day, 15 µL of Protein G Dynabeads (washed 1x with 500 µL RIPA buffer prior to use) were added to the lysates (0.45 mg total beads added) and incubated for 3 h at 4 °C with rotation. Samples were subsequently applied to a magnetic rack to precipitate beads and were washed 4x with 400 µL RIPA buffer for 10 min each at RT with rotation. Protein complexes were eluted from beads and DSP crosslinker reduced via incubation for 5 min at ≈95 °C in 35 µL loading buffer (three parts RIPA buffer to one part 4X Laemmli Sample Buffer + 10% β-mercaptoethanol (Bio-Rad)). Samples were then analyzed by Western blot as appropriate, with 10 µL immunoprecipitate loaded per lane.

2.12. Inhibition of HSP90 in Replicon Cells and During WNV and ZIKV Infection

Seeded Huh7 WNrep cells were treated with DMSO, or with 1 µM of either geldanamycin (GA; Cayman Chemical, reconstituted in DMSO), EC144 (Tocris, Bristol, UK; reconstituted in DMSO), or NITD008 (Tocris, reconstituted in DMSO) for 12 and 24 h prior to lysis for RNA and protein analysis, or for 24 h prior to fixation and staining for immunofluorescence assay.

For HSP90 inhibition during infectious WNV and ZIKV replication, seeded Vero cells were first infected with viruses at MOI = 5 for 12 h to establish logarithmic replication. Mock- or virus-infected cells were then treated with DMSO or 1 µM HSP90 inhibitors for a further 12 h prior to lysis and analysis via Western blot.

2.13. Proteasome Inhibition with MG-132

A549 cells were mock-infected or infected with WNV-TX at MOI = 5. At 18 h post-infection (hpi), cells were treated with either DMSO or 40 µM MG-132 (Sigma) for a further 6 h prior to lysis and analysis via Western blot.

2.14. Generation of CRISPR-Targeted IFNAR1$^{-/-}$ PH5CH8 Cells

For clustered regularly interspaced short palindromic repeats (CRISPR)/Cas9 targeting of IFNAR1, we generated the plasmid pRRL-MND-IFNAR1-2A-Puro by in-fusion cloning of the IFNAR1 genomic RNA (gRNA) sequence: 5′-GACCCTAGTGCTCGTCGCCGTGG-3′ into pRRL-MND-2A-Puro. PH5CH8 cells were transfected using the Amaxa 96-well Nucleofector Kit SF according to manufacturer's instructions. Then, 48h post-transfection, cells were selected in growth medium with 2 µg/mL of puromycin. Gene targeting was confirmed by T7 endonuclease I assay, loss of IFNAR1 cell surface expression by flow cytometry, and loss in the response to type I IFN treatment.

2.15. Generation of Huh7 WNrep Cells and Replicon-Cured Cells

BHK-21 cells harboring the WNV subgenomic replicon Rluc/NeoRep were kindly provided by Pei-Yong Shi (Lo et al., 2003). Rluc/NeoRep contains a dual reporter system and was derived from the parental WNV Replicon by replacing most of the structural gene region (nt 190 to 2379) with an in-frame Renilla luciferase (Rluc) reporter gene. A neomycin phospho-transferase (Neo) gene under the control of an EMCV IRES was placed just downstream from the NS5 gene in the 3′-untranslated region (UTR). Total RNA was isolated from these cells with TRIzol according to the manufacturer's instructions and stored at −80 °C. Rluc/NeoRep RNA was treated with DNaseI (Ambion, Austin, TX, USA). Huh7 cells were transfected with 1, 2, or 4 µg of DNaseI-treated Rluc/NeoRep RNA using Transmessenger Transfection Reagent (QIAGEN) according to the manufacturer's instructions. After 3 h, transfection mix was replaced with cDMEM for 16-20 h at 37 °C. Cells were then washed in 1X PBS, trypsinized, transferred to new plates in complete Dulbecco's Modified Eagle Medium (cDMEM), and allowed to recover for 48 h at 37 °C. After recovery, cDMEM was replaced with cDMEM containing G418 (G418 DMEM, 400 µg/mL) for selection of resistant colonies. Resistant colonies were transferred to new plates via colony selection discs and expanded in the presence of G418 (400 µg/mL). Approximately 11 weeks after RNA transfection, WNV Rluc/NeoRep protein expression was examined by Western blot analysis. Renilla luciferase activity of cells was also confirmed. Several distinct clones of WNV Rluc/NeoRep with variable protein and luciferase expression were recovered in Huh7 cells, with clone #8 used in these experiments.

To create a Huh7-derived control cell line for WNV Rluc/NeoRep clone #8, this line was subjected to curing with IFN. Briefly, the replicon cell line was passaged in the presence of pegylated IFNα-2b (PEG-INTRON, Schering, Berlin, Germany) at a concentration of 100 U/mL in cDMEM. After 11–25 days of maintenance in PEG-INTRON, cells were collected and analyzed for Renilla luciferase activity. When Renilla luciferase activity was below the background level of Huh7 control cells, the cured replicon cells were switched to cDMEM. Loss of the WNV replicon was confirmed by re-exposing the cells to lethal G418 to confirm death, and by RT-PCR analysis of RNA.

2.16. Quantification and Statistical Analysis

Statistical analyses were performed as indicated in figure legends, using GraphPad Prism version 8 software.

3. Results

3.1. WNV and ZIKV Broadly Inhibited JAK/STAT Signaling Following Cytokine Stimulation of Target Cells

To determine the breadth of flavivirus-mediated inhibition of JAK/STAT signaling, we first assessed acute responses to JAK/STAT-dependent cytokine treatment in cells infected with WNV or ZIKV. We conducted immunoblot analyses of cytokine-induced pY-STAT responses to IFNβ for all six STAT family members in WNV- and ZIKV-infected A549 (human lung epithelial) cells (Figure 1A,B). Mock infected cells responded to IFNβ with pY of all STATs by 30 min post-treatment. In contrast, WNV- and ZIKV-infected cells were unable to mediate pY-STATs in response to high dose IFNβ (Figure 1A,B). Immunofluorescence revealed nuclear pY-STAT1 did not accumulate in response to IFNγ in WNV- and ZIKV-infected A549 cells, though pY-STAT1 did accumulate in adjacent non-infected bystander cells (Figure 1C,D). The attenuated response to IFNγ-induced pY-STAT1 was confirmed by immunoblot of WNV-infected cells (Figure S1A). WNV-infected cells also showed inhibited pY-STAT1 and 2 in response to IFNλ3 (Figure 1E). Moreover, WNV and ZIKV infection each led to inhibition of pY-STAT1 and 3 in response to human and mouse IL6 (Figure 1F,G and Figure S1B,C), and of pY-STAT6 following IL4 treatment (Figure 1H,I). Immunofluorescence also showed that WNV-infected THP-1 macrophages failed to accumulate pY STAT1 and 3 in response to IFNβ and IL10, respectively (Figure S1D,E). Thus, WNV and ZIKV impose a broad blockade to pY-STAT mediated by several unrelated cytokines, implying one or more common features of JAK/STAT signaling could be dysregulated during flavivirus infection. Of note, background levels of pY-STAT1 and total levels of STAT1 and STAT2 were observed as being increased following virus infection in several analyses (Figure 1A,B,E,F and Figure S1A). These observations were fully expected in IFN-competent A549 cells, as IFN signaling is only blocked by flaviviruses after approximately 20–24 hpi [17]. Early antiviral signaling would then be expected induce some pY-STAT1 and ISGs (including total STAT1 and STAT2); importantly, however, neither of these were further increased upon exogenous cytokine stimulation, indicative of the virus-imposed JAK/STAT inhibition by 24hpi.

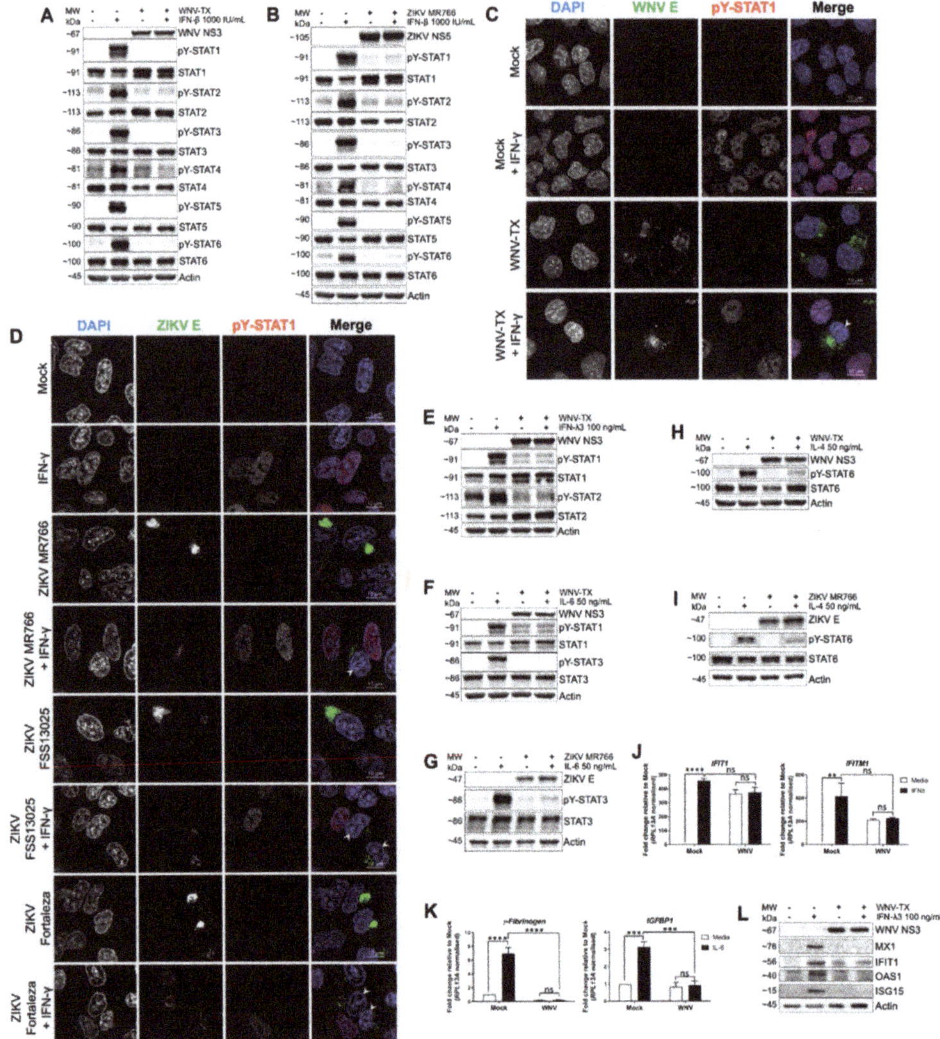

Figure 1. West Nile virus (WNV) and *Zika virus* (ZIKV) infections inhibit pY-STAT and gene expression in response to multiple cytokines. pY-STAT responses were inhibited in WNV- and ZIKV-infected A549 cells at 30 min post-treatment with (**A,B**) inferno (IFN)β, (**C,D**) 10 ng/mL IFNγ (arrow heads), (**E**) IFNλ3, (**F,G**) interleukin (IL)6, and (**H,I**) IL4. WNV-infected A549 cells did not respond to 17 h treatment with (**J**) IFNβ (*IFITM1* and *IFIT1* induction) or (**K**) IL6 (*γFibrinogen* and *IGFBP1* induction). (**L**) WNV-infected PH5CH8 IFNAR1$^{-/-}$ cells did not respond to 24 h IFNλ3 treatment via expression of the interferon stimulated genes MX1, IFIT1, OAS1, and ISG15. All infections were MOI = 5 for 24 h prior to cytokine treatment unless otherwise stated. All data represent three independent experiments or are a combination of three experiments (mean ± standard error of the mean (SEM)). See also Figure S1. **** = $p < 0.0001$, *** = $p < 0.001$, ns = $p > 0.05$. Two-way ANOVA with Tukey's posttests.

This blockade to pY-STAT corresponded to inhibition of cytokine-responsive gene induction. Analysis of endogenous interferon-induced protein with tetratricopeptide repeats 1 (*IFIT1*) and interferon-induced transmembrane protein 1 (*IFITM1*) expression in cells responding to IFNβ (Figure 1J) showed that WNV blocked induction of these ISGs in response to the cytokine (virus-induced ISG mRNA remaining from early antiviral signaling prior to establishment of JAK/STAT inhibition was not further increased). Additionally, WNV blocked induction of *γFibrinogen* and insulin-like growth factor-binding protein 1 (*IGFBP1*) in response to IL6 (Figure 1K). To assess the response of WNV-infected cells to IFNλ3, we treated human epithelial cells (hepatocytes) specifically lacking functional IFNα/β receptor expression (PH5CH8 IFNAR1$^{-/-}$ cells) and measured expression of ISG proteins MX dynamin like GTPase 1 (MX1), IFIT1, 2′5′-oligoadenylate synthetase 1 (OAS1), and ISG15. WNV infection abrogated induction of each ISG by IFNλ3 (Figure 1L).

Importantly, NF-κB signaling in response to IL1β treatment in WNV-infected A549 cells was intact, with comparable cytokine-responsive inhibitor of κB alpha (IκBα) degradation and pS-p65 induction (Figure S1F). Upstream signaling components in this non-JAK/STAT pathway were also largely unaffected by infection (Figure S1G).

3.2. Cell-Intrinsic WNV NS Proteins mediated Broad JAK/STAT Inhibition

To define which factors produced during WNV infection contributed to JAK/STAT signaling inhibition, we examined the virus replication cycle using two complementary approaches (Figure 2A–F). Huh7 hepatoma cells stably harboring a WNV replicon [31] (Figure 2A) were used to assess the JAK/STAT pathway in the context of NS protein expression and viral RNA replication, but in the absence of structural proteins. Treatment with IFNλ3 (Figure 2B) showed pY-STAT1 was reduced in WNV replicon cells compared with cells cured of replicon. Similarly, pY-STAT3 abundance was diminished in IL6-treated WNV replicon cells compared to cured controls (Figure 2C).

To assess virion impact on JAK/STAT signaling, we exposed cells to UV-inactivated WNV present in conditioned culture supernatant (Figure 2D). This approach allowed us to assess the JAK/STAT inhibitory role of viral factors such as virions, subviral particles, and NS1 protein released from infected cells [32], as well as host cytokines and metabolites produced in response to infection. Cells were first exposed to UV-inactivated WNV-conditioned media and then treated with either IL6 (Figure 2E) or IL4 (Figure 2F). Immunoblot analysis showed that cytokine treatment induced pY-STAT3 and 6 to levels comparable with control cells treated with conditioned media from mock infections. By comparison, cells exposed to supernatants containing live WNV exhibited a block of cytokine-induced pY-STAT (Figure 2E,F). Thus, extracellular factors secreted from WNV-infected cells are dispensable for the JAK/STAT signaling blockade, whereas one or more viral NS proteins are likely responsible for the broad block to pY-STAT.

To identify viral NS protein(s) responsible for JAK/STAT signaling blockade, we independently expressed each NS protein in cells and assessed their capacity to inhibit pY-STAT3 in response to IL6 (Figure 2G,I) and pY-STAT1 in response to IFNγ (Figure 2H,I) by immunofluorescence assay. As controls, we also assessed pY-STAT in cells expressing WNV capsid protein or vector alone. Expression of viral proteins themselves did not induce pY-STAT in resting cells without cytokine treatment (Figure S2). We found that among WNV NS proteins, expression of NS5 caused significant inhibition of pY-STAT accumulation in cells treated with either cytokine. Expression of NS2A, NS2B, and NS2B/3 showed significant inhibition of pY-STAT1 following treatment with IFNγ only. Expression of any individual NS protein was unable to block JAK/STAT signaling as efficiently as live WNV, suggesting viral NS proteins likely act in a complementary manner to confer broad JAK/STAT antagonism.

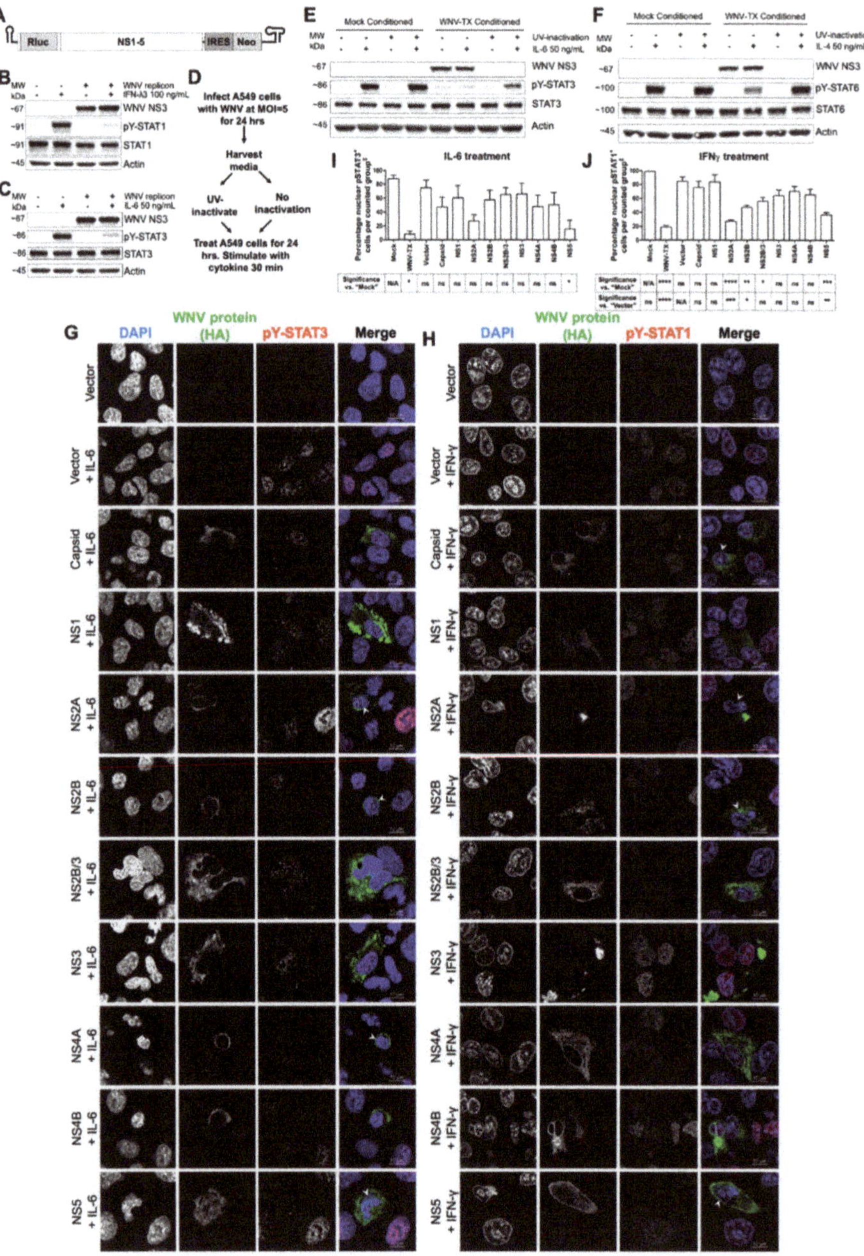

Figure 2. WNV nonstructural proteins were found to be responsible for Janus kinase (JAK)/STAT signaling inhibition intrinsically within infected cells. (**A**) Huh7 WNV replicon cells showed low pY-STAT responses to (**B**) IFNλ3 and (**C**) IL6 compared to cured controls. (**D**) Methodology to generate infectious and non-infectious WNV-conditioned media. Blockade of pY-STAT in response to (**E**) IL6 and (**F**) IL4 was only observed with infectious conditioned media. Recombinant expression of WNV C and nonstructural (NS) proteins in A549 cells showed differential inhibition of pY-STAT responses to (**G,I**) 200 ng/mL IL6 and (**H,J**) 10 ng/mL IFNγ (arrow heads), with WNV NS5 significantly responsible

for JAK/STAT inhibition across both cytokine treatments. All cytokine treatments were for 30 min. All data represent three independent experiments or are a combination of three experiments (mean ± SEM) except "Vector" in (**I**), and "Mock" and "Vector" in (**J**) ($n = 2$). See also Figure S2. **** = $p < 0.0001$, *** = $p < 0.001$, ** = $p < 0.01$, * = $p < 0.05$, ns = $p > 0.05$. One-way ANOVA with Tukey's posttests. ‡ "Counted group" = total cell population in several fields per experiment as measured by 4′,6-diamidino-2-phenylindole dihydrochloride (DAPI) for "Mock" and "Vector", all WNV E-positive cells for "WNV", and all hemagglutinin epitope tag (HA)-positive cells for WNV proteins. The red pY-STAT3 channel for images in (**G**) was enhanced equally post-acquisition across all samples to aid visual clarity.

3.3. Flavivirus Infection Lead to JAK and Cytokine Receptor Degradation by the Proteasome

Considering the fact that the broad inhibition of pY-STAT responses and gene induction by WNV and ZIKV appeared to be unrelated to the identity of the cytokine used to stimulate cells (Figure 1), we sought to determine whether infection with these viruses affected JAK proteins proximal to these events in the JAK/STAT signaling cascade. Immunoblot of WNV-infected A549 cells treated under control conditions (DMSO) demonstrated that at 24 hpi, all JAK family members (JAK1-3 and Tyk2) had reduced protein abundance compared to mock-infected cells. This loss of abundance was partially rescued by treatment with the proteasome inhibitor MG-132 for 6 h, starting at 18 hpi. MG132 led to increased JAK family protein abundance in WNV-infected cells to levels similar to mock, DMSO-treated cells (Figure 3A,B). Time course analyses demonstrated that loss of JAK1 abundance could be observed from approximately 16 hpi (Figure S1H). To assess whether JAK loss was in part due to RNA regulation, we quantified *JAK1*, *JAK2*, *JAK3*, and *Tyk2* transcripts within mock- and WNV-infected cells at 24 hpi (Figure 3C). WNV infection led to induction of JAK mRNA, and thus loss of JAK abundance is likely mediated solely by post-transcriptional processes.

To determine if loss of JAK family protein abundance was a phenotype extending beyond the virulent WNV strain we predominantly used, we evaluated the abundance of JAK1 and Tyk2 in A549 cells infected with a panel of flaviviruses, as well as control viruses including Sendai virus (SeV) and encephalomyocarditis virus (EMCV), at 36 hpi (Figure 3D,E). Compared to mock-infected cells, infection with WNV-TX, WNV-MAD, ZIKV MR766, JEV, and DENV-2 all showed a trend of variably decreased JAK1 and Tyk2 abundance (Figure 3D). Quantification of band intensity across three independent experiments demonstrated that significant decreases in both JAK1 and Tyk2 occurred during WNV-TX and ZIKV MR766 infection (Figure 3E), whereas only decrease in Tyk2 abundance was significant for JEV. Despite the trend towards decreased abundance of these JAK family proteins, quantification from WNV-MAD and DENV-2 infection (both viruses that replicate less robustly in A549 cells) demonstrated that these decreases were not significant. EMCV infection (which shuts down host translation) [33] led to significant loss of JAK1 and Tyk2. In contrast, SeV infection displayed only slight alteration in abundance of these proteins. Therefore, loss of JAK protein abundance is a feature common to several flavivirus infections.

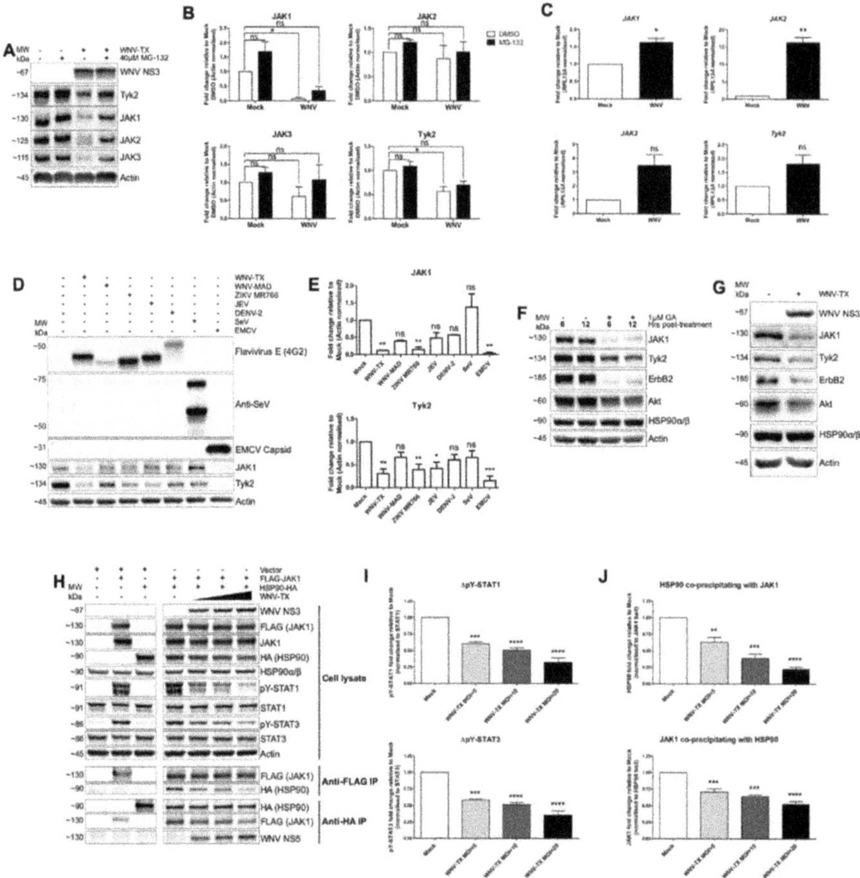

Figure 3. Flavivirus infection led to loss of abundance of heat shock protein 90 (HSP90) client kinases including JAK proteins, with interaction between HSP90 and JAK1 disrupted. (**A,B**) JAK family proteins are decreased in abundance during WNV infection. Treatment with MG-132 at 18 h post-infection (hpi) partially recovered JAK abundance in WNV-infected cells. (**C**) Transcripts of JAK1, JAK2, JAK3, and Tyk2 increased during WNV infection. (**D,E**) Infection of A549 cells with WNV-TX, ZIKV MR766, Japanese encephalitis virus (JEV), and encephalomyocarditis virus (EMCV) led to loss of Tyk2 and (for all except JEV) JAK1 by 36 hpi. Infection with less robustly infecting flaviviruses WNV-MAD and dengue virus (DENV)-2, as well as with Sendai virus (SeV), did not lead to significant protein loss. (**F**) Treatment of A549 cells with HSP90 inhibitor geldanamycin (GA) for 6 and 12 h diminished abundance of JAK1, Tyk2, erythroblastosis oncogene B2 (ErbB2), and AKR mouse strain thymoma-related protein (Akt). (**G**) HSP90 client kinases ErbB2 and Akt displayed decreased abundance in WNV-infected A549 cells in a similar manner to JAK1 and Tyk2 by 24 hpi. (**H–J**) HEK-293T cells mock- or WNV-infected (MOI = 5, 10, and 20) were co-transfected with FLAG-epitope tagged JAK1 (FLAG-JAK1) and HSP90-HA. (**H,I**) Transfection of JAK1 was sufficient for pY-STAT1 and three responses, and this pY-STAT was significantly blocked with increasing WNV. (**J**) Reciprocal HA- and FLAG-tagged co-immunoprecipitation (co-IP) showed significantly decreased interaction of co-precipitated HSP90 and JAK1 with increasing WNV. All infections were in A549 cells at MOI = 5 for 24 h prior to cytokine treatment unless stated, except SeV which was infected at 40 hemagglutination units per well. All data represent three independent experiments or are a combination of three experiments (mean ± SEM), except (**G**) ($n = 2$). **** $= p < 0.0001$, *** $= p < 0.001$, ** $= p < 0.01$, * $= p < 0.05$, ns $= p > 0.05$. Paired t-tests (**C**). One-way ANOVA with Dunnett's Multiple Comparison post-test (**I, J**). Irrelevant lanes were cropped between controls and infected samples in (**H**).

3.4. NS5 Disrupted HSP90-Client Kinase Interaction to Block JAK/STAT Signaling

To identify the mechanism WNV NS5 uses to drive JAK/STAT inhibition, we evaluated cellular processes known to regulate JAK stability and cytokine-responsive pY-STAT. As HSP90 is required to mediate the correct kinase-active conformational folding and stability of JAK proteins (and other specific client kinases) [34,35], we reasoned that viral dysregulation of HSP90 might confer reduced abundance of JAKs and other HSP90 clients within infected cells. Indeed, we found that chemical inhibition of HSP90 activity with geldanamycin (GA) led to loss in abundance of JAK1 and Tyk2, as well as other HSP90 client kinases Akt and ErbB2 at 6 and 12 h post-treatment (Figure 3F). Importantly, we were able to see an analogous loss in abundance in all four of these HSP90 client kinases at 24 hpi with WNV (Figure 3G), directly phenocopying chemical HSP90 inhibition. Thus, viral dysregulation of HSP90 might be responsible for the broad block to pY-STAT during WNV infection.

We directly assessed the effect of WNV infection on the interaction of HSP90 with JAK1 using overexpression of epitope-tagged proteins and co-immunoprecipitation (co-IP) (Figure 3H,J) from virus infected cells. HEK-293T cells were mock- or WNV-infected at MOI = 5, 10, and 20. At 2 hpi, cells were co-transfected with plasmids encoding FLAG-JAK1 and HA-HSP90α (with vector controls). Immunoblot of whole cell lysate (Figure 3H) showed that transfection of FLAG-JAK1 was sufficient to drive ligand- and receptor-independent pY-STAT1 and 3 by 24 hpi. Increased WNV MOI led to significantly decreased levels of pY-STAT (Figure 3I), without loss in abundance of ectopic, overexpressed JAK1. These results demonstrated a fundamental point—degradation of JAK1 is not required per se for pY-STAT inhibition. Degradation presumably occurs subsequent to the mechanism rendering kinase activity of JAKs defective.

Reciprocal co-IP of JAK1 and HSP90 (Figure 3J) revealed the mechanism leading to kinase-inactive JAK proteins. Increasing infection with WNV led to significantly decreased interaction between JAK1 and HSP90. Presumably, this diminished interaction prevented HSP90 from properly chaperoning JAK1 to promote an active conformational fold. Importantly, we found this decline in JAK1-HSP90 interaction occurred concomitant with increasing interaction between HSP90 and viral NS5 (Figure 3J). Thus, the precise mechanism of pY-STAT inhibition involves WNV NS5 targeting of HSP90 to disrupt its interaction with JAKs, preventing the JAK activity that drives pY-STAT (Figure 3H,I), and leading eventually to JAK proteasomal degradation (see Figure 3A,B,D,G).

3.5. HSP90 Antagonism Was Found to be Linked to Interaction with Viral NS5 at Sites of RNA Replication

To examine the context of virus-induced changes to HSP90 function, we examined localization of this chaperone within flavivirus-infected A549 cells at 24 hpi. Immunofluorescence showed HSP90 co-localized with or is adjacent to double-stranded RNA (dsRNA) within replication complexes of WNV-TX, WNV-MAD, JEV, ZIKV MR766, ZIKV Fortaleza, DENV-2, and DENV-4 (Figure 4A). High-resolution imaging showed both isoforms HSP90α and HSP90β co-localized with WNV NS5 and dsRNA (Figure 4B,C and Figure S3A,B).

We conducted analyses to interrogate the interaction of endogenous HSP90 and NS5 in WNV- and ZIKV-infected cells using reversible cross-linked IP (ReCLIP). Endogenous HSP90 bound to virus-produced WNV NS5 in reciprocal co-IP analyses (Figure 4D). Similarly, ZIKV NS5 was found in complex with endogenous HSP90 in infected cells (Figure 4E). The HSP90–NS5 interaction was specifically recovered by anti-NS5 or anti-HSP90, as the NS5-HSP90 complex was not recovered using isotype control IgG2a antibody. We confirmed specific interaction of ZIKV NS5 with HSP90 via further analysis of FLAG-tagged ZIKV NS5 in HEK-293T cells and co-IP of HSP90 (Figure S3C). ZIKV NS5 could co-IP endogenous HSP90, however, neither vector nor ZIKV capsid protein conferred HSP90 interaction.

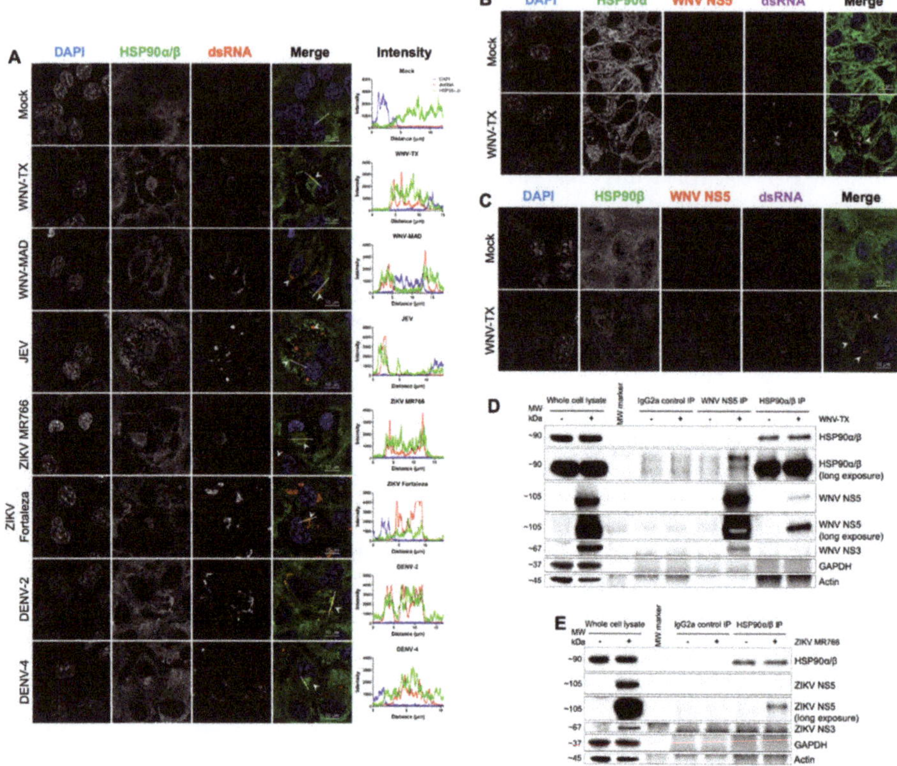

Figure 4. Flavivirus NS5 interacted with HSP90 at sites of viral RNA replication. (**A**) HSP90 co-localized with double-stranded RNA (dsRNA) at 24 hpi with WNV-TX, WNV-MAD, JEV, ZIKV MR766, ZIKV Fortaleza, DENV-2, and DENV-4 (arrow heads). WNV NS5 co-localized with both (**B**) HSP90α and (**C**) HSP90β and dsRNA in infected A549 cells (arrow heads). (**D**) ReCLIP experiments with WNV-infected A549 cells showed NS5 co-precipitation with HSP90, and HSP90 was recovered following NS5 IP. Controls showed NS3 was detected following NS5 IP (replication complexes), and neither glyceraldehyde 3-phosphate dehydrogenase (GAPDH) nor Actin were co-precipitated under any condition. The IgG2a isotype control failed to IP any analyzed protein. (**E**) ReCLIP analysis of ZIKV-infected cells showed an analogous NS5–HSP90 interaction. All infections were at MOI = 5 for 24 h (except DENV-4 where MOI = 3). All data represent three independent experiments, except (**E**) (*n* = 2).

3.6. HSP90 Activity Was not Required to Support Flavivirus RNA Replication nor to Stabilize Viral Proteins

We considered the possibility that flaviviruses may usurp HSP90 to promote viral RNA replication and viral protein function, as has been reported for flavivirus interactions with HSP70 [26–28]. To address this notion, we assessed the effect of chemical HSP90 inhibition upon the abundance of viral RNA and proteins, and on dsRNA NS protein localization within Huh7 WNV replicon cells. Compared to DMSO-treated control cells, WNV replicon cells 24 hpt with HSP90 inhibitors GA and EC144 showed stable co-localization of NS3 and NS5 by immunofluorescence analysis (Figure S4A). However, upon HSP90 inhibition in treated cells, NS1 and dsRNA shifted from co-localization with the NS proteins to distribute as bright foci directly adjacent to NS3 and NS5 (Figure S4A,B). Changes in abundance of replicon antigens was undetectable under conditions of HSP90 inhibition in treated cells. In contrast, a decrease in intensity was observed for controls treated with the flavivirus replication inhibitor NITD008.

Analysis of replicon genomic RNA (WNrep gRNA; Figure 5A) showed HSP90 inhibition via EC144 treatment significantly increased gRNA abundance by 24 h. In contrast, NITD008 treatment led to decreased gRNA. Analysis of HSP70A mRNA showed significant induction of this chaperone upon HSP90 inhibition as expected [36], but not following NITD008 treatment.

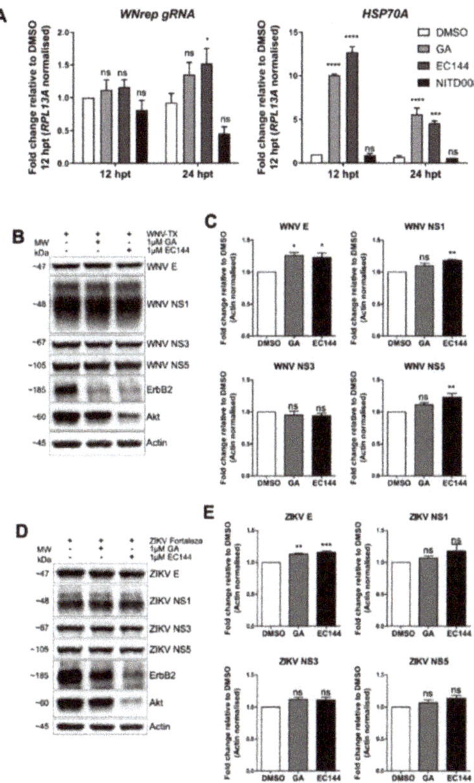

Figure 5. Flavivirus RNA replication and protein stability did not require HSP90 activity. (**A**) Huh7 WNV replicon cells were treated with HSP90 inhibitors GA and EC144, or nucleoside analogue NITD008 for 12 and 24 h to assess effects on RNA replication. Analysis of WNV replicon RNA (WNrep genomic RNA (gRNA)) revealed increased abundance following HSP90 inhibitor treatment, in contrast to reduced gRNA following NITD008. This was coincident with significant induction of HSP70A mRNA following HSP90 inhibitor treatment. Vero cells 12 hpi with (**B,C**) WNV or (**D,E**) ZIKV at MOI = 5 were treated with DMSO or HSP90 inhibitors for a further 12 h. (**B,D**) HSP90 inhibitors reduced Akt and ErbB2 abundance, and changes to viral proteins were not detected. Quantities of (**C**) WNV E, NS1, and NS5, as well as (**E**) ZIKV E were moderately yet significantly increased upon HSP90 inhibition. All data represent three independent experiments or are a combination of three experiments (mean ± SEM), with the exception of HSP70A mRNA quantification in (**C**) ($n = 2$). **** = $p < 0.0001$, *** = $p < 0.001$, ** = $p < 0.01$, * = $p < 0.05$, ns = $p > 0.05$. Two-way repeated measures ANOVA with Holm–Sidak's post-tests (**A**), and one-way ANOVA with Dunnett's multiple comparison post-tests (**C,E**).

We also measured the abundance of WNV replicon NS proteins at 12 and 24 hpt by immunoblot analysis (Figure S4C,D). Though reduced abundance of Akt and ErbB2 were found upon GA and EC144 treatment, neither inhibitor caused a decrease in replicon NS proteins. This contrasts with significant 30–50% reduction of NS proteins with NITD008. We also examined the impact of HSP90 inhibition upon protein abundance during live WNV and ZIKV infection. Vero cells 12 hpi were treated with DMSO, GA, or EC144 for a further 12 h. For both viruses, HSP90 inhibitors decreased Akt and ErbB2 abundance, though little change to viral proteins was detected (Figure 5B,D). Quantification of band intensity showed that WNV and ZIKV protein levels did not decrease during HSP90 inhibition (Figure 5C,E). Indeed, WNV E, NS1, NS5, and ZIKV E were significantly increased upon HSP90 inhibitor treatment, as similarly shown for DENV infection [37]. Taken together, these observations indicate that flavivirus replication is not dependent upon HSP90 activity.

4. Discussion

Our detailed pathway analyses revealed the fact that flavivirus-directed JAK/STAT signaling inhibition extends far beyond classical IFN responses, with antagonism of this pathway found in response to every cytokine we investigated. We proposed a mechanistic model of flavivirus JAK/STAT antagonism in which flavivirus NS5 (possibly in conjunction with other NS proteins) interacts with host HSP90 at the site of viral RNA replication. As a consequence of NS5 interaction, the HSP90-kinase client homeostasis is disrupted, leading to inappropriate chaperoning of JAK proteins, loss of their activity, protein kinase instability, and subsequent degradation via the proteasome. This process reduces abundance of JAKs, with the remaining pool of kinases lacking chaperone-supported enzymatic activity. Inactive JAKs cannot transmit pY to STAT proteins, and thus infected cells become refractory to cytokines that signal through the JAK/STAT pathway, including IFNs, proinflammatory cytokines, and immune regulatory cytokines.

HSP90 has been well described as a regulator of JAK activity and stability [34,35,38]. Inhibition of HSP90 leads to degradation of JAK1 and JAK2 to attenuate the pY-STAT1 and 2 response to IFNβ and IFNγ [34]. In Hodgkin lymphoma cells, inhibition of HSP90 led to loss of JAK1, JAK3, and Tyk2, and constitutive pY of STAT1, 3, 5, and 6 was ablated [35]. Thus, HSP90 inhibitors phenocopy JAK/STAT antagonism by flaviviruses, a similarity of function that we confirmed with side-by-side analyses. Our data revealed for the first time that flavivirus NS5 binds HSP90 to mediate broad STAT inhibition, which extends upon descriptive reports from other groups that identified JEV, DENV, and ZIKV NS5 interaction with HSP90, but did not define the relevance of this interaction [39–41]. Several DENV proteins have also been shown to bind HSP90 [37]. Our demonstration of JAK destabilization as a consequence of HSP90–NS5 interaction provides important functional insight into viral evasion of host innate immunity and inflammatory signaling. Moreover, we showed that Akt and ErbB2 levels decrease during flavivirus infection as a result of NS5-HSP90 interaction, implying that processes mediated by these and other HSP90-client kinases are likely altered during infection, possibly contributing to viral pathogenesis.

We showed that HSP90 chaperone activity is not required for flavivirus replication or protein stability but that HSP90 inhibition leads to slight enhancement of infection. This contrasts with a recent report showing activity of grp94 (an endoplasmic reticulum-resident paralog of HSP90) is required for DENV and ZIKV infection [42]. Vero cells were specifically chosen for these experiments as they lack IFN production [43,44], and thus the moderately enhanced viral growth that we observed was unlikely due to further dysregulation of antiviral signaling. Rather, we propose that induction of HSP70 likely contributes to the greater WNV and ZIKV replication seen here upon HSP90 inhibition, as HSP70 is known to enhance flavivirus replication [26–28]. Thus, flavivirus interaction with HSP90 may represent a common strategy among flaviviruses to dysregulate and suppress host IFN antiviral signaling rather than directly supporting viral replication and protein function. The broad inhibition of pY-STAT resulting from the NS5-HSP90 interaction then imparts collateral dysregulation of IFN-independent cytokine actions.

The remarkable breadth of cytokine signaling inhibition imposed by the NS5–HSP90 interaction may have consequences for flavivirus disease phenotypes outside of the context of innate antiviral defenses. In the context of an acute immune response, flavivirus-infected myeloid cells (important targets of infection) [13,45,46] would be less responsive to anti-inflammatory IL10 and IL4 signaling through STAT3 and 6, respectively, perhaps altering the immune phenotype polarization of myeloid cells and enhancing immune-mediated tissue damage. Additionally, reduced response of flavivirus-infected antigen-presenting cells to IFNγ through pY-STAT1 may affect adaptive immunity priming actions to alter T cell-mediated immunity.

Our findings of HSP90 interaction share similarities with recent reports on involvement of HSP70 in flavivirus infection and host inflammatory signaling [26–28], suggesting chaperones are key regulators of pathogenesis. Flaviviruses productively infect and transmit between diverse hosts from arthropods (ticks and mosquitoes) to vertebrates (birds and mammals) [47]. Hence, flaviviruses must exploit diverse systems to promote replication; a setting that may favor the targeting of factors conserved between reservoirs and vectors. HSP70 and HSP90 are highly conserved between these organisms, both in structure and function [48,49], and thus it is feasible that strategies developed to exploit these chaperones could be viable across diverse hosts. Indeed HSP70 inhibitors antagonize flavivirus replication in both insect and mammalian cells [26,27], and JAK/STAT inhibition by flaviviruses in mosquito cells is dependent on NS5 and involves proteasomal activity [50]. Thus, we propose flaviviruses target HSPs to facilitate infection in diverse hosts, with HSP70 and grp94 [42] activity usurped for viral replication and HSP90 activity disrupted to antagonize innate immune defense. This division of functions regarding flavivirus interactions with cytoplasmic HSPs (i.e., interaction with HSP70 to promote viral protein folding vs. targeting of HSP90 to disrupt host immunity) is striking. These chaperones normally operate in close cooperation with one another [49], and thus this distinct separation of roles warrants further investigation to discern the consequences of this division for host cell function. Whether a HSP antagonism strategy is utilized to promote pathogenesis of other virus groups also remains to be investigated. As chaperones support so many cellular functions, the unforeseen medical consequences of viral HSP antagonism are likely substantial.

Supplementary Materials: The following are available online at http://www.mdpi.com/2073-4409/9/4/899/s1: Figure S1: WNV infection inhibited STAT phosphorylation in response to multiple cytokines. Figure S2: Recombinant WNV protein expression did not induce pY-STAT3 or pY-STAT1. Figure S3: HSP90 interacted with NS5 and was localized to sites of viral dsRNA. Figure S4: HSP90 inhibition did not perturb WNV replication. Table S1: Reagents and resources used.

Author Contributions: J.A.R. conceived and performed experiments, wrote the manuscript, and secured funding. K.E.-N., E.C.D.-V., M.R.F., L.D.H., E.A.H., J.S., A.Y.L, F.W.S., B.C.K., and A.S. contributed to the design and performance of experiments. A.F., J.E.S.-B., and R.S. provided reagents, expertise, and feedback. M.G.J. conceived experiments, secured funding, and provided supervision. All authors edited and approved the final manuscript.

Funding: This research was funded by National Institutes of Health grants T32AI007509, AI145296, AI143265, AI104002, AI100625, AI127463, and AHA Award 17POST33660907 (EAH). A Perkins Coie Award for Discovery supported J.A.R.

Acknowledgments: In addition to generous gifts acknowledged in Table S1, the authors thank the members of University of Washington Immunology and the Center for Innate Immunity and Immune Disease for their helpful discussions and insight throughout this study.

Conflicts of Interest: The authors declare no conflict of interest.

References

1. Guzman, M.G.; Gubler, D.J.; Izquierdo, A.; Martinez, E.; Halstead, S. Dengue infection. *Nat. Rev. Dis. Prim.* **2016**, *2*, 16055. [CrossRef] [PubMed]

2. Roehrig, J. West Nile Virus in the United States—A Historical Perspective. *Viruses* **2013**, *5*, 3088–3108. [CrossRef] [PubMed]

3. Faria, N.R.; Azevedo, R.D.S.D.S.; Kraemer, M.U.G.; Souza, R.; Cunha, M.S.; Hill, S.C.; Theze, J.; Bonsall, M.B.; Bowden, T.A.; Rissanen, I.; et al. Zika virus in the Americas: Early epidemiological and genetic findings. *Science* **2016**, *352*, 345–349. [CrossRef] [PubMed]

4. LaZear, H.M.; Diamond, M.S. Zika Virus: New Clinical Syndromes and Its Emergence in the Western Hemisphere. *J. Virol.* **2016**, *90*, 4864–4875. [CrossRef]

5. França, G.V.A.; Schuler-Faccini, L.; Oliveira, W.; Henriques, C.M.P.; Carmo, E.H.; Pedi, V.D.; Nunes, M.L.; De Castro, M.C.; Serruya, S.; Silveira, M.F.; et al. Congenital Zika virus syndrome in Brazil: A case series of the first 1501 livebirths with complete investigation. *Lancet* **2016**, *388*, 891–897. [CrossRef]

6. Melo, A.S.D.O.; Aguiar, R.; Ribeiro, S.T.C.; Batista, A.G.M.; Ferreira, T.; Sampaio, V.V.; Moura, S.R.M.; Rabello, L.P.; Gonzaga, C.E.; Ximenes, R.; et al. Congenital Zika Virus Infection: Beyond Neonatal Microcephaly. *JAMA Neurol.* **2016**, *73*, 1407–1416. [CrossRef]

7. Walker, C.L.; Little, M.-T.E.; Roby, J.A.; Armistead, B.; Gale, M.; Rajagopal, L.; Nelson, B.R.; Ehinger, N.; Mason, B.; Nayeri, U.; et al. Zika virus and the nonmicrocephalic fetus: Why we should still worry. *Am. J. Obstet. Gynecol.* **2019**, *220*, 45–56. [CrossRef]

8. Waldorf, K.A.; Stencel-Baerenwald, J.E.; Kapur, R.P.; Studholme, C.; Boldenow, E.; Vornhagen, J.; Baldessari, A.; Dighe, M.K.; Thiel, J.; Merillat, S.; et al. Fetal brain lesions after subcutaneous inoculation of Zika virus in a pregnant nonhuman primate. *Nat. Med.* **2016**, *22*, 1256–1259. [CrossRef]

9. Waldorf, K.A.; Nelson, B.R.; Stencel-Baerenwald, J.E.; Studholme, C.; Kapur, R.P.; Armistead, B.; Walker, C.L.; Merillat, S.; Vornhagen, J.; Tisoncik-Go, J.; et al. Congenital Zika virus infection as a silent pathology with loss of neurogenic output in the fetal brain. *Nat. Med.* **2018**, *24*, 368–374. [CrossRef]

10. Loo, Y.-M.; Gale, M. Immune Signaling by RIG-I-like Receptors. *Immunity* **2011**, *34*, 680–692. [CrossRef]

11. Suthar, M.S.; Ma, D.Y.; Thomas, S.; Lund, J.M.; Zhang, N.; Daffis, S.; Rudensky, A.Y.; Bevan, M.J.; Clark, E.A.; Kaja, M.-K.; et al. IPS-1 Is Essential for the Control of West Nile Virus Infection and Immunity. *PLOS Pathog.* **2010**, *6*. [CrossRef] [PubMed]

12. Ivashkiv, L.B.; Donlin, L.T. Regulation of type I interferon responses. *Nat. Rev. Immunol.* **2013**, *14*, 36–49. [CrossRef] [PubMed]

13. Suthar, M.S.; Diamond, M.S.; Jr, M.G. West Nile virus infection and immunity. *Nat. Rev. Genet.* **2013**, *11*, 115–128. [CrossRef] [PubMed]

14. Best, S.M. The Many Faces of the Flavivirus NS5 Protein in Antagonism of Type I Interferon Signaling. *J. Virol.* **2017**, *91*, e01970–16. [CrossRef] [PubMed]

15. Diamond, M.S. Mechanisms of Evasion of the Type I Interferon Antiviral Response by Flaviviruses. *J. Interf. Cytokine Res.* **2009**, *29*, 521–530. [CrossRef]

16. Evans, J.D.; Seeger, C. Differential Effects of Mutations in NS4B on West Nile Virus Replication and Inhibition of Interferon Signaling. *J. Virol.* **2007**, *81*, 11809–11816. [CrossRef]

17. Keller, B.C.; Fredericksen, B.L.; Samuel, M.A.; Mock, R.E.; Mason, P.W.; Diamond, M.S.; Gale, M., Jr. Resistance to alpha/beta interferon is a determinant of West Nile virus replication fitness and virulence. *J. Virol.* **2006**, *80*, 9424–9434. [CrossRef]

18. Laurent-Rolle, M.; Boer, E.; Lubick, K.J.; Wolfinbarger, J.B.; Carmody, A.B.; Rockx, B.; Liu, W.; Ashour, J.; Shupert, W.L.; Holbrook, M.R.; et al. The NS5 Protein of the Virulent West Nile Virus NY99 Strain Is a Potent Antagonist of Type I Interferon-Mediated JAK-STAT Signaling. *J. Virol.* **2010**, *84*, 3503–3515. [CrossRef]

19. Daffis, S.; LaZear, H.M.; Liu, W.J.; Audsley, M.; Engle, M.; Khromykh, A.A.; Diamond, M.S. The Naturally Attenuated Kunjin Strain of West Nile Virus Shows Enhanced Sensitivity to the Host Type I Interferon Response. *J. Virol.* **2011**, *85*, 5664–5668. [CrossRef]

20. Ambrose, R.L.; MacKenzie, J.M. West Nile Virus Differentially Modulates the Unfolded Protein Response To Facilitate Replication and Immune Evasion. *J. Virol.* **2010**, *85*, 2723–2732. [CrossRef]

21. Liu, W.J.; Wang, X.J.; Mokhonov, V.; Shi, P.-Y.; Randall, R.; Khromykh, A.A. Inhibition of Interferon Signaling by the New York 99 Strain and Kunjin Subtype of West Nile Virus Involves Blockage of STAT1 and STAT2 Activation by Nonstructural Proteins. *J. Virol.* **2005**, *79*, 1934–1942. [CrossRef] [PubMed]

22. Muñoz-Jordán, J.L.; Laurent-Rolle, M.; Ashour, J.; Martínez-Sobrido, L.; Ashok, M.; Lipkin, W.I.; García-Sastre, A. Inhibition of Alpha/Beta Interferon Signaling by the NS4B Protein of Flaviviruses. *J. Virol.* **2005**, *79*, 8004–8013. [CrossRef] [PubMed]

23. Schuessler, A.; Funk, A.; LaZear, H.M.; Cooper, D.; Torres, S.; Daffis, S.; Jha, B.K.; Kumagai, Y.; Takeuchi, O.; Hertzog, P.; et al. West Nile Virus Noncoding Subgenomic RNA Contributes to Viral Evasion of the Type I Interferon-Mediated Antiviral Response. *J. Virol.* **2012**, *86*, 5708–5718. [CrossRef] [PubMed]

24. Shuai, K.; Liu, B. Regulation of JAK–STAT signalling in the immune system. *Nat. Rev. Immunol.* **2003**, *3*, 900–911. [CrossRef]

25. Bauer, S.; Kerr, B.J.; Patterson, P.H. The neuropoietic cytokine family in development, plasticity, disease and injury. *Nat. Rev. Neurosci.* **2007**, *8*, 221–232. [CrossRef]

26. Taguwa, S.; Maringer, K.; Li, X.; Bernal-Rubio, D.; Rauch, J.N.; Gestwicki, J.E.; Andino, R.; Fernández-Sesma, A.; Frydman, J. Defining Hsp70 Subnetworks in Dengue Virus Replication Reveals Key Vulnerability in Flavivirus Infection. *Cell* **2015**, *163*, 1108–1123. [CrossRef]

27. Taguwa, S.; Yeh, M.-T.; Rainbolt, T.K.; Nayak, A.; Shao, H.; Gestwicki, J.E.; Andino, R.; Frydman, J. Zika Virus Dependence on Host Hsp70 Provides a Protective Strategy against Infection and Disease. *Cell Rep.* **2019**, *26*, 906–920. [CrossRef]

28. Pujhari, S.; Brustolin, M.; Macias, V.; Nissly, R.; Nomura, M.; Kuchipudi, S.; Rasgon, J.L. Heat shock protein 70 (Hsp70) mediates Zika virus entry, replication, and egress from host cells. *Emerg. Microbes Infect.* **2019**, *8*, 8–16. [CrossRef]

29. Kumar, A.; Hou, S.; Airo, A.M.; Limonta, D.; Mancinelli, V.; Branton, W.; Power, C.; Hobman, T.C. Zika virus inhibits type-I interferon production and downstream signaling. *EMBO Rep.* **2016**, *17*, 1766–1775. [CrossRef]

30. Smith, A.L.; Friedman, D.B.; Yu, H.; Carnahan, R.H.; Reynolds, A.B. ReCLIP (Reversible Cross-Link Immuno-Precipitation): An Efficient Method for Interrogation of Labile Protein Complexes. *PLoS ONE* **2011**, *6*, e16206. [CrossRef]

31. Wang, C.; Gale, M.; Keller, B.; Huang, H.; Brown, M.S.; Goldstein, J.L.; Ye, J. Identification of FBL2 As a Geranylgeranylated Cellular Protein Required for Hepatitis C Virus RNA Replication. *Mol. Cell* **2005**, *18*, 425–434. [CrossRef] [PubMed]

32. Muller, D.A.; Young, P. The flavivirus NS1 protein: Molecular and structural biology, immunology, role in pathogenesis and application as a diagnostic biomarker. *Antivir. Res.* **2013**, *98*, 192–208. [CrossRef] [PubMed]

33. Jen, G.; Detjen, B.M.; Thach, R.E. Shutoff of HeLa cell protein synthesis by encephalomyocarditis virus and poliovirus: A comparative study. *J. Virol.* **1980**, *35*, 150–156. [CrossRef] [PubMed]

34. Shang, L.; Tomasi, T.B. The Heat Shock Protein 90-CDC37 Chaperone Complex Is Required for Signaling by Types I and II Interferons. *J. Boil. Chem.* **2005**, *281*, 1876–1884. [CrossRef] [PubMed]

35. Schoof, N.; Von Bonin, F.; Truemper, L.; Kube, D. HSP90 is essential for Jak-STAT signaling in classical Hodgkin lymphoma cells. *Cell Commun. Signal.* **2009**, *7*, 17. [CrossRef] [PubMed]

36. Chang, Y.-S.; Lee, L.-C.; Sun, F.-C.; Chao, C.-C.; Fu, H.-W.; Lai, Y.-K. Involvement of calcium in the differential induction of heat shock protein 70 by heat shock protein 90 inhibitors, geldanamycin and radicicol, in human non-small cell lung cancer H460 cells. *J. Cell. Biochem.* **2005**, *97*, 156–165. [CrossRef]

37. Srisutthisamphan, K.; Jirakanwisal, K.; Ramphan, S.; Tongluan, N.; Kuadkitkan, A.; Smith, D.R. Hsp90 interacts with multiple dengue virus 2 proteins. *Sci. Rep.* **2018**, *8*, 4308. [CrossRef]

38. Bocchini, C.E.; Kasembeli, M.M.; Roh, S.-H.; Tweardy, D.J. Contribution of chaperones to STAT pathway signaling. *JAK-STAT* **2014**, *3*, e970459. [CrossRef]

39. Lu, C.-Y.; Chang, Y.-C.; Hua, C.-H.; Chuang, C.; Huang, S.-H.; Kung, S.-H.; Hour, M.-J.; Lin, C.-W. Tubacin, an HDAC6 Selective Inhibitor, Reduces the Replication of the Japanese Encephalitis Virus via the Decrease of Viral RNA Synthesis. *Int. J. Mol. Sci.* **2017**, *18*, 954. [CrossRef]

40. Carpp, L.N.; Rogers, R.S.; Moritz, R.L.; Aitchison, J.D. Quantitative Proteomic Analysis of Host-virus Interactions Reveals a Role for Golgi Brefeldin A Resistance Factor 1 (GBF1) in Dengue Infection. *Mol. Cell. Proteom.* **2014**, *13*, 2836–2854. [CrossRef]

41. Kovanich, D.; Saisawang, C.; Sittipaisankul, P.; Ramphan, S.; Kalpongnukul, N.; Somparn, P.; Pisitkun, T.; Smith, D.R. Analysis of the Zika and Japanese Encephalitis Virus NS5 Interactomes. *J. Proteome Res.* **2019**, *18*, 3203–3218. [CrossRef] [PubMed]

42. Rothan, H.A.; Zhong, Y.; Sanborn, M.A.; Teoh, T.C.; Ruan, J.; Yusof, R.; Hang, J.; Henderson, M.J.; Fang, S. Small molecule grp94 inhibitors with antiviral activity against Dengue and Zika virus. *Antiviral Res.* **2019**, 104590. [CrossRef] [PubMed]

43. Emeny, J.M.; Morgan, M.J. Regulation of the Interferon System: Evidence that Vero Cells have a Genetic Defect in Interferon Production. *J. Gen. Virol.* **1979**, *43*, 247–252. [CrossRef] [PubMed]

44. Desmyter, J.; Melnick, J.L.; Rawls, W.E. Defectiveness of Interferon Production and of Rubella Virus Interference in a Line of African Green Monkey Kidney Cells (Vero). *J. Virol.* **1968**, *2*, 955–961. [CrossRef]

45. Michlmayr, D.; Andrade, P.; Gonzalez, K.; Balmaseda, A.; Harris, E. CD14+CD16+ monocytes are the main target of Zika virus infection in peripheral blood mononuclear cells in a paediatric study in Nicaragua. *Nat. Microbiol.* **2017**, *2*, 1462–1470. [CrossRef]

46. Balsitis, S.J.; Flores, D.; Coloma, J.; Beatty, P.R.; Alava, A.; McKerrow, J.H.; Castro, G.; Harris, E. Tropism of dengue virus in mice and humans defined by viral nonstructural protein 3-specific immunostaining. *Am. J. Trop. Med. Hyg.* **2009**, *80*, 416–424. [CrossRef]

47. MacKenzie, J.S.; Gubler, D.J.; Petersen, L.R. Emerging flaviviruses: The spread and resurgence of Japanese encephalitis, West Nile and dengue viruses. *Nat. Med.* **2004**, *10*, S98–S109. [CrossRef]

48. Robert, J. Evolution of heat shock protein and immunity. *Dev. Comp. Immunol.* **2003**, *27*, 449–464. [CrossRef]

49. Schopf, F.H.; Biebl, M.M.; Buchner, J. The HSP90 chaperone machinery. *Nat. Rev. Mol. Cell Boil.* **2017**, *18*, 345–360. [CrossRef]

50. Paradkar, P.; Duchemin, J.-B.; Rodriguez-Andres, J.; Trinidad, L.; Walker, P.J. Cullin4 Is Pro-Viral during West Nile Virus Infection of Culex Mosquitoes. *PLoS Pathog.* **2015**, *11*. [CrossRef]

Article

Network of Interactions between ZIKA Virus Non-Structural Proteins and Human Host Proteins

Volha A. Golubeva [1,†], Thales C. Nepomuceno [1,2,†], Giuliana de Gregoriis [2], Rafael D. Mesquita [3], Xueli Li [1], Sweta Dash [1,4], Patrícia P. Garcez [5], Guilherme Suarez-Kurtz [2], Victoria Izumi [6], John Koomen [7], Marcelo A. Carvalho [2,8,*] and Alvaro N. A. Monteiro [1,*]

[1] Cancer Epidemiology Program, H. Lee Moffitt Cancer Center and Research Institute, Tampa, FL 33612, USA; golubeva.3@osu.edu (V.A.G.); thales.cn@gmail.com (T.C.N.); xueli.li@moffitt.org (X.L.); sweta.dash@moffitt.org (S.D.)

[2] Divisão de Pesquisa Clínica, Instituto Nacional de Câncer, Rio de Janeiro 20230-130, Brazil; gregoriis@gmail.com (G.d.G.); Kurtz@inca.gov.br (G.S.-K.)

[3] Departamento de Bioquímica, Instituto de Química, Federal University of Rio de Janeiro, Rio de Janeiro 21941-909, Brazil; rdmesquita@iq.ufrj.br

[4] Cancer Biology PhD Program, University of South Florida, Tampa, FL 33612, USA

[5] Institute of Biomedical Science, Federal University of Rio de Janeiro, Rio de Janeiro 20230-130, Brazil; ppgarcez@gmail.com

[6] Proteomics and Metabolomics Core, H. Lee Moffitt Cancer Center and Research Institute, Tampa, FL 33612, USA; victoria.izumi@moffitt.org

[7] Chemical Biology and Molecular Medicine Program, H. Lee Moffitt Cancer Center and Research Institute, Tampa, FL 33612, USA; john.koomen@moffitt.org

[8] Instituto Federal do Rio de Janeiro-IFRJ, Rio de Janeiro 20270-021, Brazil

[*] Correspondence: marcelo.carvalho@ifrj.edu.br (M.A.C.); alvaro.monteiro@moffitt.org (A.N.A.M.); Tel.: +55-21-2566-7774 (M.A.C.); +813-7456321 (A.N.A.M.)

[†] These authors contributed equally to this work.

Received: 3 September 2019; Accepted: 1 January 2020; Published: 8 January 2020

Abstract: The Zika virus (ZIKV) is a mosquito-borne Flavivirus and can be transmitted through an infected mosquito bite or through human-to-human interaction by sexual activity, blood transfusion, breastfeeding, or perinatal exposure. After the 2015–2016 outbreak in Brazil, a strong link between ZIKV infection and microcephaly emerged. ZIKV specifically targets human neural progenitor cells, suggesting that proteins encoded by ZIKV bind and inactivate host cell proteins, leading to microcephaly. Here, we present a systematic annotation of interactions between human proteins and the seven non-structural ZIKV proteins corresponding to a Brazilian isolate. The interaction network was generated by combining tandem-affinity purification followed by mass spectrometry with yeast two-hybrid screens. We identified 150 human proteins, involved in distinct biological processes, as interactors to ZIKV non-structural proteins. Our interacting network is composed of proteins that have been previously associated with microcephaly in human genetic disorders and/or animal models. Further, we show that the protein inhibitor of activated STAT1 (PIAS1) interacts with NS5 and modulates its stability. This study builds on previously published interacting networks of ZIKV and genes related to autosomal recessive primary microcephaly to generate a catalog of human cellular targets of ZIKV proteins implicated in processes related to microcephaly in humans. Collectively, these data can be used as a resource for future characterization of ZIKV infection biology and help create a basis for the discovery of drugs that may disrupt the interaction and reduce the health damage to the fetus.

Keywords: ZIKV; protein–protein interaction; non-structural viral proteins; network

---

1. Introduction

Zika virus (ZIKV) is a neurotropic arthropod-borne virus belonging to Flaviviridae family, along with other Flaviviruses capable of infecting central nervous system, such as West Nile Virus, St. Louis Encephalitis Virus, and Japanese Encephalitis Virus. It is commonly transmitted though the bite of an infected *Aedes aegypti* mosquito. Importantly, besides the mosquito bites, human-to-human modes of transmission have also been documented, including sexual activity, blood transfusions, and mother to fetus [1].

Since its first confirmed human infection in the 1960s, there were three documented Zika virus (ZIKV) outbreaks worldwide. The first two occurred in Micronesia and French Polynesia in 2007 and 2013, respectively. The most recent one (2015–2016) started in the northeastern region of Brazil and rapidly spread through South America, the Caribbean, and Mexico. By July 2016, locally transmitted cases of Zika infection were first reported in the United States (Florida). According to the World Health Organization (WHO), 73 different countries had reported ZIKV infections by February of 2016 [2,3]. According to the Centers for Disease Control & Prevention, there have been no recorded local transmissions of the Zika virus in the continental United States in 2018 and 2019. However, with the globally increasing rate of travelling and the historical ability of viruses to acquire genetically modified virulence, the search for effective methods of Zika prevention and control remains important.

ZIKV infections in adults have been associated with neurological conditions such as Guillain-Barré syndrome, acute flaccid paralysis, and meningoencephalitis [4–7]. The Brazilian outbreak was the first time that ZIKV infection (presented in pregnant women) was correlated to congenital microcephaly in newborns [8,9]. Both in vitro and in vivo models have demonstrated that ZIKV has a tropism toward human neural progenitor cells [10–12]. In these cells, ZIKV infection is followed by apoptosis, corroborating the hypothesis of ZIKV as the etiological agent of these neurological disorders [4,5,10–12]. Further, independent studies have shown that the microcephaly and neural development-associated phenotypes is not a distinct feature of the Asian lineage [12–16]. However, the precise molecular mechanism(s) underlying these ZIKV-related manifestations is not understood.

ZIKV is a Baltimore class IV arbovirus from the Flaviviridae family. The ZIKV genome encodes a polyprotein that is processed by both viral and host proteases into ten proteins. Three of them (the capsid, pre-membrane, and envelope) are responsible for the structural organization of the virus. The other seven are non-structural (NS) proteins (NS1, NS2A NS2B, NS3, NS4A, NS4B, and NS5) responsible for regulatory function, viral replication, and subvert host responses [17].

The identification of virus–host protein–protein interaction is essential to better understand viral pathogenesis and to identify cellular mechanisms that could be pharmacologically targeted [18]. To gain further insight into the ZIKV pathogenesis, we generated a virus–host protein–protein interaction network focused on the interactions mediated by the non-structural proteins encoded by the Brazilian ZIKV genotype. Here, we present a network composed of proteins related to neuron projection development, microcephaly-associated disorders, and by protein complexes linked to replication and infection of other members of the Flaviviridae family. In addition, we integrate our dataset with previously published ZIKV protein interaction networks, highlighting common and unique protein interaction partners [19–21]. In addition, we show a PIAS1-dependent control of NS5 protein stability. Taken together, these data can be used as a resource to improve the understanding of the ZIKV pathogenesis and identify putative pharmacological targets for future treatment approaches.

2. Materials and Methods

2.1. cDNA Constructs

We generated the cDNA of seven individual NS proteins (NS1, NS2A, NS3, NS2B, NS4A, NS4B, and NS5) corresponding to a ZIKV Brazilian isolate from the state of Pernambuco (GenBank AMD16557.1) [22]. We used the virus isolate as template for polymerase chain reaction (PCR) to generate cDNAs for NS1, NS2A, NS2B, NS4A, and NS4B. We were unable to obtain the correct product corresponding to NS3 and NS5 cDNAs using the virus isolate. We generated the Brazilian genotype cDNAs for NS3 and NS5 using nucleotide substitutions introduced by site-directed mutagenesis on the Asian lineage (PRVABC59 strain) cDNAs for NS3 and NS5 (pLV_Zika_Flag_NS3 and pLV_Zika_NS5_Flag plasmids; Addgene #79634 and #79639, respectively).

ZIKV cDNAs coding for NS proteins were cloned into pGBKT7 (Clontech) in frame with the GAL4 DNA binding domain (DBD) for yeast-two hybrid (Y2H) assays, and into pNTAP (Agilent) in frame with the Streptavidin-binding peptide (SBP) and Calmodulin-binding peptide (CBP) epitope tags for the tandem affinity purification coupled to mass spectrometry (TAP-MS) assays. The glutathione-S-transferase (GST)-tagged baits used for Y2H validations were generated by subcloning the cDNA from the isolated pGADT7 (Clontech) plasmid into the pDEST27 vector using Gateway recombination cloning according to the manufacturer's instructions (ThermoFisher).

To validate Y2H interactions, recovered Y2H plasmids containing prey cDNAs were amplified by PCR using primers containing attb sites. PCR products were cloned into pDONR221 for Gateway recombination cloning (Invitrogen) and subsequently into pDEST27, to produce an N-terminal GST fusion.

PIAS1 cDNA (NM 001320687.1) was obtained via PCR amplification using a human leukocyte cDNA library and was cloned into pGEX-6P1 using *EcoRI* and *SalI* sites. For expression in mammalian cells, PIAS1 cDNA was subcloned into the pEBG vector using *Bam*HI and *Not*I sites. All constructs were confirmed by Sanger sequencing.

2.2. Y2H Library Screening

To identify direct human brain protein targets of ZIKV NS proteins, we used the MATCHMAKER Gold Y2H system (Clontech). Seven ZIKV viral proteins (NS1, NS2A, NS2B, NS3, NS4A, NS4B, and NS5) were transformed into the Saccharomyces cerevisiae strain Y2HGold (Clontech) alone or co-transformed together with an empty pGADT7 vector and tested for auto-activation and toxicity (defined by low transformation efficiency, small colony phenotype, or inability to grow in liquid culture), as previously described by our group [23].

All bait proteins were expressed in Y2HGold and did not induce toxic effects on the yeast cell cycle or survival (Figure S3A,B). Y2HGold transformants expressing each bait were mated to Y187 strain expressing a pre-transformed human brain normalized cDNA library (Matchmaker® Gold Yeast Two-Hybrid System; Cat.no. 630486; Clontech) for 20 h. The mated cultures were then plated on quadruple dropout medium (SD -Trp/-Leu/-His/-Ade) and incubated for 8–12 days (NS5 was screened twice). For every screen, more than 1×10^6 transformants were screened (Table S1). Yeast miniprep DNA was used to recover pGADT7 fusions from each positive clone (Clontech Yeast Plasmid Isolation Kit), amplified by KOD polymerase chain reaction (PCR) and Sanger sequenced using a T7 primer. Out of frame clones were discarded and in-frame clones were kept for further analysis (Table S2).

2.3. Validation of Y2H Interactions

Protein–protein interactions identified in Y2H screens were validated by expression in human embryonic kidney (HEK) 293FT cells and protein pulldowns. HEK293FT cells were co-transfected with pDEST27 containing prey fusions to GST, and pNTAP containing bait fusions to SBP and CBP. Cells were collected after 24 h and lysed in 1% 3-[(3-Cholamidopropyl)dimethylammonio]-1-Propanesulfonate (CHAPS) lysis buffer (1% CHAPS, 150 mM NaCl, 10 mM 4-(2-hydroxyethyl)-1-piperazineethanesulfonic acid (HEPES) (pH 7.4) with protease and phosphatase inhibitors). Whole cell lysates were subjected to affinity purification of the TAP-tagged NS constructs using streptavidin-conjugated agarose beads, which were washed four times with 1% CHAPS lysis buffer, and then analyzed by Western blot using anti-GST (GE27–4577-01; Sigma Aldrich) and anti-CBP tag antibodies (GenScript; Cat.no. A00635).

2.4. Tandem Affinity Purification Coupled to Mass Spectrometry

HEK293FT cells were transfected using the calcium phosphate method with the SBP-CBP-tagged NS or control (Green Fluorescent Protein; GFP) vectors (Figure 1A). HEK cells have been previously used as a model for characterizing host–pathogen protein–protein interactions (PPIs) [20,24,25].

About 1×10^8 cells were used for the purification of the protein complexes using the InterPlay TAP purification kit (Stratagene) as described previously [23].

A nanoflow ultra-high-performance liquid chromatograph (RSLC, Dionex) coupled to an electrospray bench top orbitrap mass spectrometer (Q-Exactive plus, Thermo) was used for liquid chromatography-tandem mass spectrometry (LC-MS/MS) peptide sequencing experiments. Samples were loaded onto a pre-column and washed for 8 min with aqueous 2% acetonitrile and 0.04% trifluoroacetic acid. Trapped peptides were eluted onto an analytical column (C18 PepMap100, Thermo) and separated using a 90-min gradient delivered at 300 nl/min. Sixteen tandem mass spectra were collected in a data-dependent manner following each survey scan using a 15 s exclusion for previously sampled peptide peaks (QExactive, Thermo).

2.5. Analysis of Proteomics Data

We used Scaffold (Version 4.8.5) to obtain the original samples report of all TAP-MS based peptide and protein identifications. Peptide identifications were retained if they satisfied a minimum of 95.0% threshold. Protein identifications were accepted if they met greater than 50.0% threshold with a minimum of two identified peptides.

To further analyze the original Scaffold mass spectrometry data, we used APOSTL, an integrative Galaxy pipeline for affinity proteomics data [26]. The following global cutoffs were applied to 7996 interactions and generated a list of 88 high confidence interactions: SaintScore cutoff: 0.5; FoldChange cutoff: 0; normalized spectral abundance factor (NSAF) score cutoff: 0.0000025.

APOSTL also interrogates the CRAPome database (http://crapome.org/), which contains common contaminants in affinity purification–mass spectrometry data [27]. Seventeen proteins displayed a CRAPome score >90% and were candidates to be called non-specific interactors. However, two hits were plausible as they were previously implicated in microcephaly (RAB18 and NEDD1), and two hits were found to be associated with ZIKV NS proteins in a previously published independent study (AHCYL1 and GET4) [20]. Moreover, only 2 out of 17 displayed multiple interactions, suggesting that the other 15 proteins do not constitute non-specific interactors in our assay. Therefore, we decided to retain all proteins, but have indicated the high CRAPome score in Figure 2 when appropriate.

2.6. Generation of the Microcephaly-Associated Protein–Protein Interaction Network (PIN)

We generated a microcephaly-associated PIN by searching National Center for Biotechnology Information (NCBI) ENTREZ Gene using [microcephaly] AND [Homo sapiens] as a query. This exercise led to 277 genes, which were manually curated to remove those without an Online Mendelian Inheritance in Man (OMIM) designation (i.e., pseudogenes and partially characterized loci) with a final tally of 261 genes. These genes were used as input to BisoGenet [28], which adds edges between the input nodes, to generate a microcephaly-associated network with 370 interactions (Table S13). BisoGenet settings were 'input nodes only' (Methods) and checking 'protein–protein interactions' only leaving 'Protein DNA interaction' and 'microRNA silencing interaction' unchecked. Significant interaction clusters were identified using ClusterONE (Version 1.0) [29] using the following settings: 'minimum size' = 5; 'minimum density' = 0.5; 'edge weights' = unweighted. Gene ontology (GO) enrichment of clusters was done using BINGO [30] as a Cytoscape plug-in.

2.7. Network Generation and GO Analysis

Network graphics were generated with Cytoscape version 3.7.1 [31]. Each NS integrated dataset was analyzed using WebGestalt to determine the enrichment of GO terms. For each bait set (all proteins that interact with each NS bait), the number of genes in the set that was scored for a term was obtained. The number was then divided by the number of genes in the GO database for that term to obtain an enrichment ratio. Enrichment ratios were log2-transformed to depict increase and decrease changes as numerically equal, but with an opposite sign. To allow for log transformation, enrichment ratios with a 0.0 value were replaced by half of the lowest non-zero value in the complete set. Bait sets were clustered with Cluster 3.0 using the Correlation (uncentered) metric of similarity with no filtering, and the clustering method chosen was complete linkage. It was visualized using Java TreeView v 1.1.6r4 [32].

2.8. Mitocheck Analysis and Clustering

The Mitocheck phenotype database (20,921 genes), which scores 14 mitosis-related phenotypes in a binary form (presence = 1; absence = 0) from RNA interference screens, was downloaded from http://www.mitocheck.org/ as a tab-delimited file, in which genes are represented in rows and phenotypes in columns.

The enrichment and clustering (for each bait set) were performed as described above. Further, we also deconvoluted the integrated dataset to genes that were positive to at least one of the 14 phenotypes (Table S9). These new data sets were clustered with Cluster 3.0 using the Correlation (uncentered) metric of similarity with no filtering, but log2-transformed to depict increase and decrease changes as numerically equal, but with an opposite sign. The clustering method chosen was complete linkage. It was visualized using Java TreeView v 1.1.6r4 [32].

2.9. GST Pulldown Assay

HEK293FT cells were transfected using Polyethylemine (PEI; Polysciences Inc.) as previously described [33]. GST pulldown assays were performed by incubating for 16h at 4 °C Glutathione Sepharose 4B (GE Healthcare) with whole cell lysates prepared 48 h after transfection. The resin was extensively washed with 'mild' Radioimmunoprecipitation Assay (RIPA) buffer (50 mM Tris-Cl pH 7.4, 150 mM NaCl, 1 mM Ethylenediaminetetraacetic acid, 1% NP40, and 2.5 mM Dithiothreitol), boiled in loading buffer, and analyzed by Western blotting using anti-GST (Santa Cruz Biotechnology; B-4) and anti-GFP (Millipore; MAB3580) antibodies.

2.10. Protein Stability Assay

HEK293FT cells were transfected with expression vectors containing GST-tagged PIAS1 and GFP- or CBP-tagged NS5 (empty expression vectors were used as negative controls) and treated 24 h later with 10 µg/mL of cyclohexamide (or Dimethylsulfoxide) for varying lengths of time. Whole cell lysates were analyzed by Western blotting using anti-GST, anti-CBP, or anti-GFP and anti-β-actin (Santa Cruz Biotechnology; C4).

3. Results

3.1. Yeast-Two Hybrid Screenings

To build the first pair-wise protein–protein interaction database of ZIKV NS proteins encoded by the Brazilian genotype, we performed stringent yeast-two hybrid (Y2H) screenings using ZIKV NS proteins as baits to screen a normalized human brain cDNA library (Figure 1A and Table S1). The Y2H screenings generated a protein–protein interaction network (PIN) ranging from 1 (NS2A) to 56 (NS3) interactions, totaling 99 unique protein hits and 109 bait–prey interactions (Figure 1A,B and Table S2).

To validate our Y2H PIN, a subset of 38 bait–prey interactions (38.4% of the Y2H PIN) was tested for interaction in human HEK293FT cells. All NS coding sequences were cloned in a eukaryotic expression vector in frame with the streptavidin and calmodulin binding peptides (SBP and CBP, respectively). Candidate interactor cDNAs were expressed in frame with glutathione-S-transferase (GST). SBP pulldown assays were performed against GST-tagged preys in HEK293FT cells, and 76.3% of interactions were confirmed (Figure 1C). As some true interactors might not validate in these specific conditions, Figure 1C retains all interactions, with validated ones indicated. However, further analysis was conducted, retaining only the validated hits.

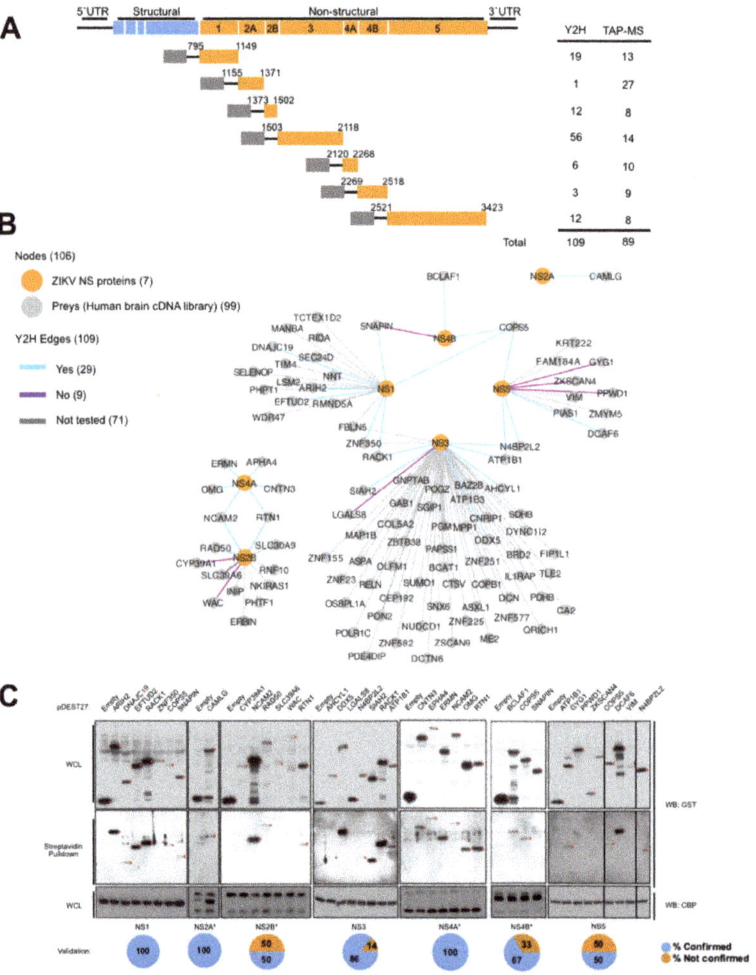

Figure 1. Yeast two-hybrid (Y2H) protein–protein interaction network (PIN). (**A**) Schematic representation of the ZIKA virus (ZIKV) genome and non-structural (NS) constructs used on our protein–protein interaction screenings. Grey boxes represent the GAL4 DNA binding domain (DBD) and the streptavidin binding protein (SBP)–calmodulin binding protein (CBP) tag used on the yeast two-hybrid (Y2H) and tandem affinity purification (TAP) assays, respectively. The number of hits identified by each assay and bait is summarized on the right. UTR, untranslated region. (**B**) Network of the interactions identified by Y2H screens. ZIKV NS nodes are colored in orange, and gray nodes indicate other proteins that were recovered as preys. Edge colors represent results from validation. Edges to all preys are shown and 'not validated' (No), 'validated' (Yes), and 'not tested interactions' (Not tested), and are denoted by blue, purple, and gray edges, respectively. The color legend is depicted on the upper left-hand corner. (**C**) Streptavidin pulldown of TAP-tagged NS constructs from 293FT whole cell lysates followed by Western blotting with the indicated antibodies. The individual percentage of hits validated by bait is depicted on the right. Red dots indicate the expected band size. Red asterisk for Streptavidin pulldown assay.

3.2. Tandem Affinity Purification Followed by Mass Spectrometry

To further characterize the NS-mediated protein interactions, we expressed all baits as fusions to SBP and CBP in HEK293FT cells (Figure 1A). We then performed tandem affinity purification coupled to mass spectrometry (TAP-MS), which resulted in a high-confidence PIN with interactions ranging from 8 (NS2B and NS5) to 27 (NS2A), totaling 62 unique protein hits and 89 bait–prey interactions (Figures 1A and 2A,B; Table S3).

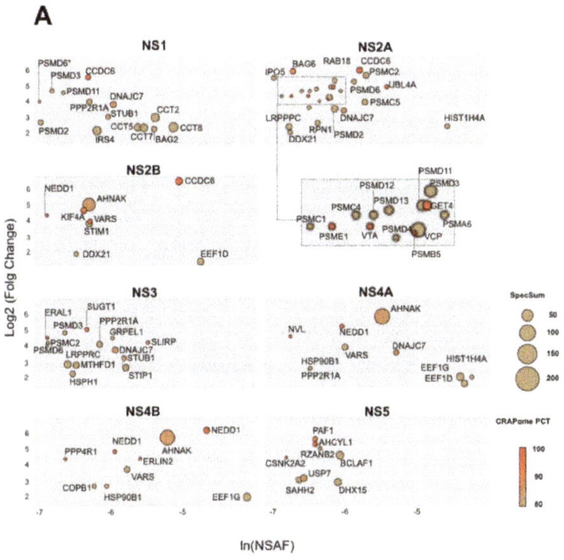

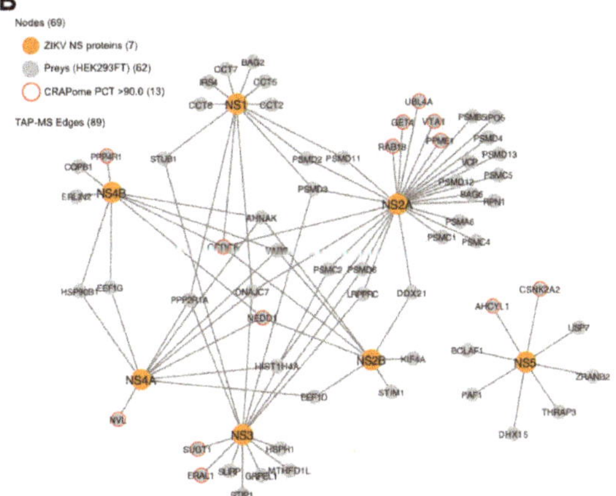

Figure 2. TAP-MS PIN. (**A**) Protein interaction profile of TAP-MS screenings plotted based on normalized spectral abundance factor (NSAF) [34] (X axis) and specificity based on fold change of spectral counts (Y axis) between TAP-tagged NS proteins and TAP-tagged GFP (negative control). Node size denotes the spectral sum (Spec sum) obtained for each protein. Node color denotes CRAPome PCT score according to the scale. (**B**) Network of the interactions identified by TAP-MS screens. Orange nodes represent ZIKV NS proteins (baits) and gray nodes represent human proteins (preys). Gray nodes with a red circle indicate a prey with high CRAPome score.

3.3. Merged ZIKV PIN

We combined the final Y2H and TAP-MS networks to generate the Merged ZIKV Network containing 157 nodes (including the baits) with 189 interactions between NS proteins and human host proteins (Figure 3A) (Table S4). Only 3 out of 151 hits were common to both Y2H and MS-based networks (BCLAF1, AHCYL1, and COPB1), indicating a limited overlap between methods.

3.4. Gene Ontology

Gene ontology (GO) enrichment analysis of the Merged ZIKV PIN identified a subset of proteins mainly involved in 13 non-redundant biological processes (Figure S1). Among the hits identified, 13 are members of the proteasome complex (11 unique to the 26S subunit) and five members of the chaperonin-containing TCP1 (CCT) complexes (8.7% and 3.3% of our PIN, respectively) (Figure S1 and Table S5). GO enrichment analysis of cellular components revealed an enrichment of peptidase complex, chaperone complex, and ficolin-1-rich granules (Table S6).

Next, we applied unsupervised clustering of bait sets according to their GO enrichment ratios for biological processes and cellular components (Figure 3B,C and Tables S7 and S8). Interestingly, protein bait sets were clustered into two major groups (NS1, NS2A and NS3 versus NS2B, NS4A, NS4B, and NS5).

3.5. Phenoclusters

Autosomal recessive primary microcephaly (MCPH) development is intrinsically associated with impaired mitosis [35]. Therefore, we used data from the Mitocheck project database (http://www.mitocheck.org/) [36,37] to determine the enrichment (or depletion) ratio of our bait sets for each mitotic phenotype scored in Mitocheck (Table S9). We then used the enrichment ratios to cluster bait sets according to their functional similarities (Figure 3D). Bait sets clustered around two large components according to their involvement in mitotic processes. One cluster (NS1, NS2A, and NS3 bait sets) presented enrichment of mitotic phenotypes, while the second (NS2B, NS4A, NS4B, and NS5) did not, suggesting that NS1, NS2A, and NS3 are more likely to disrupt cellular mitotic processes (Figure 3D).

Finally, to identify individual preys more likely to be involved in mitotic processes, we clustered all preys according to their Mitocheck enrichment rations (Figure S2) and identified a cluster of nine proteins (CEP192, FAM184A, PAPSS1, EFTUD2, ZNF155, BAG6, SELENOP, KIF4A, and PHPT1) with phenotypes consistent with centrosomal abnormalities (Table S10). This analysis reflected the clustering pattern for NS1, NS2A, and NS3 bait sets obtained when clustering for GO biological processes and cellular components (Figure 3B–D).

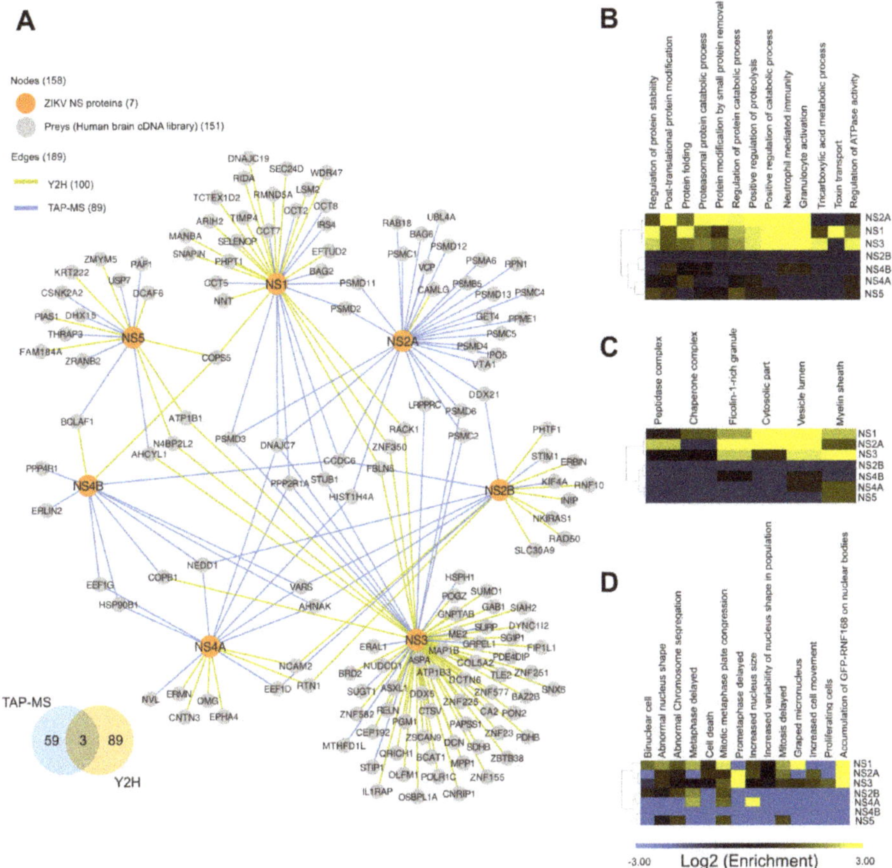

Figure 3. Merged ZIKV PIN. (**A**) Network of the interactions identified by Y2H and TAP-MS screens. The color legend is depicted on the upper left-hand corner. (**B,C**) Clustering of bait sets according to gene ontology (GO) enrichment ratio for biological processes (**B**) and cellular component (**C**). (**D**) Phenoclusters (clustering of bait sets according to enrichment or depletion of Mitocheck phenotype classes. Clustering and visualization were performed using Cluster v3.0 software and TreeView v1.1.6r4, respectively.

3.6. Integration with Other ZIKV PINs

Our work builds on three previous physical interaction networks of host and ZIKV proteins [19–21]. We integrated our PIN with the published networks to evaluate the level of overlap between the four PINs (Figure 4A; Table S11). No common hit was shared by all four PINs and pair-wise overlaps ranged from 1 to 50 hits, suggesting that the ZIKV–host protein interacting network is still far from reaching saturation (Figure 4A,B).

We identified five highly internally connected clusters among the integrated PIN (Figure 4C). All five clusters contained components of the ZIKV PIN from this study. Four of them were enriched in proteins involved in the following: (a) anaphase promoting complex (APC)-dependent proteasomal ubiquitin-dependent protein catabolic process (GO31145); (b) protein amino acid N-linked glycosylation via asparagine (GO18279); (c) protein folding (GO6457); (d) regulation of transcription (GO45449); and (e) histone H2B ubiquitination (GO33523).

Finally, 14 unique nodes of our merged ZIKV PIN (9.3% of the data set) have been shown to be important for proper replication of different *Flavivirus* (Table S12) [38–40], suggesting that our network also contains proteins that could explain the mechanisms of ZIKV replication and help identify therapeutic targets.

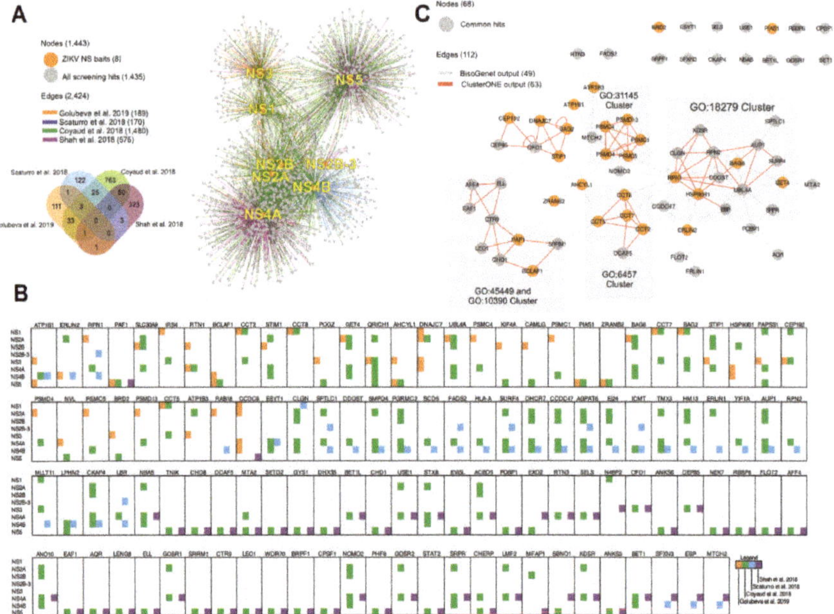

Figure 4. Integration of ZIKV PINs. (**A**) Network of the interactions identified by this study, Scaturro et al. 2018, Shah et al. 2018, and Coyaud et al. 2018. The color legend is depicted on the upper left-hand corner. (**B**) Graphic representation of common preys according to the bait and study in which they were identified. (**C**) Clustering of bait sets according to overlapping protein complexes among the integrated network using ClusterONE (Version 1.0) [29]. Gene ontologies of these networks were obtained by the Cytoscape plugin, BINGO [30]. GO accession numbers represent the following biological process: GO31145—anaphase promoting complex (APC)-dependent proteasomal ubiquitin-dependent protein catabolic process, GO18279—protein amino acid N-linked glycosylation via asparagine, GO6457—protein folding, GO45449—regulation of transcription, and GO33523—histone H2B ubiquitination. Proteins identified in this study are represented as orange nodes.

3.7. Integration with a Microcephaly-Associated Network

We generated a new network composed of microcephaly-associated genes/proteins (Table S13) and our merged ZIKV PIN using BisoGenet to impute known interactions between nodes in this network (see Experimental Procedures) (Figure 5A). Four highly cohesive (i.e., highly connected internally, but only sparsely with the rest of the network) clusters emerged (Figure 5B). Of those, only three contained components of the ZIKV PIN: anaphase promoting complex (APC)-dependent proteasomal ubiquitin-dependent protein catabolic process (GO31145), centrosome duplication (GO7099), and COPI coating of Golgi vesicle (GO48205). Finally, we identified four nodes common to both PINs (CEP192, ASXL1, VARS, and EFTUD2). A similarly limited overlap between the microcephaly-associated and merged ZIKV PIN was also obtained with the three other previously determined ZIKV PINs (Figure 5C).

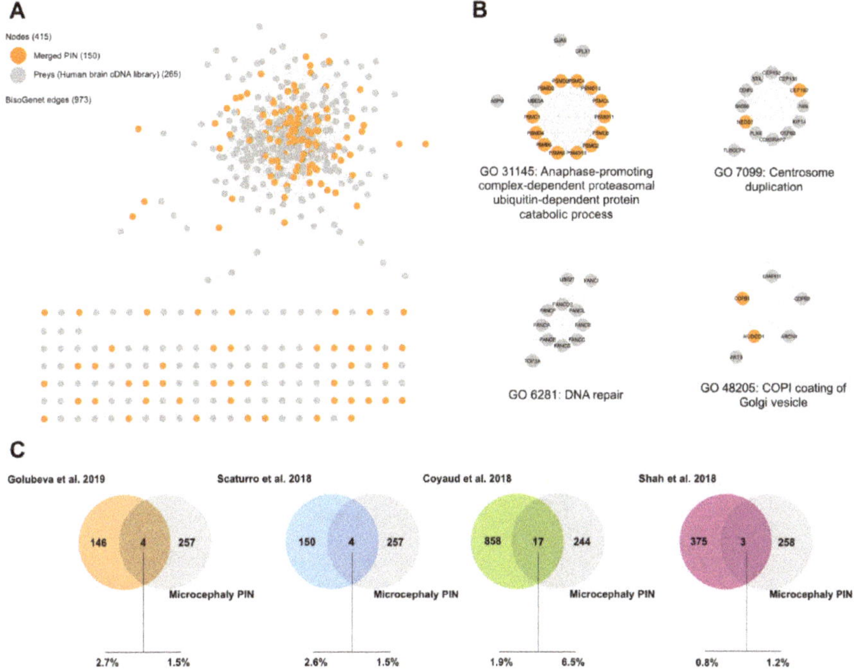

Figure 5. Integration of Microcephaly-associated PIN with the merged ZIKV PIN. (**A**) Network of the interactions identified by this study, and proteins related to the Microcephalic phenotype (see Experimental Procedures). (**B**) Clustering of bait sets according to overlapping protein complexes among the integrated network using ClusterONE (Version 1.0) [29]. Gene ontology of these networks were obtained by the Cytoscape plugin, BINGO [30]. Proteins identified in this study are represented as orange nodes. (**C**) Venn diagrams represent the overlap between the Microcephaly-associated PIN and the individual ZIKV PINs.

3.8. PIAS1 Modulates NS5 Protein Stability

The Y2H screening identified PIAS1 (protein inhibitor of activated STAT1) as an interacting partner of NS5. We confirmed this interaction using a GST-pulldown assay in HEK293FT cells co-transfected with GST-tagged PIAS1 and GFP-tagged NS5 (Figure 6A).

PIAS1 is an E3 SUMO-protein ligase implicated in the maintenance of protein stability [41,42]. Curiously, SUMOylation of the DENV NS5 has been linked to its stabilization, thus stimulating viral replication in human cells [43]. This observation prompted us to evaluate the impact of PIAS1 overexpressing on ZIKV NS5 protein stability. HEK293FT cells were transfected with GST-tagged PIAS1 and GFP- or CBP-tagged NS5 constructs and treated with cyclohexamide (CHX) to inhibit de novo protein synthesis. NS5 half-life was evaluated in a time course (Figure 6B). PIAS1 overexpressed is correlated with a decrease in NS5 protein stability, as seen 12 h after CHX treatment (compare lines 4 and 8). PIAS1 modulates steady state levels of NS5 even at lower expression levels of PIAS1 (Figure 6C) and levels of NS5 decrease in a PIAS1 dose-dependent manner (Figure 6D). Taken together, our data suggest that PIAS1 is involved in modulating the stability of ZIKV NS5.

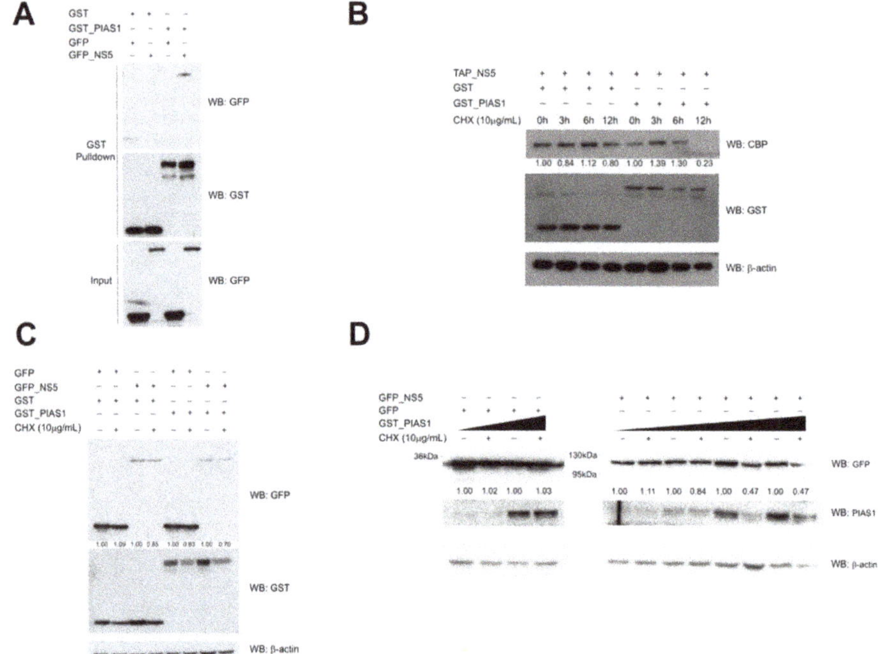

Figure 6. PIAS1 interacts with and modulates ZIKV NS5 protein stability. (**A**) GST pulldown was conducted using whole cell lysates of HEK293FT cells transfected with GST-tagged PIAS1 and GFP-tagged NS5 (empty vectors were used as negative controls). Input represents 10% of the lysate used in the GST pulldown assay. (**B**) HEK293FT cells transfected with GST-tagged constructs (empty vector or PIAS1) and pNTAP NS5 and treated with 10 µg/mL cyclohexamide (CHX) for the indicated time points. (**C**) HEK203FT cells transfected with GST and GFP-tagged constructs. Cells were treated, 24 h after transfections, with 10 µg/mL cyclohexamide (CHX) for 12 h. (**D**) HEK293FT cells were transfected with GFP-tagged constructs (empty vector or NS5) and different amounts of GST-tagged PIAS1 cDNA (0.5 µg, 1 µg, 2 µg, or 4 µg). pQCXIH was used to normalize the amount of transfected DNA. At 20 h post-transfection, cells were treated with 10 µg/mL cyclohexamide (CHX) for 12 h. + and – signs indicate presence or absence of the reagent indicated on the left, respectively.

4. Discussion

In humans, ZIKV infection was correlated with congenital microcephaly in newborns and with other neurological conditions in adults [4–7,10–12]. Still, little is known about the molecular mechanism of ZIKV infection and how it relates to neurological disorders. Here, we present a human host protein–ZIKV (Brazilian genotype) NS protein interaction network. This network was obtained by a combination of yeast two-hybrid (Y2H) screens and tandem affinity purification coupled to mass spectrometry (TAP-MS). Y2H screens primarily reveal direct pair-wise interactions and are capable of detecting transient interactions, while TAP-MS will reveal proteins engaged in stable complexes, which will eventually result in the identification of indirect interactions [44–47]. The use of both methods results in a comprehensive panorama of ZIKV protein–protein interactions.

The merged network combining two complementary methods (Y2H and TAP-MS) contains 157 nodes and 189 interactions with a limited overlap between the two methods, consistent with other previously determined PIN [23,46,47]. Further, the subset of Y2H interactions validated in human cells displayed a false positive rate of ~24% (9/38) averaged across all seven ZIKV NS proteins, as judged by confirmation interaction experiments using SBP pulldown assays. This is in line with other published Y2H screens [23,48–51]. These results suggest that this PIN contains high-confidence interactions.

Twenty-nine human proteins interacted with more than one ZIKV protein (19.3% of all hits). Similar relatively high levels of promiscuity of human proteins in relation to their viral interactors were also found in previous studies. Scaturro et al. [19] and Coyaud et al. [20] had 10.5% and 36% of all hits interacting to more than one ZIKV protein, respectively. Although Shah et al. [21] had a much lower (0.3% of all hits) rate of human proteins interacting to more than one ZIKV protein, a high level of promiscuity of human proteins is also apparent across studies, where the same human proteins are often found interacting with distinct ZIKV proteins. For example, all human proteins shared between Shah et al. [21] and Scaturro et al. [19], or between Shah et al. and this study, were found to interact with different ZIKV NS proteins. In addition, comparisons across other studies showed consistently high levels of discordance in bait interactions (Figure 4B). These data suggest that different ZIKV NS proteins have common targets in the human proteome. However, it is unclear why different studies detected exclusive interactions with different baits. It is conceivable that several ZIKV NS proteins interact with large protein complexes, such as the 26S subunit of the proteasome complex, via different targets; furthermore, differences in the biology of the cells providing the proteome (i.e., levels of protein expression and formation of specific protein complexes), the biochemical methods, or the filtering criteria for significant interactions may also determine which interactions are robust enough to result in detection.

Several aspects could account for the low level of overlapping between studies. Although every study used a different cell line, three of them used derivatives of HEK293 cells (HEK 293T, HEK 293FT, HEK 293 T-rex) and one used SK-N-BE2 neuroblastoma cells [19]. Yet, low overlap was also encountered in pair-wise comparison between studies with HEK cells (Shah et al. and Golubeva et al.; Shah et al. and Coyaud et al.). Conversely, a slightly higher overlap can be found between studies with HEK cells (Golubeva et al. and Coyaud et al.), as well as with studies with different cell lines (Coyaud et al. and Scaturro et al.). These observations suggest that differences in cell lines are unlikely to explain the low overlap.

In addition to affinity purification followed by mass spectrometry used in every study, complementary methods were also used such as BioID (a proximity-dependent labeling approach) [20] and the yeast two-hybrid (this study) that could partly explain the differences between studies. Other small differences in methods (expression vectors, affinity tags) and preys (all viral proteins versus only non-structural proteins; amino acid sequence differences between virus strains used) could also contribute to the differences across studies. Alternatively, some could represent spurious interactions detected as a result of the overexpression of the baits; however, all four studies used stringent cut-off measures and validation experiments, ensuring that the number of false positive results is likely to be low. Furthermore, the limited number of proteins in our dataset with high CRAPome [27] scores indicating consistent recovery in affinity proteomics as non-specific background also suggests that the differences across studies are unlikely to be explained by a large number of false positives. We propose that the low overlap among these studies suggests that they have not reached saturation and other ZIKV protein-interacting host proteins are still to be discovered.

Further, we identified multiple components of CCT (chaperonin containing TCP1 or TriC-TCP-1 ring complex) complex as targets. This complex plays a role in trafficking of telomerase and small Cajal body (CB) RNAs through the proper folding of the telomerase cofactor, TCAB1 [52]. CBs are transcription-dependent nuclear compartments and play a critical role in neuron biology through snRNP and snoRNP assembly [53]. Interestingly, Coyaud et al. [20] demonstrated that ZIKV NS5 expression leads to an increase in the absolute number of CBs per cell, but to a reduction of the volume of these CBs, suggesting that NS5 expression could lead to CB fragmentation. Our data point to the interaction of NS1 with multiple components of the CCT complex, suggesting that NS1 could also play a role in CB stability and in neural disorders. Additionally, it has already been shown that the Dengue virus (DENV) infection occurs in an NS1/CCT-dependent manner [54].

Centrosomal abnormalities lead to impaired mitosis, which is a hallmark of MCPH. In fact, our data set presents multiple proteins related to phenotypes associated with impaired mitosis (Figure 3B,C). Furthermore, our PIN shares 24 (16% of all unique hits) known interaction partners of 14 (out of 18) MCPH loci plus CEP63 (Table S14).

In that context, CEP192 (identified as an NS3 interaction partner by Y2H) plays a central role in the initial steps of centriole duplications through the interaction and recruitment of CEP152 (MCPH9) and PLK4, respectively, which is necessary for the proper recruitment of SAS6 (MCPH14), STIL (MCPH7), and CENPJ (MCPH6) [55–60]. Our data suggest that NS3 could interfere with centriole duplication and, consequently, could be important for the ZIKV-associated microcephaly phenotype. Furthermore, GO enrichment analysis and Mitocheck phenoclusters suggest that NS1, NS2A, and NS3 target host factors are implicated in mitotic phenotypes.

In humans, viral infection activates the type-I interferon (IFN-I) signaling leading to STAT1/2 activation. The ZIKV5 protein acts as an antagonist of the IFN-I pathway by stimulating STAT2 (but not STAT1) degradation [61]. STAT1 activity is modulated by PIAS1, which has been implicated in herpes simplex viral replication [62]. We identified PIAS1 as an interacting partner of NS5 and showed that overexpression of PIAS1 results in a shorter NS5 protein half-life. Our data suggest that PIAS1 can modulate the levels of ZIKV NS5, but it is unclear the extent to which this modulation may affect ZIKV replication. Interestingly, a CRISPR/Cas9 screening revealed that PIAS1-depleted cells are more sensitive to ZIKV infection-dependent lethality [38]. Collectively, these data suggest that PIAS1 might play an important role in ZIKV biology by modulating NS5 protein levels.

In summary, the data presented here together with three previously published studies [19–21] provide a valuable resource to dissect the mechanistic underpinnings of central nervous system perturbations caused by ZIKV infection and to identify potential pharmacological targets. A small number of overlapping hits across different studies suggest that the screens are still far from reaching saturation.

Supplementary Materials: The following are available online at http://www.mdpi.com/2073-4409/9/1/153/s1, Figure S1: GO enrichment of the merged network; Figure S2: Phenoclusters of individual preys; Figure S3: Y2H bait expression, control transformations, and matings; Table S1. Yeast two-hybrid screening data; Table S2. Yeast two-hybrid hits; Table S3. TAP-MS hits - APOSTL output; Table S4. Merged PIN (Y2H + TAP/MS); Table S5. GO (Cellular component) enrichment membership; Table S6. GO (Biological Process) enrichment membership; Table S7. Bait-specific GO (Biological Process) enrichment ratios; Table S8. Bait-specific GO (Cellular component) enrichment ratios; Table S9. Phenoclusters of bait sets; Table S10. Phenoclusters of individual preys; Table S11. Integrated ZIKV PIN; Table S12. Flavivirus replication factors (functional screens) intersection with Merged ZIKV PIN; Table S13. Microcephaly-associated genes; Table S14. Merged ZIKV PIN and MCPH subnetwork.

Author Contributions: V.A.G., T.C.N., R.D.M., G.S.-K., M.A.C. and A.N.A.M. data curation; V.A.G. and T.C.N. formal analysis; V.A.G., T.C.N., G.S.-K., M.A.C. and A.N.A.M. supervision; V.A.G., T.C.N. and X.L. validation; V.A.G., T.C.N., G.d.G., S.D., X.L., P.P.G., V.I. and J.K. investigation; V.A.G., T.C.N., G.d.G., X.L., P.P.G., V.I. and J.K. methodology; V.A.G. and T.C.N. writing—original draft; M.A.C. and A.N.A.M. conceptualization; M.A.C. and A.N.A.M. writing—review and editing; M.A.C. and A.N.A.M. resources; M.A.C. and A.N.A.M. funding acquisition. All authors have read and agreed to the published version of the manuscript.

Funding: This work was supported by Florida Department of Health Zika Research Grant Initiative pilot award 7ZK29 and in part by the Proteomics and Metabolomics Core at the Moffitt Cancer Center through its NCI CCSG grant (P30-CA76292), and by Moffitt's Center for Immunization and Infection Research in Cancer (CIIRC). T.C.N. is a Fulbright Scholar.

Acknowledgments: We thank the individuals and families who have generously donated their time, samples, and information to facilitate research on ZIKV. We also thank Brent Kuenzi for help with proteomic data analysis.

Conflicts of Interest: The authors declare that they have no conflicts of interest related to the contents of this article.

References

1. Lessler, J.; Chaisson, L.H.; Kucirka, L.M.; Bi, Q.; Grantz, K.; Salje, H.; Carcelen, A.C.; Ott, C.T.; Sheffield, J.S.; Ferguson, N.M.; et al. Assessing the global threat from zika virus. *Science* **2016**, *353*, aaf8160. [CrossRef] [PubMed]

2. Kindhauser, M.K.; Allen, T.; Frank, V.; Santhana, R.S.; Dye, C. Zika: The origin and spread of a mosquito-borne virus. *Bull. World Health Organ.* **2016**, *94*, 675–686C. [CrossRef]

3. Grubaugh, N.D.; Faria, N.R.; Andersen, K.G.; Pybus, O.G. Genomic insights into zika virus emergence and spread. *Cell* **2018**, *172*, 1160–1162. [CrossRef] [PubMed]

4. Carteaux, G.; Maquart, M.; Bedet, A.; Contou, D.; Brugieres, P.; Fourati, S.; Cleret de Langavant, L.; de Broucker, T.; Brun-Buisson, C.; Leparc-Goffart, I.; et al. Zika virus associated with meningoencephalitis. *N. Engl. J. Med.* **2016**, *374*, 1595–1596. [CrossRef] [PubMed]

5. Parra, B.; Lizarazo, J.; Jimenez-Arango, J.A.; Zea-Vera, A.F.; Gonzalez-Manrique, G.; Vargas, J.; Angarita, J.A.; Zuniga, G.; Lopez-Gonzalez, R.; Beltran, C.L.; et al. Guillain-barre syndrome associated with zika virus infection in colombia. *N. Engl. J. Med.* **2016**, *375*, 1513–1523. [CrossRef] [PubMed]

6. Craig, A.T.; Butler, M.T.; Pastore, R.; Paterson, B.J.; Durrheim, D.N. Acute flaccid paralysis incidence and zika virus surveillance, pacific islands. *Bull. World Health Organ.* **2017**, *95*, 69–75. [CrossRef]

7. Miner, J.J.; Diamond, M.S. Zika virus pathogenesis and tissue tropism. *Cell Host Microbe* **2017**, *21*, 134–142. [CrossRef]

8. Werner, H.; Fazecas, T.; Guedes, B.; Lopes Dos Santos, J.; Daltro, P.; Tonni, G.; Campbell, S.; Araujo Junior, E. Intrauterine zika virus infection and microcephaly: Correlation of perinatal imaging and three-dimensional virtual physical models. *Ultrasound Obst. Gyn.* **2016**, *47*, 657–660. [CrossRef]

9. Wang, J.N.; Ling, F. Zika virus infection and microcephaly: Evidence for a causal link. *Int. J. Environ. Res. Public Health* **2016**, *13*, 1031. [CrossRef]

10. Garcez, P.P.; Loiola, E.C.; Madeiro da Costa, R.; Higa, L.M.; Trindade, P.; Delvecchio, R.; Nascimento, J.M.; Brindeiro, R.; Tanuri, A.; Rehen, S.K. Zika virus impairs growth in human neurospheres and brain organoids. *Science* **2016**, *352*, 816–818. [CrossRef]

11. Tang, H.; Hammack, C.; Ogden, S.C.; Wen, Z.; Qian, X.; Li, Y.; Yao, B.; Shin, J.; Zhang, F.; Lee, E.M.; et al. Zika virus infects human cortical neural progenitors and attenuates their growth. *Cell Stem Cell* **2016**, *18*, 587–590. [CrossRef] [PubMed]

12. Cugola, F.R.; Fernandes, I.R.; Russo, F.B.; Freitas, B.C.; Dias, J.L.; Guimaraes, K.P.; Benazzato, C.; Almeida, N.; Pignatari, G.C.; Romero, S.; et al. The brazilian zika virus strain causes birth defects in experimental models. *Nature* **2016**, *534*, 267–271. [CrossRef] [PubMed]

13. Besnard, M.; Eyrolle-Guignot, D.; Guillemette-Artur, P.; Lastere, S.; Bost-Bezeaud, F.; Marcelis, L.; Abadie, V.; Garel, C.; Moutard, M.L.; Jouannic, J.M.; et al. Congenital cerebral malformations and dysfunction in fetuses and newborns following the 2013 to 2014 zika virus epidemic in french polynesia. *Euro Surveillance* **2016**, *21*. [CrossRef] [PubMed]

14. Goodfellow, F.T.; Willard, K.A.; Wu, X.; Scoville, S.; Stice, S.L.; Brindley, M.A. Strain-dependent consequences of zika virus infection and differential impact on neural development. *Viruses* **2018**, *10*, 550. [CrossRef] [PubMed]

15. Jaeger, A.S.; Murrieta, R.A.; Goren, L.R.; Crooks, C.M.; Moriarty, R.V.; Weiler, A.M.; Rybarczyk, S.; Semler, M.R.; Huffman, C.; Mejia, A.; et al. Zika viruses of african and asian lineages cause fetal harm in a mouse model of vertical transmission. *PLoS Negl. Trop. Dis.* **2019**, *13*, e0007343. [CrossRef] [PubMed]

16. Udenze, D.; Trus, I.; Berube, N.; Gerdts, V.; Karniychuk, U. The african strain of zika virus causes more severe in utero infection than asian strain in a porcine fetal transmission model. *Emerg. Microbes Infect.* **2019**, *8*, 1098–1107. [CrossRef]

17. Hasan, S.S.; Sevvana, M.; Kuhn, R.J.; Rossmann, M.G. Structural biology of zika virus and other flaviviruses. *Nat. Struct. Mol. Biol.* **2018**, *25*, 13–20. [CrossRef]

18. Ramage, H.; Cherry, S. Virus-host interactions: From unbiased genetic screens to function. *Annu. Rev. Virol.* **2015**, *2*, 497–524. [CrossRef]

19. Scaturro, P.; Stukalov, A.; Haas, D.A.; Cortese, M.; Draganova, K.; Plaszczyca, A.; Bartenschlager, R.; Gotz, M.; Pichlmair, A. An orthogonal proteomic survey uncovers novel zika virus host factors. *Nature* **2018**, *561*, 253–257. [CrossRef]

20. Coyaud, E.; Ranadheera, C.; Cheng, D.; Goncalves, J.; Dyakov, B.J.A.; Laurent, E.M.N.; St-Germain, J.; Pelletier, L.; Gingras, A.C.; Brumell, J.H.; et al. Global interactomics uncovers extensive organellar targeting by zika virus. *Mol. Cell Proteomics* **2018**, *17*, 2242–2255. [CrossRef]

21. Shah, P.S.; Link, N.; Jang, G.M.; Sharp, P.P.; Zhu, T.; Swaney, D.L.; Johnson, J.R.; Von Dollen, J.; Ramage, H.R.; Satkamp, L.; et al. Comparative flavivirus-host protein interaction mapping reveals mechanisms of dengue and zika virus pathogenesis. *Cell* **2018**, *175*, 1931–1945. [CrossRef] [PubMed]

22. Calvet, G.; Aguiar, R.S.; Melo, A.S.; Sampaio, S.A.; de Filippis, I.; Fabri, A.; Araujo, E.S.; de Sequeira, P.C.; de Mendonca, M.C.; de Oliveira, L.; et al. Detection and sequencing of zika virus from amniotic fluid of fetuses with microcephaly in brazil: A case study. *Lancet Infect. Dis.* **2016**, *16*, 653–660. [CrossRef]

23. Woods, N.T.; Mesquita, R.D.; Sweet, M.; Carvalho, M.A.; Li, X.; Liu, Y.; Nguyen, H.; Thomas, C.E.; Iversen, E.S., Jr.; Marsillac, S.; et al. Charting the landscape of tandem brct domain-mediated protein interactions. *Sci. Signal.* **2012**, *5*, rs6. [CrossRef] [PubMed]

24. Graham, F.L.; Smiley, J.; Russell, W.C.; Nairn, R. Characteristics of a human cell line transformed by DNA from human adenovirus type 5. *J. Gen. Virol.* **1977**, *36*, 59–74. [CrossRef]

25. Eckhardt, M.; Zhang, W.; Gross, A.M.; Von Dollen, J.; Johnson, J.R.; Franks-Skiba, K.E.; Swaney, D.L.; Johnson, T.L.; Jang, G.M.; Shah, P.S.; et al. Multiple routes to oncogenesis are promoted by the human papillomavirus-host protein network. *Cancer. Discov.* **2018**, *8*, 1474–1489. [CrossRef]

26. Kuenzi, B.M.; Borne, A.L.; Li, J.; Haura, E.B.; Eschrich, S.A.; Koomen, J.M.; Rix, U.; Stewart, P.A. Apostl: An interactive galaxy pipeline for reproducible analysis of affinity proteomics data. *J. Proteome Res.* **2016**, *15*, 4747–4754. [CrossRef]

27. Mellacheruvu, D.; Wright, Z.; Couzens, A.L.; Lambert, J.P.; St-Denis, N.A.; Li, T.; Miteva, Y.V.; Hauri, S.; Sardiu, M.E.; Low, T.Y.; et al. The crapome: A contaminant repository for affinity purification-mass spectrometry data. *Nat. Methods* **2013**, *10*, 730–736. [CrossRef]

28. Martin, A.; Ochagavia, M.E.; Rabasa, L.C.; Miranda, J.; Fernandez-de-Cossio, J.; Bringas, R. Bisogenet: A new tool for gene network building, visualization and analysis. *BMC Bioinform.* **2010**, *11*, 91. [CrossRef]

29. Nepusz, T.; Yu, H.; Paccanaro, A. Detecting overlapping protein complexes in protein-protein interaction networks. *Nat. Methods* **2012**, *9*, 471–472. [CrossRef]

30. Maere, S.; Heymans, K.; Kuiper, M. Bingo: A cytoscape plugin to assess overrepresentation of gene ontology categories in biological networks. *Bioinformatics* **2005**, *21*, 3448–3449. [CrossRef]

31. Shannon, P.; Markiel, A.; Ozier, O.; Baliga, N.S.; Wang, J.T.; Ramage, D.; Amin, N.; Schwikowski, B.; Ideker, T. Cytoscape: A software environment for integrated models of biomolecular interaction networks. *Genome Res.* **2003**, *13*, 2498–2504. [CrossRef] [PubMed]

32. Saldanha, A.J. Java treeview–extensible visualization of microarray data. *Bioinformatics* **2004**, *20*, 3246–3248. [CrossRef] [PubMed]

33. Longo, P.A.; Kavran, J.M.; Kim, M.S.; Leahy, D.J. Transient mammalian cell transfection with polyethylenimine (pei). *Methods Enzymol.* **2013**, *529*, 227–240. [PubMed]

34. Zybailov, B.L.; Florens, L.; Washburn, M.P. Quantitative shotgun proteomics using a protease with broad specificity and normalized spectral abundance factors. *Mol. Biosyst.* **2007**, *3*, 354–360. [CrossRef] [PubMed]

35. Naveed, M.; Kazmi, S.K.; Amin, M.; Asif, Z.; Islam, U.; Shahid, K.; Tehreem, S. Comprehensive review on the molecular genetics of autosomal recessive primary microcephaly (mcph). *Genet. Res.* **2018**, *100*, e7. [CrossRef] [PubMed]

36. Neumann, B.; Walter, T.; Heriche, J.K.; Bulkescher, J.; Erfle, H.; Conrad, C.; Rogers, P.; Poser, I.; Held, M.; Liebel, U.; et al. Phenotypic profiling of the human genome by time-lapse microscopy reveals cell division genes. *Nature* **2010**, *464*, 721–727. [CrossRef]

37. Gudjonsson, T.; Altmeyer, M.; Savic, V.; Toledo, L.; Dinant, C.; Grofte, M.; Bartkova, J.; Poulsen, M.; Oka, Y.; Bekker-Jensen, S.; et al. Trip12 and ubr5 suppress spreading of chromatin ubiquitylation at damaged chromosomes. *Cell* **2012**, *150*, 697–709. [CrossRef]

38. Savidis, G.; McDougall, W.M.; Meraner, P.; Perreira, J.M.; Portmann, J.M.; Trincucci, G.; John, S.P.; Aker, A.M.; Renzette, N.; Robbins, D.R.; et al. Identification of zika virus and dengue virus dependency factors using functional genomics. *Cell Rep.* **2016**, *16*, 232–246. [CrossRef]

39. Marceau, C.D.; Puschnik, A.S.; Majzoub, K.; Ooi, Y.S.; Brewer, S.M.; Fuchs, G.; Swaminathan, K.; Mata, M.A.; Elias, J.E.; Sarnow, P.; et al. Genetic dissection of flaviviridae host factors through genome-scale crispr screens. *Nature* **2016**, *535*, 159–163. [CrossRef]

40. Ma, H.; Dang, Y.; Wu, Y.; Jia, G.; Anaya, E.; Zhang, J.; Abraham, S.; Choi, J.G.; Shi, G.; Qi, L.; et al. A crispr-based screen identifies genes essential for west-nile-virus-induced cell death. *Cell Rep.* **2015**, *12*, 673–683. [CrossRef]

41. Liu, Y.; Zhang, Y.D.; Guo, L.; Huang, H.Y.; Zhu, H.; Huang, J.X.; Liu, Y.; Zhou, S.R.; Dang, Y.J.; Li, X.; et al. Protein inhibitor of activated stat 1 (pias1) is identified as the sumo e3 ligase of ccaat/enhancer-binding protein beta (c/ebpbeta) during adipogenesis. *Mol. Cell. Biol.* **2013**, *33*, 4606–4617. [CrossRef] [PubMed]

42. Rabellino, A.; Carter, B.; Konstantinidou, G.; Wu, S.Y.; Rimessi, A.; Byers, L.A.; Heymach, J.V.; Girard, L.; Chiang, C.M.; Teruya-Feldstein, J.; et al. The sumo e3-ligase pias1 regulates the tumor suppressor pml and its oncogenic counterpart pml-rara. *Cancer Res.* **2012**, *72*, 2275–2284. [CrossRef] [PubMed]

43. Su, C.I.; Tseng, C.H.; Yu, C.Y.; Lai, M.M.C. Sumo modification stabilizes dengue virus nonstructural protein 5 to support virus replication. *J. Virol.* **2016**, *90*, 4308–4319. [CrossRef] [PubMed]

44. Golemis, E.A.; Gyuris, J.; Brent, R. Two-hybrid system/interaction traps. In *Current protocols in Molecular Biology*; Ausubel, F.M., Brent, R., Kingston, R., Moore, D., Seidman, J., Smith, J.A., Struhl, K., Eds.; John Wiley & Sons: New York, NY, USA, 1994; pp. 13.14.11–13.14.17.

45. Aebersold, R.; Mann, M. Mass spectrometry-based proteomics. *Nature* **2003**, *422*, 198–207. [CrossRef]

46. Yu, H.; Braun, P.; Yildirim, M.A.; Lemmens, I.; Venkatesan, K.; Sahalie, J.; Hirozane-Kishikawa, T.; Gebreab, F.; Li, N.; Simonis, N.; et al. High-quality binary protein interaction map of the yeast interactome network. *Science* **2008**, *322*, 104–110. [CrossRef]

47. Rajagopala, S.V.; Sikorski, P.; Caufield, J.H.; Tovchigrechko, A.; Uetz, P. Studying protein complexes by the yeast two-hybrid system. *Methods* **2012**, *58*, 392–399. [CrossRef]

48. Sprinzak, E.; Sattath, S.; Margalit, H. How reliable are experimental protein-protein interaction data? *J. Mol. Biol.* **2003**, *327*, 919–923. [CrossRef]

49. Ito, T.; Chiba, T.; Ozawa, R.; Yoshida, M.; Hattori, M.; Sakaki, Y. A comprehensive two-hybrid analysis to explore the yeast protein interactome. *Proc. Nat. Acad. Sci. USA* **2001**, *98*, 4569–4574. [CrossRef]

50. Giot, L.; Bader, J.S.; Brouwer, C.; Chaudhuri, A.; Kuang, B.; Li, Y.; Hao, Y.L.; Ooi, C.E.; Godwin, B.; Vitols, E.; et al. A protein interaction map of drosophila melanogaster. *Science* **2003**, *302*, 1727–1736. [CrossRef]

51. Li, S.; Armstrong, C.M.; Bertin, N.; Ge, H.; Milstein, S.; Boxem, M.; Vidalain, P.O.; Han, J.D.; Chesneau, A.; Hao, T.; et al. A map of the interactome network of the metazoan *C. elegans*. *Science* **2004**, *303*, 540–543. [CrossRef]

52. Freund, A.; Zhong, F.L.; Venteicher, A.S.; Meng, Z.; Veenstra, T.D.; Frydman, J.; Artandi, S.E. Proteostatic control of telomerase function through tric-mediated folding of tcab1. *Cell* **2014**, *159*, 1389–1403. [CrossRef] [PubMed]

53. Lafarga, M.; Tapia, O.; Romero, A.M.; Berciano, M.T. Cajal bodies in neurons. *RNA Biol.* **2017**, *14*, 712–725. [CrossRef] [PubMed]

54. Hafirassou, M.L.; Meertens, L.; Umana-Diaz, C.; Labeau, A.; Dejarnac, O.; Bonnet-Madin, L.; Kummerer, B.M.; Delaugerre, C.; Roingeard, P.; Vidalain, P.O.; et al. A global interactome map of the dengue virus ns1 identifies virus restriction and dependency host factors. *Cell Rep.* **2017**, *21*, 3900–3913. [CrossRef] [PubMed]

55. Gomez-Ferreria, M.A.; Rath, U.; Buster, D.W.; Chanda, S.K.; Caldwell, J.S.; Rines, D.R.; Sharp, D.J. Human cep192 is required for mitotic centrosome and spindle assembly. *Curr. Biol.* **2007**, *17*, 1960–1966. [CrossRef] [PubMed]

56. Zhu, F.; Lawo, S.; Bird, A.; Pinchev, D.; Ralph, A.; Richter, C.; Muller-Reichert, T.; Kittler, R.; Hyman, A.A.; Pelletier, L. The mammalian spd-2 ortholog cep192 regulates centrosome biogenesis. *Curr. Biol.* **2008**, *18*, 136–141. [CrossRef] [PubMed]

57. Joukov, V.; Walter, J.C.; De Nicolo, A. The cep192-organized aurora a-plk1 cascade is essential for centrosome cycle and bipolar spindle assembly. *Mol. Cell* **2014**, *55*, 578–591. [CrossRef]

58. Firat-Karalar, E.N.; Rauniyar, N.; Yates, J.R., 3rd; Stearns, T. Proximity interactions among centrosome components identify regulators of centriole duplication. *Curr. Biol.* **2014**, *24*, 664–670. [CrossRef]

59. Gupta, G.D.; Coyaud, E.; Goncalves, J.; Mojarad, B.A.; Liu, Y.; Wu, Q.; Gheiratmand, L.; Comartin, D.; Tkach, J.M.; Cheung, S.W.; et al. A dynamic protein interaction landscape of the human centrosome-cilium interface. *Cell* **2015**, *163*, 1484–1499. [CrossRef]

60. Barbelanne, M.; Tsang, W.Y. Molecular and cellular basis of autosomal recessive primary microcephaly. *Biomed. Res. Int.* **2014**, *2014*, 547986. [CrossRef]

61. Grant, A.; Ponia, S.S.; Tripathi, S.; Balasubramaniam, V.; Miorin, L.; Sourisseau, M.; Schwarz, M.C.; Sanchez-Seco, M.P.; Evans, M.J.; Best, S.M.; et al. Zika virus targets human stat2 to inhibit type i interferon signaling. *Cell Host Microbe* **2016**, *19*, 882–890. [CrossRef]
62. Brown, J.R.; Conn, K.L.; Wasson, P.; Charman, M.; Tong, L.; Grant, K.; McFarlane, S.; Boutell, C. Sumo ligase protein inhibitor of activated stat1 (pias1) is a constituent promyelocytic leukemia nuclear body protein that contributes to the intrinsic antiviral immune response to herpes simplex virus 1. *J. Virol.* **2016**, *90*, 5939–5952. [CrossRef] [PubMed]

Article

Participation of Extracellular Vesicles from Zika-Virus-Infected Mosquito Cells in the Modification of Naïve Cells' Behavior by Mediating Cell-to-Cell Transmission of Viral Elements

Pedro Pablo Martínez-Rojas [1], Elizabeth Quiroz-García [1], Verónica Monroy-Martínez [1], Lourdes Teresa Agredano-Moreno [2], Luis Felipe Jiménez-García [2] and Blanca H. Ruiz-Ordaz [1,*]

[1] Departamento de Biología Molecular y Biotecnología, Instituto de Investigaciones Biomédicas, Universidad Nacional Autónoma de México, Av. Universidad 3000, Ciudad Universitaria, Coyoacán, Ciudad de México 04510, México; pedropablo.martinezrojas@gmail.com (P.P.M.-R.); geliqg@gmail.com (E.Q.-G.); vmonroy@iibiomedicas.unam.mx (V.M.-M.)

[2] Departamento de Biología Celular, Facultad de Ciencias, Universidad Nacional Autónoma de México, Av. Universidad 3000, Ciudad Universitaria, Coyoacán, Ciudad de México 04510, México; agredano-moreno@ciencias.unam.mx (L.T.A.-M.); luisfelipe_jimenez@ciencias.unam.mx (L.F.J.-G.)

* Correspondence: bhro@unam.mx or bhro@iibiomedicas.unam.mx; Tel.: +521-55-56228931

Received: 16 October 2019; Accepted: 31 December 2019; Published: 4 January 2020

Abstract: To date, no safe vaccine or antivirals for Zika virus (ZIKV) infection have been found. The pathogenesis of severe Zika, where host and viral factors participate, remains unclear. For the control of Zika, it is important to understand how ZIKV interacts with different host cells. Knowledge of the targeted cellular pathways which allow ZIKV to productively replicate and/or establish prolonged viral persistence contributes to novel vaccines and therapies. Monocytes and endothelial vascular cells are the main ZIKV targets. During the infection process, cells are capable of releasing extracellular vesicles (EVs). EVs are mediators of intercellular communication. We found that mosquito EVs released from ZIKV-infected (C6/36) cells carry viral RNA and ZIKV-E protein and are able to infect and activate naïve mosquito and mammalian cells. ZIKV C6/36 EVs promote the differentiation of naïve monocytes and induce a pro-inflammatory state with tumor necrosis factor-alpha (TNF-α) mRNA expression. ZIKV C6/36 EVs participate in endothelial vascular cell damage by inducing coagulation (TF) and inflammation (PAR-1) receptors at the endothelial surface of the cell membranes and promote a pro-inflammatory state with increased endothelial permeability. These data suggest that ZIKV C6/36 EVs may contribute to the pathogenesis of ZIKV infection in human hosts.

Keywords: extracellular vesicles; Zika virus; cellular communication; C6/36 cells; human monocytes; endothelial vascular cells

---

1. Introduction

Zika virus (ZIKV) is an emerging arthropod-borne *Flavivirus*, transmitted mainly by mosquitoes of the genus *Aedes*, but the ZIKV infection could also be produced by sexual contact or vertical transmission from mother to child [1,2]. ZIKV was first isolated in 1947 from the blood of a sentinel Rhesus monkey No. 766, stationed in the Zika forest in Uganda. Again, in 1948, ZIKV was isolated in the same forest from a pool of *Aedes (Ae) africanus* mosquitoes. Thereafter, serological and entomological data indicated that ZIKV circulates actively in East and West Africa and South-East Asia. In 2007, ZIKV caused an outbreak of relatively mild disease characterized by rash, arthralgia, and conjunctivitis on Yap Island in the Southwestern Pacific Ocean. This was the first time that the virus was detected outside of Africa [3].

Later, a ZIKV epidemic in Brazil was present in 2015 and spread rapidly throughout South and Central America in 2016. The Pan American Health Organization (PAHO) has received reports of more than 7.5×10^5 cases of Zika in 84 cities or territories in America [3,4]. The ZIKV infection during pregnancy can cause fetal loss, microcephaly, and other brain abnormalities that are classified as congenital Zika syndrome [5,6]. Further, severe forms of encephalopathies, meningoencephalitis, myelitis, uveitis, autoimmunity (Guillain-Barré syndrome), and severe thrombocytopenia have been associated with ZIKV infection [7,8]. The pathogenic mechanisms that give rise to severe forms of Zika are still unclear, and to date, no safe vaccine or specific antiviral treatments for ZIKV infection are available [9]. A rapid and successful expansion of ZIKV has occurred due to the high virulence of circulating strains, immunologically susceptible populations, and the wide distribution of its vectors [10,11].

Ae. aegypti and *Ae. albopictus* mosquitoes are the primary vectors of several *Flavivirus* such as ZIKV and dengue virus (DENV) [12]. Female mosquitoes acquire the virus from an infected host during feeding, it undergoes replication in the gut and disseminates to the salivary glands, and the virus is released into the saliva, where it is transmitted to the host during subsequent feeding [13,14]. Cime et al. (2015) reported that *Ae. aegypti* saliva plays an important role during DENV transmission to the host cells. Likewise, they detected an enhanced viral infection of mammalian cells in the presence of mosquito salivary gland extract [15]. However, the mechanisms in the transmission of *Flavivirus* from vector to host are not entirely understood [16]. In human hosts, monocytes, macrophages, endothelial vascular cells, and central nervous system cells are identified as main ZIKV target cells [17–19]. During differentiation or activation, cells release extracellular vesicles (EVs) [20]. EVs are considered crucial mediators of intercellular communication and play a role in the pathophysiology of inflammation-associated disorders [21].

EVs are a heterogeneous group of particles naturally released by the cells, delimited by a lipid bilayer, and cannot replicate. The classification proposed by the International Society of Extracellular Vesicles (ISEV) has established that EVs can be distinguished by their biogenesis. Vesicles are derived from the plasma membrane (microparticles [MPs]) and are also derived from endosomal maturation (exosomes). Further, they differ in size, where the MPs (> 200 nm) are grouped as large EVs (lEVs), and the exosomes (< 200 nm) are grouped as small EVs (sEVs) [22]. These EVs can be identified by the presence of different membrane markers (phosphatidylserine [PS] in lEVs or tetraspanins in sEVs) or by their internal content, since they transport active biomolecules (proteins and different types of RNA) capable of modifying the response of the cells with which they interact [22,23].

Small EVs are formed as intraluminal vesicles within multivesicular bodies during the endosome maturation process and released into the extracellular space through highly specialized cellular secretory pathways [24]. During the infectious process by some RNA viruses such as flaviviruses, the viral replication cycle and the biogenesis of sEVs can converge, so different viral components (antigens, genomes, or complete viruses) can be part of the internal content, being potential vehicles for viral transmission, evasion of the host's immune response, and the enhancement of pathophysiological processes by promoting the spread of the pathogen to immunologically privileged sites [25,26]. Therefore, sEVs are considered a new, alternative mechanism that is efficient for viral spread [27]. Large EVs are formed by cytoskeleton rearrangement and released from the plasma membrane after the cell activation process [24]. In blood circulation, MPs facilitate cell–cell interaction and induce different responses associated with inflammation, thrombosis, or vascular dysfunction [28]. Virus-infected cells secrete lEVs that may contain viral proteins and RNAs [29]. Little is known about the EV participation function in the vector–human host interaction during the flaviviruses transmission-infection processes. Recently, Vora et al. (2018) reported that DENV-infected mosquito cells release EVs that contain infectious DENV RNA and proteins, favoring viral transmission from the vector to human keratinocytes and endothelial cells (ECs) [16]. Likewise, Reyes-Ruiz et al. (2019) reported that sEVs from DENV-infected mosquito cells have homologous proteins to human CD9 tetraspanin, containing virion-like particles inside them [30].

To date, the participation of EVs released from *Aedes* mosquito cells during the ZIKV infection process has not been described. This study aimed to evaluate the potential role of EVs from ZIKV-infected C6/36 cells in viral-element, cell-to-cell transmission to the main host's target cells (monocytes and endothelial vascular cells) as well in the naïve cellular behavior modification. EVs from ZIKV-infected mosquito cells were then isolated by differential ultracentrifugation, characterized by nanoparticle tracking analysis, identified by transmission electron microscopy, and subject to phosphatidylserine (MPs) and tetraspanin CD63 (exosomes) detection by cytofluorometry assays. The isolated sEVs were purified by using paramagnetic beads coated with anti-CD63 antibodies, thus demonstrating their endosomal origin. The possible modification of cellular behavior mediated by ZIKV C6/36 EVs was evaluated by using different cell activation assays. Our results support that EVs (small and large) from ZIKV-infected *Aedes* mosquito cells modify host cells responses, which could be implicated in the pathogenic mechanisms associated with the progression to severe forms of the disease.

2. Materials and Methods

2.1. Cell Cultures and Zika Virus Strain

Larvae lysate cells (C6/36) from mosquitos *Aedes albopictus* (ATCC CRL-1660, USA), kidney epithelial cells (Vero) from monkeys *Cercopithecus aethiops* (ATCC CCL-81), human monocytes (THP-1) from peripheral blood (ATCC TIB-202), and human endothelial cells (HMEC-1) from the dermal microvasculature (ATCC CRL-3243) were used in this study. The C6/36 cells were maintained in Leibovitz L15 medium (Biowest, Riverside, MO, USA) supplemented with 10% (*v/v*) fetal bovine serum (FBS; Biowest, Nuaillé, France), 10% tryptose phosphate broth (DIFCO, Lawrence, KS, USA), 1% 200 mM L-glutamine (Biowest), and 1% antibiotic solution (10,000 U/mL penicillin, 10 mg/mL streptomycin, and 25 µg/mL amphotericin B; Biological Industries, Cromwell, CT, USA) and were incubated at 28 °C without CO_2 (Lab-Line Ambi-Hi-Low Chamber, Lab-Line Instruments Inc., Melrose Park, IL, USA). The Vero and THP-1 cells were maintained in Dulbecco's Modified Eagle medium (DMEM, Biowest) and Roswell Park Memorial Institute (RPMI) 1640 medium (Biowest), respectively. Both media were supplemented with 10% Fetal Bovine Serum (FBS), 1% 200 mM L-glutamine, and 1% antibiotic solution. The HMEC-1 cells were maintained in MCDB-131 medium (Sigma-Aldrich, St. Louis, MO, USA) supplemented with 10% FBS, 1% 200 mM L-glutamine, 10 ng/mL epidermal growth factor (Sigma-Aldrich), 1 µg/mL hydrocortisone (Sigma-Aldrich), and 1% antibiotic solution. The Vero, THP-1, and HMEC-1 cells were incubated at 37 °C in 5% CO_2 (Series 8000 WJ CO_2 Incubator, Thermo Fisher Scientific, Waltham, MA, USA). Dr. Amadou A. Sall, from Institut Pasteur Dakar, kindly provided the reference strain ZIKV MR766 (Genebank Accession HQ234498.1), with the following passage history: 146 × in suckling mouse, 1 × C6/36 cells, and 1 × Vero cells. Additionally, in our laboratory, it was passaged twice in C6/36 cells.

2.2. ZIKV Propagation and Titration

The ZIKV MR766 strain with passage (P) number 3 was propagated in C6/36 cells and titrated in confluent Vero monolayers and was used in all experiments. The ZIKV titer was determined by the lytic plaque assay as follows: Vero cells were seeded into a 24-well culture plate (Corning, Corning, NY, USA) and incubated at 37 °C in 5% CO_2 until confluence. Each culture well was inoculated with 400 µL of diluted ZIKV (serial log (10-fold) dilutions in serum-free medium with dilution factors from 10^{-1} up to 10^{-22}) in duplicate and incubated for 2 h at 37 °C in 5% CO_2. The viral inoculum was removed, and each well was washed with Phosphate Buffer Saline (PBS) 1× (pH 7.4). The monolayers were overlaid with 1 mL of DMEM medium containing 1% carboxymethylcellulose (Sigma, USA) and 2.5% FBS and afterwards incubated at 37 °C in 5% CO_2. The plaque formation began to be observed at the 7th day post-infection (PI). Cells were fixed with 96% methanol (J.T.Baker, Fisher Scientific, Allentown, PA, USA) on the 14th day PI and stained with 1% crystal violet (Sigma) for 15 min. The titer was calculated and expressed as plaque-forming units (PFU) per milliliter (mL), according to the following formula: PFU/mL = $N/(V \times D)$, where N corresponds to the average number of plaques counted, V is the inoculated volume of the viral dilution, and D is the less concentrated dilution from which the plaques were counted.

2.3. Preparation of Fetal Bovine Serum Depleted of Extracellular Vesicles (EVs)

The FBS was collected in sterile conical tubes (Corning) and centrifuged (GH3.8 rotor, Beckman GPR Centrifuge; Beckman Coulter, Inc., Brea, CA, USA) at 900× g for 10 min and filtered using a 0.22 μm pore (Millipore, Burlington, MA, USA). The samples were then centrifuged (SW28 rotor, Beckman XL-90 Ultracentrifuge, Beckman Coulter, Inc.) at 120,000× g for 18 h at 4 °C [22] and stored at 2 °C until use.

2.4. ZIKV Infection Assay

Cell cultures (mosquito C6/36, human monocytes, or endothelial vascular cells) were seeded in 12-well plate culture (Corning) until confluence. They were then infected with ZIKV at a multiplicity of infection (MOI) of 1, as previously described [31–33] and incubated for 2 h at 37 °C in 5% CO_2. After removal of the viral inoculum, cells were maintained in media supplemented with 5% EV-depleted FBS and incubated for 24, 48, 72, 96, and 120 h. Cytopathic effect formation was observed by light field microscopy (Olympus IX71 inverted microscope; Olympus Corp. Miami, FL, USA). Images were taken with a digital camera (Olympus DP72) attached to the microscope and analyzed with ImageJ software version 1.50i (Wayne Rasband, National Institutes of Health, Bethesda, MA, USA). The cell infection was evaluated by ZIKV envelope (E) protein detection as described below.

2.5. ZIKV Envelope (E) Protein Detection in ZIKV-Infected Cells by Cytofluorometry (FACS)

The ZIKV E protein detection at the cell membrane's surface was performed as follows: The different cells were removed from the cell culture plates (C6/36, THP-1, or HMEC-1 were scrapped and homogenized by vigorous pipetting) in sterile 1.5 mL microcentrifuge tubes (Labcon, Petaluma, CA, USA), centrifuged (Eppendorf Centrifuge 5415 R; Eppendorf International, Hamburg, Germany) at 550× g for 10 min at 4 °C, and separated from the medium. The cell pellets were fixed with 2% paraformaldehyde (Sigma) for 5 min at 4 °C, blocked for nonspecific binding sites with 2% bovine serum albumin (BSA; Biowest) for 30 min at RT, and washed with 0.5% BSA. The samples were stained with mouse anti-ZIKV E protein clone 1,413,267 antibody (Catalog #CABT-B8528, CD Creative Diagnostics, New York, NY, USA) at a 1:300 dilution in 0.5% BSA and incubated overnight at 4 °C with constant 1000 rpm agitation (Vibrax VXR basic; IKA, Wilmington, NC, USA). After washing with 0.5% BSA and centrifugation, the Alexa Fluor 555-conjugated anti-mouse IgG (H + L), a highly cross-adsorbed secondary antibody (Catalog #A-31570, Thermo Fisher Scientific), was added at a 1:500 dilution in 0.5% BSA, incubated for 2 h at room temperature (RT) with constant 1000 rpm agitation, washed with 0.5% BSA, and centrifugated. The samples were suspended in 300 μL of 0.5% BSA and analyzed by the FACSCalibur flow cytometer (BD Biosciences, San Jose, CA, USA) with CellQuest software. The mock cells were treated in the same way as the infected cells (the mock FACS average values were rested from the infected FACS values).

2.6. ZIKV Envelope (E) Protein Detection in ZIKV-Infected Cells by Immunofluorescence (IF)

To confirm the ZIKV E protein presence at the cell membrane surface level, an IF assay was performed as follows: The mosquito C6/36 cells were seeded on an 8-well separation chamber slide system (Lab-Tek II; Thermo Fisher Scientific), incubated until confluence, and infected as described above. The cells were fixed with 2% paraformaldehyde for 5 min at 4 °C, blocked for nonspecific binding sites with 2% BSA for 30 min at RT, and washed with 0.5% BSA. The cells were stained with mouse anti-ZIKV E protein antibody at a 1:300 dilution in 0.5% BSA and incubated overnight at 4 °C with constant 1000 rpm agitation. After washing with 0.5% BSA, the Alexa Fluor 555-conjugated anti-mouse IgG was added at a 1:500 dilution in 0.5% BSA, incubated for 2 h at RT, and washed with 0.5% BSA. The separation chamber was then withdrawn, the slide was covered with mounting medium with DAPI (FluoroQuest; AAT Bioquest, Sunnyvale, CA, USA), and a coverslip (Corning) was placed. The samples were observed via fluorescence microscopy (Olympus IX71 inverted microscope; Olympus Corp.), and the images were analyzed with ImageJ software.

2.7. Mosquito C6/36 EVs Isolation from the Cell Culture Medium by Ultracentrifugation

Mosquito C6/36 cells were seeded in cell culture flasks T75 (Corning). The EV isolation from the culture media of the mock and ZIKV-infected cell culture flasks were performed by ultracentrifugation (Figure S1). Briefly, C6/36 cells culture media (50 mL) were collected in sterile conical tubes and centrifuged (GH3.8 rotor, Beckman GPR Centrifuge, Beckman Coulter, Inc.) at 900× g for 10 min at 4 °C. The viable cell pellet was discarded. The supernatant was transferred to sterile conical tubes and centrifuged at 2000× g for 10 min at 4 °C. The debris pellet was discarded. The supernatant was transferred to 25 × 89 mm centrifuge tubes (Beckman Coulter, Inc.) and centrifuged (SW28 rotor, Beckman XL-90 Centrifuge) at 10,000× g for 35 min at 4 °C. The lEVs pellet was suspended in 1 mL of PBS at 4 °C and used immediately or stored at −72 °C (Revco ULT1786, Thermo Fisher Scientific). The supernatant was transferred to centrifuge tubes and centrifuged at 120,000× g for 70 min at 4 °C. The supernatant was then discarded. The sEVs and contaminant protein pellet was washed in 5 mL of PBS at 4 °C and incubated for 30 min at RT with constant 100 rpm agitation. The suspension was filtered with a 0.22 μm pore filter, and 25 mL of PBS was added. The samples were transferred to centrifuge tubes and centrifuged at 120,000× g for 70 min at 4 °C. The last supernatant obtained in this process was separated and identified as non-EV ZIKV SNT (used as a control in the EV stimulation assays) and stored at −72 °C. The sEVs pellet was suspended in 1 mL of PBS at 4 °C and used immediately or stored at −72 °C [22,34].

2.8. Characterization of EVs from Mosquito C6/36 Cells by Nanoparticle Tracking Analysis (NTA)

The characterization of EVs was performed by the detection of reflected light emitted by the Brownian motion of nanoparticles suspended in solution by nanoparticle tracking analysis (NTA) with the help of the NanoSight NS300 equipment and Malvern Instruments software. The optimal detection conditions were previously established (Figure S2A), and quantitative controls with 100 and 200 nm polystyrene microspheres (NTA4088 and NTA4089, Malvern Panalytical Products, Mexico City, Mexico) were used (Figure S2B). The nanoparticle concentration values (particles/mL) and the size (nm) were determined for each measurement. The ZIKV virions were detected (Figure S2C) to identify their presence in the EV samples from ZIKV-infected mosquito cells. In a parallel assay, the nanoparticles present in PBS and in FBS-EV-depleted were quantified to rest the number of nanoparticles obtained in the mock and ZIKV-infected C6/36 EV isolates.

2.9. C6/36 lEVs Phosphatidylserine (PS)+ Detection by an Annexin-V Binding Assay

Phosphatidylserine (PS) is located on the cytoplasmic surface of the cell plasmatic membrane, and, during the lEVs biogenesis, PS is translocated from the inner to the outer leaflet of the membrane, exposing the PS that can be detected by the Annexin-V binding assay [35]. The lEVs samples (50 μL) from C6/36 cells were suspended in 200 μL of the Annexin-V binding buffer 1× (Catalog #556454, BD Pharmingen, BD Biosciences, San Jose, CA, USA) that contained fluorescein isothiocyanate (FITC)-conjugated Annexin-V (Catalog #640906, BioLegend, San Diego, CA, USA) at a 1:200 dilution and were incubated for 20 min at RT with constant 100 rpm agitation. After a wash with 250 μL of PBS and centrifugation at 10,000× g for 35 min at 4 °C, samples were suspended in 300 μL of PBS. Polystyrene microspheres of a 1 μm diameter size (Polysciences, Inc., Warrington, PA, USA) were used as a FACS calibration control. The samples were analyzed by a FACSCalibur flow cytometer.

2.10. Mosquito C6/36 Cell Tetraspanin CD63-Like Protein Detection by FACS

Tetraspanin-like proteins in arthropod cells, including mosquito C6/36 cells, have been described previously [16,30,36–38]. Briefly, the C6/36 cells (mock and ZIKV-infected cells) were collected from the cell culture plates in sterile 1.5 mL microcentrifuge tubes, centrifuged at 550× *g* for 10 min at 4 °C, and separated from the medium. The cell pellets were fixed with 2% paraformaldehyde for 5 min at 4 °C, permeabilized with 0.1% Triton X-100 (Sigma) for 5 min at 4 °C, blocked for nonspecific binding sites with 2% BSA for 30 min at RT, and washed with 0.5% BSA. The cells were stained with the phycoerythrin (PE)-conjugated mouse anti-human CD63 antibody (Catalog #557305, BD Pharmingen) at a 1:20 dilution in 0.5% BSA and incubated for 1 h at RT with constant 1000 rpm agitation. The samples were suspended in 300 µL of 0.5% BSA and analyzed by using the FACSCalibur flow cytometer. The mouse IgG1 kappa (P3.6.2.8.1) antibody (Catalog #14-4714-82, eBioscience, San Diego, CA, USA) was used as an isotype control. Isotype FACS average values were rested from the mock and the infected cells FACS values.

2.11. Mosquito C6/36 Cells Tetraspanin CD63-Like Protein Detection by Immunofluorescence Assay

To confirm the CD63-like protein presence at the cells membrane's surface and in the cytosol, the immunofluorescence (IF) assay was performed as follows: The C6/36 cells were seeded on an 8-well separation chamber slide system, incubated until confluence, and infected as described above. The cells were fixed with 2% paraformaldehyde for 5 min at 4 °C, permeabilized with 0.1% Triton X-100 (Sigma) for 5 min at 4 °C, blocked for nonspecific binding sites with 2% BSA for 30 min at RT, and washed with 0.5% BSA. The cells were stained with the PE-conjugated mouse anti-human CD63 antibody at a 1:20 dilution in 0.5% BSA and incubated for 1 h at RT. The mouse IgG1 kappa antibody was used as an isotype control. Finally, the mock and the ZIKV-infected cells were stained with mouse anti-ZIKV E protein antibody at a 1:300 dilution in 0.5% BSA and incubated overnight at 4 °C with constant 1000 rpm agitation. After a wash with 0.5% BSA, the FITC-conjugated anti-mouse IgG (H + L), a highly cross-adsorbed secondary antibody (Catalog #AP308F, Merck, Kennersburg, NJ, USA), was added at a 1:500 dilution in 0.5% BSA, incubated for 2 h at RT with constant 100 rpm agitation, and washed with 0.5% BSA. The separation chamber was withdrawn, the slide was covered with mounting medium with DAPI, and a coverslip was placed. The samples were observed by fluorescence microscopy, and the images were analyzed with ImageJ software.

2.12. C6/36 sEVs CD63+ Detection (FACS) by Coupling to Anti-CD63-Coated Paramagnetic Nanobeads

To identify the presence of the tetraspanin CD63 (sEVs marker) on the sEVs membrane's surface, the C6/36 sEVs isolates were coupled with anti-CD63-coated paramagnetic nanobeads (Catalog #10606D; Invitrogen, Thermo Fisher Scientific) to be detected by cytofluorometry (Figure S3). Briefly, in sterile round-bottom microcentrifuge tubes (Labcon), 100 µL of the sEVs suspension and 20 µL of the paramagnetic nanobeads were added. The samples were incubated for 24 h at 4 °C with constant 1000 rpm agitation. After a wash with 300 µL with PBS, magnetic separation of the bead-coupled sEVs from the matrix suspension was performed using a DynaMag-2 magnet (Life Technologies, Thermo Fisher Scientific), and the supernatant was discarded. The bead-coupled sEVs were suspended in 300 µL of 0.5% BSA. For CD63 detection by FACS, the PE-conjugated mouse anti-human CD63 antibody was used as described above. The samples were analyzed using a FACSCalibur flow cytometer. The paramagnetic nanobeads were treated in the same way as the bead-coupled sEVs (the paramagnetic bead FACS average values of the bead-coupled sEVs were rested from mock and ZIKV-infected cells FACS values).

2.13. C6/36 EVs Morphological Characterization by Transmission Electron Microscopy (TEM)

The C6/36 EV isolates were fixed with a 1:1 mixture of 2.5% glutaraldehyde (Electron Microscopy Sciences [EMS], Hatfield, PA, USA) and 4% paraformaldehyde (Sigma) for 2 h at RT and washed three times (5 min each) with PBS. After fixation, the samples were incubated with 2% osmium tetroxide (Alfa Aesar, Thermo Fisher Scientific) for 90 min at RT. The fixed pellets were washed three times with PBS and dehydrated in an ascending graded series of ethanol (30, 50, 70, 80, 90, and 96%), including three passes (5 min each) in absolute ethanol (J.T.Baker) at RT. Three passes (5 min each) in propylene oxide (Sigma-Aldrich) at RT were then performed. The samples were placed in a 1:1 mixture of propylene oxide/epoxy resin for 18 h at RT and embedded in pure epoxy resin (EMS) at 60 °C for

48 h. Ultrathin sections (40–50 nm thick) were obtained in an ultramicrotome (Leica EM UC7, Leica Microsystems, Buffalo Grove, IL, USA) and mounted on copper grids (EMS) covered with formvar (EMS). The sections were contrasted with uranyl acetate (Merck, Kennerworth Fort, NJ, USA) for 30 min and lead citrate (EMS) for 10 min at RT. The preparations were observed with a transmission electron microscope (JEM1010 model; JEOL, Peabody, MA, USA) operating at 80 kV. The images were captured with a CCD300-RC camera (DAGE-MTI, Michigan City, IN, USA) adapted to the microscope and analyzed with ImageJ software.

2.14. ZIKV E Protein Detection in lEVs Isolates from ZIKV-Infected C6/36 Cells by FACS

The ZIKV E protein detection at the lEVs membrane's surface was performed as follows: The lEVs isolates were centrifuged at 10,000× *g* for 35 min at 4 °C, and the supernatant was discarded. The lEVs pellets were fixed with 2% paraformaldehyde for 5 min at 4 °C, blocked for unspecific binding sites with 2% BSA for 30 min at RT, and washed with 0.5% BSA. The lEVs were stained with mouse anti-ZIKV E protein antibody at a 1:300 dilution in 0.5% BSA and incubated overnight at 4 °C with constant 1000 rpm agitation. After washing with 0.5% BSA and centrifugation, the FITC-conjugated anti-mouse IgG secondary antibody was added at a 1:500 dilution in 0.5% BSA, incubated for 2 h at RT with constant 1000 rpm agitation, washed with 0.5% BSA, and centrifugated. The samples were suspended in 300 µL of 0.5% BSA and analyzed by the FACSCalibur flow cytometer with CellQuest software. The lEVs from mock cells were treated in the same way as those from the infected cells (the lEVs mock C6/36 FACS average values were rested of the lEVs ZIKV C6/36 FACS values).

2.15. ZIKV E Protein Detection in sEVs Isolates from ZIKV-Infected C6/36 Cells by FACS

The ZIKV E protein detection at the sEVs membrane's surface was performed as follows: The sEVs isolates were coupled with anti-CD63-coated paramagnetic beads as described above. The bead-coupled sEVs were fixed with 2% paraformaldehyde for 5 min at 4 °C, blocked for nonspecific binding sites with 2% BSA for 30 min at RT, and washed with 0.5% BSA. The bead-coupled sEVs were stained with mouse anti-ZIKV E protein antibody at a 1:300 dilution in 0.5% BSA and incubated overnight at 4 °C with constant 1000 rpm agitation. After washing with 0.5% BSA and recovery by magnetic separation, the Alexa Fluor 555-conjugated anti-mouse IgG secondary antibody was added at a 1:500 dilution in 0.5% BSA, incubated for 2 h at RT with constant 1000 rpm agitation, washed with 0.5% BSA, and recovered by magnetic separation. The samples were suspended in 300 µL of 0.5% BSA and analyzed by the FACSCalibur flow cytometer with CellQuest software. The EVs from mock cells were treated in the same way as those from the infected cells (the sEVs mock C6/36 FACS average values were subtracted from the sEVs ZIKV C6/36 FACS values).

2.16. RNA Extraction and Purification

The RNA extraction (from ZIKV stock, C6/36 EVs isolates, or cells [C6/36, THP-1, HMEC-1] pellets) was performed with the centrifugation protocol using the QIAamp RNA Mini kit (Qiagen, Hilden, Germany) according to the manufacturer's recommendations. Briefly, the samples were first lysed under highly denaturing conditions using the lysis buffer. Next, 70% ethanol was added. Preparations were mixed by pulse/vortexing (Lab-Line Vortex Mixer, Alpha Multiservices, Conroe, TX, USA) for 15 s and incubated for 15 min at RT. The entire volume of the preparations was loaded onto the QIAamp Mini spin columns placed in separation tubes to promote the RNA binding to the columns' membranes. The tubes were centrifuged (Eppendorf 5415 C Centrifuge) at 6000× *g* for 2 min at RT at every step. The contaminants were washed in two steps using the Absorber Waste 1 (AW1) and AW2 buffers that were added onto the columns. The tubes were centrifuged at 6000× *g* for 2 min at RT at each step. Finally, eluent buffer AVE (RNase-free water with 0.4% sodium azide) was added to the columns, and they were centrifugated at 16,000× *g* for 2 min at RT. The RNA filtrates were collected at 4 °C. The RNA was quantified using a NanoDrop ND1000 Spectrophotometer (Thermo Fisher Scientific) with ND-1000 software version 3.5.2. The RT-PCR protocols were performed immediately, as described below, or samples were stored at −72 °C.

2.17. ZIKV Inactivation on Viral Stock Samples and ZIKV-Infected C6/36 EVs Isolates

To inactivate the ZIKV virions, viral stock samples were irradiated at 1200 µJ (× 100) in three consecutive cycles on a UV Stratalinker 1800 (Stratagene, San Diego, CA, USA), and genomic RNA was degraded by RNase A activity assays (Figure S4A). Briefly, the total RNA in the samples was quantified, and a concentration of 10 µg/mL of RNAse A (DNase and Protease-free; Thermo Fisher Scientific) was added. The mixtures were then incubated for 1 h, 30 min, and 15 min at 37 °C with 5% CO_2. A 1:1 proportion of RiboLock RNase Inhibitor (40 U/µL; Thermo Fisher Scientific) was used for 15 min at 37 °C with 5% CO_2 to inhibit RNase A activity. The RNA degradation pattern was visualized on 2% ethidium bromide-stained (Sigma) 1.2% agarose gel (Invitrogen) using a Typhoon FLA 9500 scanner (GE Healthcare, Chicago, IL, USA) with GE control software version 1.0. The images were analyzed with ImageJ software. The inactivated ZIKV (iZIKV) was evaluated by the lytic plaque assay. The mosquito ZIKV-infected C6/36 EV isolates were irradiated at 1200 µJ (× 100) in three consecutive cycles and treated with RNase A at the best condition of incubation (Figure S4B). The samples were used immediately or held at 4 °C. These treatments were identified as EVs (lEVs or sEVs) ZIKV C6/36 (RNase + UV) and were quantified by NTA, the RNA extraction was performed for ZIKV RNA detection by RT-PCR (as describe below), and the lytic plaque assay was performed to evaluate their plaque formation ability in Vero cells (as described above).

2.18. ZIKV RNA Detection by Polymerase Chain Reaction with Reverse Transcriptase (RT-PCR)

The ZIKV RNA detection in a single step by the RT-PCR reaction has been previously described [39,40]. The master mix was prepared according to the specifications given by the OneStep RT-PCR kit (Qiagen). The primers ZIKV FW [5′-GCTGGDGCRGACACHGGRACT-3′] (Mfg. ID 275853243, Integrated DNA Technologies [IDT], San Diego, CA, USA)] and ZIKV RV [5′-RTCYACYGCCATYTGGRCTG-3′] (Mfg. ID 275853246, IDT, USA)] developed by Faye et al. [40] were used. The reaction was performed on a GeneAmp PCR System 2400 (Applied Biosystems, Foster City, CA, USA) with the following conditions: pre-PCR at 50 °C for 40 min and 95 °C for 15 min; 35 cycles at 94 °C for 30 s, 45 °C for 30 s, and 72 °C for 1 min; final elongation at 72 °C for 7 min. The amplified cDNA (amplicon of 364 bp) corresponding to the more specific genome region for the ZIKV E protein, which has no cross reaction with other *Flavivirus* [40], was visualized in 2% ethidium bromide-stained 1.2% agarose gel using a Typhoon FLA 9500 scanner with GE control software. The images were analyzed with ImageJ software.

2.19. Quantification of the Total Protein from C6/36 EVs Isolates by Micro BCA Protein Assay

The protein quantification from the C6/36 EVs isolates was performed according to the specification given by the Micro BCA Protein Assay kit (Thermo-Fisher Scientific). The calibration curve standards were performed, in dilutions of 1:2, from a concentrated solution of 2 mg/mL of BSA (Thermo-Fisher Scientific). The blanks, standards, and C6/36 EVs samples (150 µL) were added in triplicate to flat bottom 96-well microplates (Corning) following the addition of 150 µL of the kit work reagent mixture. The plate was covered with sealing tape and incubated at 37 °C for 2 h. The absorbances were measured at 562 nm on the Multiskan Ascent spectrophotometer (Thermo Labsystems) with the Ascent software version 2.6. The average of the 562 nm absorbances of the blank samples was rested from the 562 nm reading of each standard and C6/36 EVs samples. The standard curve was used to determine the protein concentration (mg/mL) of each C6/36 EVs sample.

2.20. C6/36 EVs Stimulation Assays on Naïve Vero, C6/36, THP-1, and HMEC-1 Cells

Naïve Vero cells were seeded in 24-well culture plates and incubated until confluence. The cells were inoculated with 400 µL of C6/36 EVs isolates in serial log (10-fold) dilutions (in serum-free medium with dilution factors from 10^{-1} up to 10^{-22}) in duplicate and incubated for 2 h. The C6/36 EVs inoculum was removed, and each well was washed with PBS. The monolayers were overlaid with 1 mL of DMEM medium containing 1% carboxymethyl cellulose and 2.5% EV-depleted FBS. The rest of the lytic plaque assay methodology was performed as described above.

Naïve cells (C6/36, THP-1, or HMEC-1) were seeded in 12-well culture plates and incubated until confluence (C6/36 and HMEC-1 cells): For each condition, a strip of 4 plates were used and 2.5×10^5 cells/well were added, to collect 1.0×10^6 cells at the end of the assay (Figure S5). The following conditions were applied: 0.10 mg of protein in 250 µL/well of EV isolates from mock C6/36 cells (lEVs and sEVs, separately), 0.10 mg of protein in 250 µL/well of EV isolates from ZIKV-infected C6/36 cells (lEVs and sEVs, separately), 0.10 mg of protein in 250 µL/well of EV isolates from ZIKV-infected C6/36 cells RNase A + UV-treated (lEVs and sEVs, separately), and 0.10 mg of protein in 250 µL/well of non-EV ZIKV SNT. The mock cells and ZIKV-infected cells (MOI 1) were used as negative and positive controls, respectively. All conditions were added with 250 µL of non-supplemented media and incubated for 2 h; afterward, 1.0 mL of supplemented media with 5% EV-depleted FBS was added and incubated according to the best period of time established in the ZIKV infection assay: 48 h (C6/36 cells), 72 h (HMEC-1), or 96 h (THP-1 cells). The cell cultures for each condition (in a row of 4 plates) were collected (by scrapping and homogenization by vigorous pipetting) in sterile 1.5 mL microcentrifuge tubes (1.0×10^6 cells). For each condition, the ZIKV E protein was detected, and the cytopathic effects were observed via light field microscopy. The monocytes (CD11b, CD14, and CD16) and the endothelial vascular cells (CD142 [Tissue Factor, TF], PAR-1, and CD54 [ICAM-1]) were immunophenotyped. The detection of tumor necrosis factor-alpha (TNF-α) mRNA expression by RT-PCR was performed as well, as described below.

2.21. Monocyte and Vascular Endothelial Cell Immunophenotyping

The monocytes and endothelial vascular cells from EV stimulation assays were fixed and blocked for nonspecific binding sites as described above. The human monocytes were immunophenotyped, separately, with the mouse PE-conjugated anti-human CD14 antibody (Catalog #325606, BioLegend), the mouse anti-human CD16 antibody (Catalog #555404, BD Pharmingen), and the mouse PE-conjugated anti-human CD11b antibody (Catalog #301306, BioLegend). Likewise, the ECs were immunophenotyped with the mouse FITC-conjugated anti-human CD142 (TF) antibody (Catalog #13133-MM05-F, Sino Biological, USA), the mouse anti-Protease Activated Receptor (PAR-1) antibody (Catalog #sc-13503, Santa Cruz Biotechnology, USA), and the mouse FITC-conjugated anti-human CD54 (ICAM-1) antibody (Catalog # 35-0549-T025, Tonbo Biosciences, USA). For primary antibodies, the Alexa Fluor 555-conjugated secondary antibody was added at a 1:500 dilution in 0.5% BSA. The samples were analyzed by the FACS Calibur flow cytometer.

2.22. Monocytes and Vascular Endothelial Cells TNF-α mRNA Expression by RT-PCR

The RNA extraction was performed from cells collected after EV stimulation assays for the TNF-α mRNA detection by RT-PCR. The master mix was prepared according to the specifications given by the OneStep RT-PCR kit (Qiagen), as described above. The primers TNF-α FW [5′-ACAAGCCTG-TAGCCCATGTT-3′ (Mfg. ID 110182256, IDT, USA)], TNF-α RV [5′-AAAGTAGACCTGCCC-AGACT-3′ (Mfg. ID 170166847, IDT, USA)], GAPDH housekeeping FW [5′-CCATGTTCGTCATGG-GTGTGAACCA-3′ (Mfg. ID 110179057, IDT, USA)], and GAPDH housekeeping RV [5′-GCCAGT-AGAGGCAGGGATGATGTTC-3′ (Mfg. ID 110179058, IDT, USA)] were used. The reaction was performed on a GeneAmp PCR System 2400 with the following conditions: pre-PCR at 45 °C for 60 min and 95 °C for 15 min; 35 cycles at 94 °C for 30 s, 60 °C for 30 s, and 72 °C for 30 s; final elongation at 72 °C for 10 min. The amplicons [TNF-α: 600 bp and glyceraldehyde-3-phosphate dehydrogenase (GAPDH: 294 bp)] were visualized on 2% ethidium bromide-stained 1.2% agarose gel using a Typhoon FLA 9500 scanner with GE control software. The images were analyzed with ImageJ software.

2.23. Endothelial Vascular Cells Permeability Assay

The endothelial vascular barrier integrity after the stimulation with EVs from naïve or ZIKV-infected C6/36 cells was evaluated by a Transwell assay. Briefly, sterile polycarbonate tissue culture-treated Transwell inserts (12 mm) with a 0.4 µm microporous membrane pore size (Corning) were used. The HMEC-1 cells (3.5×10^5) were seeded in the inserts' upper chambers; meanwhile, the lower chambers were filled with MCDB-131 medium supplemented with 5% EVs-depleted FBS. The cultures were incubated for 48 h at 37 °C with 5% CO_2 to reach confluence. The EVs samples (0.10 mg of protein in 250 µL) from mock C6/36 and ZIKV C6/36 (RNase + UV-treated or untreated), as described above, were applied in the upper chambers (Figure S6). The inserts were incubated for 2 h; afterward, 100 µL of the medium supplemented with 5% EVs-depleted FBS were added in the upper chambers and incubated for 48 h at 37 °C with 5% CO_2. The mock HMEC-1 cells, non-EV ZIKV SNT (0.10 mg of protein in 250 µL), and ZIKV-infected HMEC-1 (MOI 1) were used as controls. To determine the cellular permeability degree, FITC (40 kDa)-Dextran (Catalog #60842-46-8, Sigma-Aldrich) was diluted to 1:60 in MCDB-131 medium supplemented with 5% EV-depleted FBS; afterward, 250 µL was added to each upper chamber. An empty insert (without cells) with FITC-Dextran was identified as the no-cell control (NCC), corresponding to 100% of permeability. The inserts were incubated for 1 h at 37 °C with 5% CO_2. The fluorescence emitted by the FITC-Dextran solution that passes through the cell monolayer to the lower chamber was measured as follows: 100 µL of each lower chamber medium was separated and diluted to 1:50; afterward, the media were passed to a black 96-well plate (Merck) that was analyzed at 492/520 nm in a Synergy H4 hybrid multi-mode microplate reader (BioTek Instruments Inc., USA) with the Gen5 software version 2.09. The permeability percentage (p%) was calculated according to the following formula: p% = (Abs InsertX/Abs NCC) × 100, where *Abs InsertX* corresponds to the absorbance of each different EV stimulus condition or control, and *Abs NCC* corresponds to the absorbance of the no-cell control.

2.24. Statistical Analysis

Quantitative data were obtained by three independent assays. The FACS data analysis was performed with FlowJo software version 10.6.1 (BD Biosciences). The bars graphs were obtained by GraphPad Prism software version 8.1.1 (GraphPad Software Inc., USA). Values were expressed as the mean ± standard deviation (SD) and evaluated using an unpaired Student's t-test with Welch's correction (for the data to compare two means, assuming unequal SDs). The statistical significance was recognized as *, +, ¡, #, !,/, ~, or ° when $p < 0.05$, **, ++, ¡¡, ##, !!,//, ~~, or °° when $p < 0.01$, and ***, +++, ¡¡¡, ###, !!!,///, ~~~, or °°° when $p < 0.0001$.

3. Results

3.1. ZIKV Infects C6/36 Mosquito Cells

The ZIKV infection of C6/36 mosquito cells was evaluated by the presence of the ZIKV E protein (FACS) at the cell membrane surface at 24, 48, 72, 96, and 120 h PI. We found that, when using a multiplicity of infection (MOI) of 1 at 48 h PI, the mosquito cells had high levels (38.60 ± 1.05%) of ZIKV E protein at the membrane surface level (Figure 1A).

Likewise, a high percentage (40%) of ZIKV-infected C6/36 cells at 48 h PI was present (Figure 1B), when comparing the E protein levels with the mock cell, and these data were statistically significant ($p < 0.0001$). After 48 h PI, less fluorescent cells were observed in the microscopy assay (Figure S7). It could be that, after main viral infection time, most syncytia structures were in development. The ZIKV-infected C6/36 cultures (48 h PI) developed a more cytopathic effect with the formation of syncytium structures but without the detachment of the cell monolayer, showing increased viral E protein-positive cells determined by fluorescence microscopy (Figure S7). Since the optimal cell infections (activation) were obtained in these conditions (MOI 1, 48 h PI), they were used for all subsequent infection experiments. Recently, it was shown that, during the virus infection process, infected cells are activated and produce different subtypes of EVs, which may have different functions when interacting with other cells, modifying naïve cellular behavior [41]. Therefore, we evaluated whether ZIKV-infected C6/36 mosquito cells could release large and/or small EVs. EVs produced from other arbovirus-infected cells were able to mediate the cell-to-cell communication between vector-host cells [36].

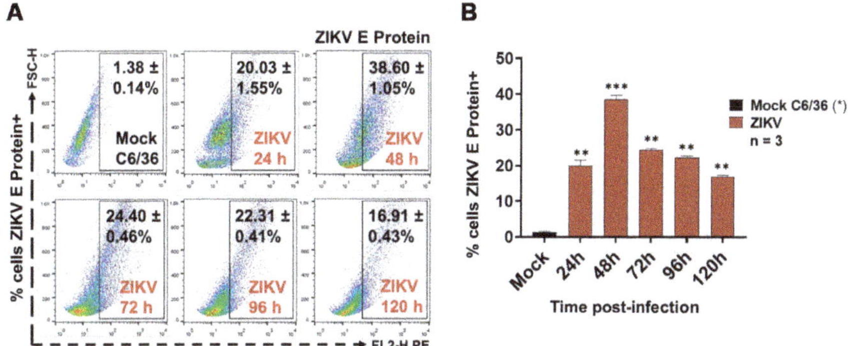

Figure 1. Zika virus (ZIKV) (multiplicity of infection (MOI) 1) infects C6/36 cells. (**A**) ZIKV envelope (E) protein detection at 24, 48, 72, 96, and 120 h post-infection (PI) by FACS assay. Dot plots are the representative mean ± standard deviation (SD) of the positive cells from three independent experiments. (**B**) ZIKV-infected cells percentages obtained by FACS. The ZIKV E protein levels were compared (by an unpaired Student's t-test) with the mock C6/36 (*) value. Statistical significance was recognized as ** when $p < 0.01$, and *** when $p < 0.0001$.

3.2. ZIKV-Infected C6/36 Cells Release Large EV Phosphatidylserine+ ZIKV E Protein+

Large EVs were developed by the outward shedding of the cell plasma membrane during cell activation and have a size greater than 200 nm, which can be identified by an Annexin-V binding assay (by FACS). This assay exposes phosphatidylserine (PS) on the outer plasma membrane leaflet [42,43]. The lEVs released from mosquito ZIKV-infected cells and mock cells were isolated from culture supernatants as described above. The characterization of all large EVs was performed by nanoparticle tracking analysis (NTA) by using the NanoSight NS300 equipment and Malvern Instruments software. The different experimental conditions for EVs detection were first established (Figure S2A), and quantitative controls with 100 and 200 nm polystyrene microspheres (NTA4088 and NTA4089, Malvern) were used (Figure S2B). Nanoparticles present in PBS and EVs-depleted FBS were also quantified to correct the number of the isolated EVs. The ZIKV virions were detected (as a peak of size of 63.5 ± 8.1 nm) to identify its presence in the EVs isolates from ZIKV-infected mosquito cells (Figure S2C).

For the NanoSight 300, a camera level of 12.0 was used with a detection limit of 2.5, a temperature of 20 °C, samples diluted in 1.0 mL of PBS, reading periods of 30 s, and three consecutive repetitions. The NTA from lEVs isolates of ZIKV-infected cells showed a concentration of $2.92 \times 10^{10} \pm 3.55 \times 10^{9}$ particles/mL with an average value of size of 319.3 ± 11.5 nm. Compared with NTA from lEVs isolates of mock cells ($2.14 \times 10^{9} \pm 4.10 \times 10^{8}$ particles/mL with an average value of size of 268.9 ± 8.2 nm), the concentration of nanoparticles from lEVs isolates from infected cells were 13.6-fold higher (Figure 2A).

In parallel, we observed that, during ZIKV infection (MOI 1, 72h PI) of the C6/36 cells (Figure 2B), the percentage of lEVs PS+ released to the cell culture supernatant (60.50 ± 0.87%) was 1.98-fold higher ($p < 0.001$) than lEVs PS+ from uninfected (30.60 ± 0.96%) mosquito cells (Figure 2C). The fluorescence emitted by Annexin-V binding was proportional to the PS presence: The ZIKV-infected C6/36 lEVs mean fluorescence intensity (MFI) was compared with that of the microbead control and mock C6/36 lEVs values (Figure S8A). The Annexin-V binding MFI in the lEVs from ZIKV-infected cultures was 1.45-fold higher than that of the lEVs from the mock cultures.

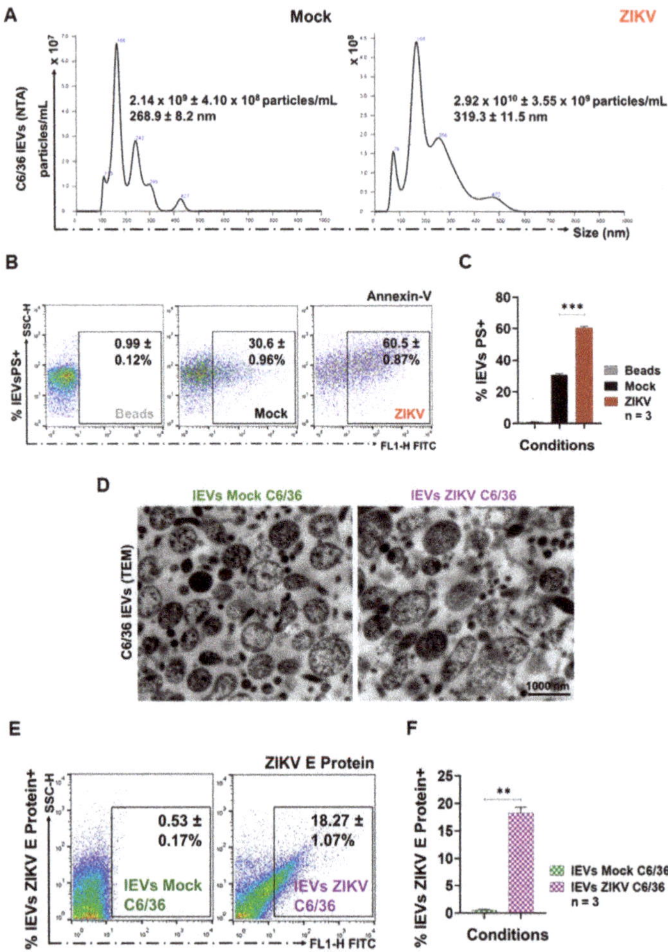

Figure 2. ZIKV-infected C6/36 cells issue the large extracellular vesicles (lEVs) phosphatidylserine (PS)+. (**A**) Nanoparticles tracking analysis (NTA) of the purified lEVs isolates from the mock and ZIKV-infected C6/36 cells. Histograms are the representative mean ± SD of the nanoparticle's concentration (particles/mL) and the size (nm) from three independent experiments. (**B**) PS detection by the Annexin-V binding assay. The fluorescence emitted by Annexin-V binding is proportional to PS levels. FACS dot plots are the representative mean ± SD of the lEVs PS+ from three independent experiments. (**C**) lEVs PS+ percentages obtained by FACS. The PS levels were compared (by an unpaired Student's t-test) with the mock lEVs (*) value. (**D**) Transmission electron microscopy (TEM) images from the mock C6/36 and the ZIKV-infected C6/36 lEVs (1000 nm scale). (**E**) ZIKV E protein detection on the lEVs surface by FACS assay. Dot plots are the representative mean ± SD of the lEVs ZIKV E protein+ from three independent experiments. (**F**) Percentages of the lEVs ZIKV E protein+ by FACS. The ZIKV E protein levels were compared (by an unpaired Student's t-test) with the mock C6/36 (*) value. Statistical significance was recognized as ** when $p < 0.01$, and *** when $p < 0.0001$.

The lEVs samples were also characterized by transmission electron microscopy (TEM): The analysis of images from the mock C6/36 lEVs and ZIKV-infected C6/36 lEVs (Figure 2D) showed different EV populations, which were heterogeneous in shape, with sizes up to 200 nm (1000 nm scale) and a defined membrane in a proper structural resolution. These data agree with other reports, which demonstrated that cells secrete EVs as a heterogeneous population with different sizes and shapes [30,44]. We did not identify viral particles inside lEVs TEMs from ZIKV-infected cells, so we evaluated the ZIKV E protein presence on the membrane's surface of lEVs. We found 18.27 ± 1.27% of positive ZIKV E protein+ lEVs ($p < 0.01$) compared with the lEVs from the mock C6/36 cells (Figure 2E,F). The ZIKV E protein (MFI) in the lEVs from ZIKV-infected cells was 18.6-fold higher than that of the lEVs from the mock cultures (Figure S8B).

3.3. ZIKV-Infected C6/36 Cells Release Small EV CD63-Like+ ZIKV E Protein+

First, the presence of CD63-like tetraspanin at the membrane surface of C6/36 cells was evaluated as well. The point time of the CD63 membrane decreased (CD63 internalization) to select the best condition for the identification of the sEVs marker (inside cells) and, likewise, the time for the optimal isolation of sEVs CD63-like+ [45]. The tetraspanin detection (FACS) was evaluated at 24, 48, 72, 96, and 120 h PI (Figure 3A). We observed that naïve C6/36 cells constitutively contain high levels of the CD63-like tetraspanin (Figure 3A,B) at the cell membrane surface (55.67 ± 2.30%). Nevertheless, in ZIKV-infected cultures, an increase in the CD63 percentage (79.40 ± 2.78%) was present at 24 h PI (1.4-fold higher), while the highest percentage (88.73 ± 1.35%) of the CD63+ cells were obtained at 48 h PI (1.6-fold higher compared with mock cells). The increased tetraspanin values were significant ($p < 0.0001$). We found that CD63 levels were decreased at 72–96 h PI (Figure 3A,B). These data suggest that CD63-tetraspanin internalization may occur after 48 h PI. Therefore, the sEVs biogenesis in mosquito cells could take place between 48 and 72 h. The sEVs CD63+ were then isolated at 72 h PI. The presence of the CD63 tetraspanin was also evaluated (red) by fluorescence microscopy (100×) in mock and ZIKV-infected C6/36 cultures (green for ZIKV E protein). The presence of the CD63 tetraspanin inside C6/36 cells suggests the endosomal nature of small EVs (Figure S9A).

The NTA from sEVs isolates of ZIKV-infected cells showed a concentration of $3.17 \times 10^{11} \pm 5.62 \times 10^{10}$ particles/mL with an average value of size of 125.5 ± 1.6 nm. Compared with the NTA from sEVs isolates of mock cells ($2.39 \times 10^{10} \pm 4.41 \times 10^{9}$ particles/mL with an average value of size of 107.8 ± 3.1 nm), the concentration of nanoparticles from sEVs isolates from infected cells were 13.3-fold higher (Figure 3C).

The sEVs isolates were then identified by positive selection, using paramagnetic nanobeads coated with anti-CD63 antibodies (Figure S3). Previously, we determined whether ZIKV viral particles could cross-react with the paramagnetic nanobeads (Figure S9B,C). We found that ZIKV did not couple to the nanobeads, so the sEVs CD63+ detection is free of viral particles. The sEVs CD63+ percentage from ZIKV-infected cultures (19.32 ± 0.93%) was 1.7-fold higher than the sEVs CD63+ from the mock cultures (11.08 ± 0.34%), showing a significant difference ($p < 0.01$) (Figure 3D,E). The MFI values (obtained by FACS) were proportional to the sEVs CD63+ presence (Figure S9D). We found that the sEVs CD63+ from ZIKV-infected cells showed MFI values 1.3-fold higher than the sEVs CD63+ from the mock cultures.

The TEM images (Figure 3F) from the mock and ZIKV-infected C6/36 sEVs (500 nm scale) show a heterogeneous population of sEVs [46] in terms of their size (fewer than 200 nm in diameter), shape, and content; this population is also well defined by a bilipid membrane. We did not identify viral particles inside sEVs TEMs from ZIKV-infected cells, so we evaluated the ZIKV E protein presence on the membrane's surface of sEVs. We found 31.19 ± 0.28% of positive ZIKV E protein+ sEVs coupled with paramagnetic nanobeads ($p < 0.001$) compared with the sEVs from the mock C6/36 cells (Figure 3G,H). The ZIKV E protein (MFI) in the sEVs from ZIKV-infected cells was 4.03-fold higher than that of the sEVs from the mock cultures (Figure S9E).

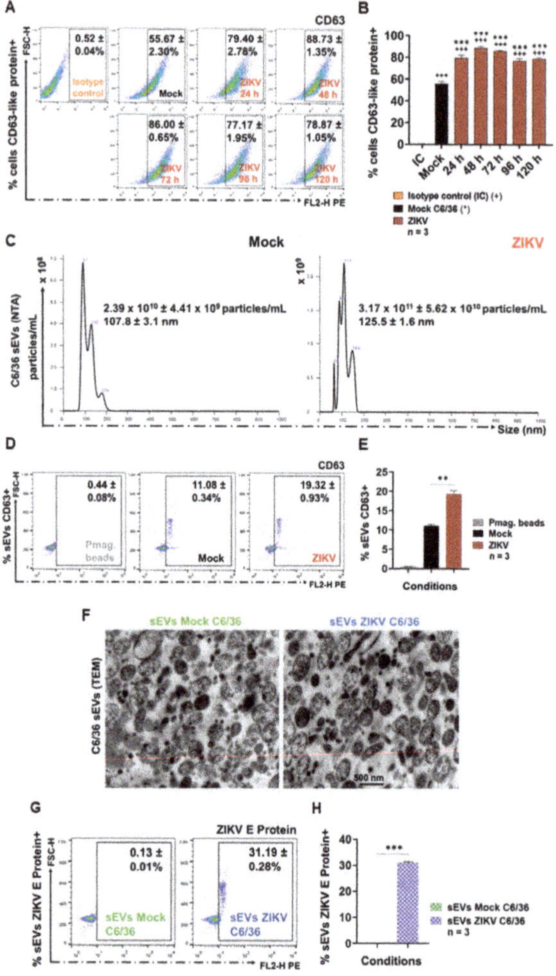

Figure 3. ZIKV-infected C6/36 cells issue small EVs (sEVs) CD63+. (**A**) CD63-like detection at 24, 48, 72, 96, and 120 h PI by FACS. Dot plots are the representative mean ± SD of positive cells from three independent experiments. Isotype IgG1 antibody was used as negative control. (**B**) Cells CD63-like+ percentages obtained by FACS. The levels of CD63-like protein were compared (by an unpaired Student's t-test) with the isotype control (+) and the mock cells (*) values. (**C**) NTA of the purified sEVs isolates from the mock and the ZIKV-infected C6/36 cells. Histograms are the representative mean ± SD of the nanoparticle's concentration (particles/mL) and the size (nm) from three independent experiments. (**D**) sEVs CD63+ coupled with paramagnetic bead detection by FACS. Dot plots are the representative mean ± SD of the sEVs CD63+ from three independent experiments. (**E**) sEVs CD63+ percentages obtained by FACS. The CD63 levels were compared (by an unpaired Student's t-test) with the mock sEVs CD63+ (*) values. (**F**) Transmission electron microscopy (TEM) images from the mock and the ZIKV-infected C6/36 sEVs (500 nm scale). (**G**) ZIKV E protein detection on the sEVs coupled with paramagnetic beads by FACS assay. Dot plots are the representative mean ± SD of the positive sEVs ZIKV E protein+ from three independent experiments. (**H**) Percentages of the sEVs ZIKV E protein+ by FACS. The ZIKV E protein levels were compared (by an unpaired Student's t-test) with the sEVs mock C6/36 (*) values. Statistical significance was recognized as ++ or ** when $p < 0.01$, and +++ or *** when $p < 0.0001$.

3.4. ZIKV C6/36 EVs, after ZIKV Inactivation, Carry Viral RNA, Reproduce Lytic Plaque Formation on Vero Cells, and Favor Infection in Naïve Mosquito Cells

We determined whether ZIKV mosquito EVs participate during the infectious process as was recently reported for other arthropod-borne flaviviruses [16,30,36]. It was then determined whether ZIKV C6/36 EVs contain viral elements as RNA or E protein and whether they support the infection of naïve cells.

First, we determined whether ZIKV could be inactivated by RNAse A activity and UV radiation at 1200 μJ (× 100) in three consecutive cycles (Figure S4A). We evaluated different conditions for incubation times and found that the complete RNA degradation from ZIKV viral stock occurred for 1 h at 37 °C (Figure 4A). The infection capability of inactivated ZIKV (iZIKV) was evaluated by a lytic plaque assay (Figure 4B) and found no lytic plaque formation after the inactivation process.

We then proceeded to inactivate free ZIKV virions in C6/36 mosquito EV isolates (small/large). These samples were irradiated at 1200 μJ (×100) in three consecutive cycles by using an UV Stratalinker (Figure S4B). In addition, the possible presence of ZIKV genomic RNA in the EVs samples, as a possible precipitation product during the EVs isolation process, was eliminated with RNase A, as described above: By using 10 μg/mL of RNase A (DNase and Protease-free) added to the different EV isolates for 1 h at 37 °C, we observed total RNA degradation in all samples (sEVs and lEVs) (Figure 4A).

To evaluate the integrity of the EVs (sEVs and lEVs) after the ZIKV inactivation process, we proceed to quantified them by NTA. The NTA from lEVs ZIKV C6/36 (RNase A + UV) showed a concentration of $2.48 \times 10^{10} \pm 4.07 \times 10^9$ particles/mL with an average size value of 304.1 ± 10.9 nm, while the NTA from sEVs ZIKV C6/36 (RNase A + UV) showed a concentration of $2.49 \times 10^{11} \pm 2.29 \times 10^{10}$ particles/mL with an average size value of 150.9 ± 5.5 nm (Figure 4C). The NTA histograms showed the same patterns of the NTA histograms shown in Figures 2A and 3C, so the EVs' integrity is preserved. In Figure 4C, the peak containing particles of nearly 50 nm, compatible with ZIKV, was substantially reduced.

Next, to evaluate the possible ZIKV genomic RNA presence inside small and large EVs (RNase A + UV-treated), all samples were processed for RNA extraction and purification by using the QIAamp RNA Mini kit, according to the manufacturer's instructions. The samples were first lysed under highly denaturing conditions using the lysis buffer. The ZIKV-RNA amplification was performed by RT-PCR, according to the specifications given by the OneStep RT-PCR kit (see Materials and Methods). The amplified cDNA corresponded to the specific 364 bp E-amplicon for the ZIKV [40] envelope protein, which was visualized on 2% ethidium bromide-stained 1.2% agarose gel (Figure 4D). These findings suggest that both small and large EVs from ZIKV-infected mosquito cells may carry viral RNA after RNase A + UV treatment.

As a result of the significance of these data, we also evaluated the possible mammalian naïve cell infection via small/large ZIKV C6/36 EVs. With this aim, first, we used the epithelial cells (Vero) from the monkey *Cercopithecus aethiops* (used as gold standard cells for infection assay) [32] to perform plaque assays (as described above) in the presence of small and large ZIKV C6/36 EVs (also RNase A + UV-treated and untreated samples). The presence of lytic plaques in ZIKV-infected Vero cells was present in high or undetermined amounts at different dilutions (Figure 4E). Importantly, lytic plaques were also observed in cultures in the presence of small and large ZIKV C6/36 EVs isolates in a more concentrated amount, which were also formed in high or undetermined quantities. However, plaque formation was not detected in the negative control named non-EVs ZIKV SNT (the final supernatant obtained in the last centrifugation during the isolation of EVs from ZIKV-infected C6/36 cell culture media). These data suggest the ZIKV infection of monkey epithelial cells via lEVs/sEVs released from ZIKV-infected C636 mosquito cells.

Therefore, different EVs stimulation assays were performed using naïve C6/36 cells in the presence of ZIKV (MOI 1), mock C6/36 EVs, or ZIKV C6/36 small/large EVs (including RNase A + UV-treated and untreated isolates) (Figure S5). The following conditions were applied: 0.10 mg of protein in 250 μL/well of EVs isolates from mock C6/36 cells (lEVs and sEVs, separately), 0.10 mg of protein in 250 μL/well of EVs isolates from ZIKV-infected C6/36 cells (lEVs and sEVs, separately), 0.10 mg of protein in 250 μL/well of EVs isolates from ZIKV-infected C6/36 cells RNase A + UV-treated (lEVs and sEVs, separately), and 0.10 mg of protein in 250 µL/well of non-EVs ZIKV SNT. The mock cells and ZIKV-infected cells (MOI 1) were used as negative and positive controls, respectively. We also evaluated the iZIKV infection capability on C6/36 cells (Figure S11A,B), and we found that iZIKV did not infect naïve C6/36 cells after 48 h of incubation, because the ZIKV E protein was undetectable.

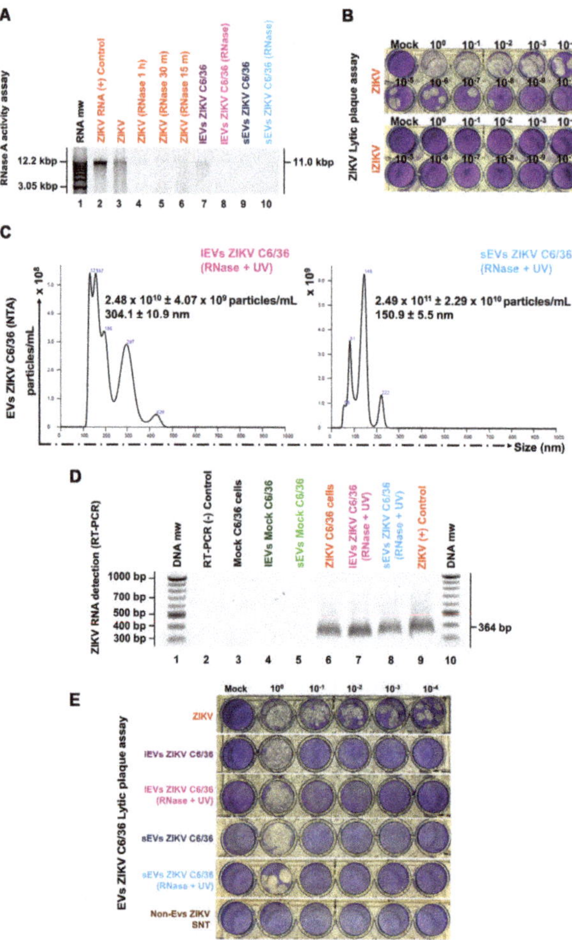

Figure 4. EVs from ZIKV-infected C6/36 cells, after ZIKV inactivation by RNase A activity assay and UV radiation, carry viral RNA and favor the mammalian cell infection. (**A**) RNase A activity assay. RNase A (10 µg/mL) was added to the purified ZIKV RNA and incubated for 1 h, 30 min, and 15 min at 37 °C with 5% CO_2. Additionally, the RNase A activity (1 h incubation) was evaluated in C6/36 EVs isolates. The RNA degradation pattern was visualized on 2% ethidium bromide-stained 1.2% agarose gel. (**B**) Inactivated ZIKV (iZIKV) titration by a lytic plaque assay. (**C**) NTA of the EV isolates (from ZIKV-infected cells) treated with RNase A and UV. Histograms are the representative mean ± SD of the nanoparticle's concentration (particles/mL) and the size (nm) from three independent experiments. (**D**) ZIKV RNA detection (RT-PCR) in ZIKV-infected C6/36 EVs (RNase A + UV-treated) samples. The ZIKV amplicon (364 bp from E genome conserved region) was visualized on 2% ethidium bromide-stained 1.2% agarose gel. (**E**) Evaluation of the ZIKV-infected C6/36 EVs (RNase A + UV-treated and untreated samples) in the viral transmission to naïve Vero cells by a lytic plaque assay.

The samples were incubated over 48 h (the best infection time) and evaluated for possible viral infection by means of ZIKV E protein detection using a FACS assay. The ZIKV E protein presence was detected in 44.39 ± 0.69% of ZIKV-infected C6/36 cells (Figure 5A–C), showing statistical significance ($p < 0.0001$) in relation to the mock cells (*) (34.7-fold higher). In naïve C6/36 cells stimulated with ZIKV C6/36 lEVs, the E protein was found in 38.24 ± 1.15% of cells and 28.44 ± 1.37% of cells stimulated with ZIKV C6/36 lEVs (RNase A + UV-untreated and treated) (#,!), showing statistical significance ($p < 0.0001$) compared with the mock C6/36 cells and stimuli in the presence of mock C6/36 EVs (Figure 5A,B). Similarly, in naïve mosquito cells cultured with ZIKV C6/36 sEVs, the viral E protein was found in 39.31 ± 0.58% of cells and 33.00 ± 0.29% of cells stimulated with ZIKV C6/36 sEVs (RNase A + UV-untreated and treated) (/, ~), showing statistical significance ($p < 0.0001$) compared with the mock cells and naïve cells stimulated with EVs from mock C6/36 cells (Figure 5A–C). The present data suggest that ZIKV C6/36 (large and small) EVs could favor the infection of mosquito naïve cells.

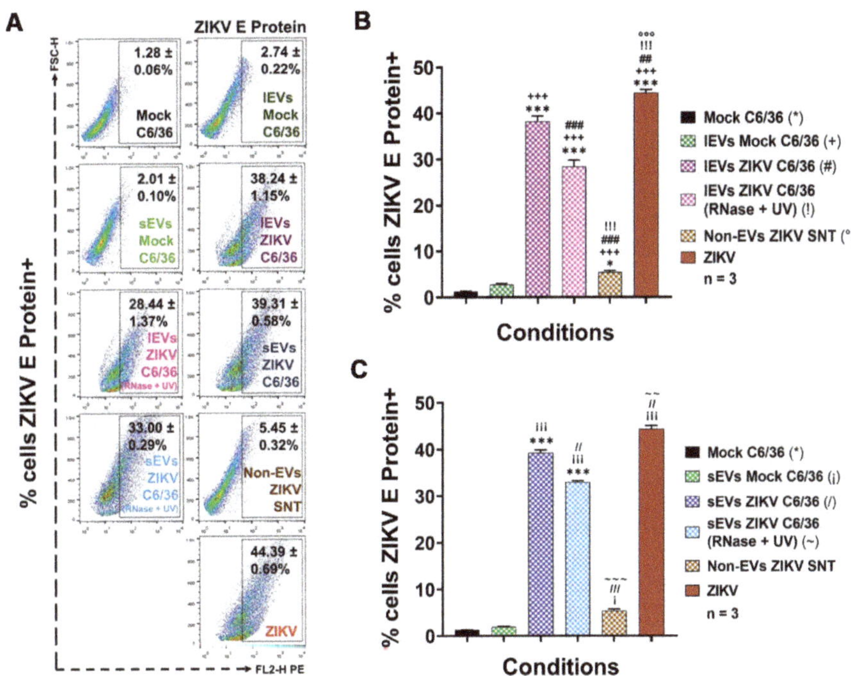

Figure 5. ZIKV E protein is present on the membrane's surface of naïve C6/36 cells after the stimulus with ZIKV-infected C6/36 EVs. (**A**) ZIKV E protein detection at different EV stimuli conditions (FACS assay). Dot plots are the representative mean ± SD of the positive cells from three independent experiments. (**B**) Percentages of ZIKV E protein+ cells (FACS) after the lEVs stimuli. The levels of the ZIKV E protein were compared (by an unpaired Student's t-test) between all conditions' values. Statistical significance was recognized as *, +, #, !, or ° when $p < 0.05$, **, ++, ##, !!, or °° when $p < 0.01$, and ***, +++, ###, !!!, or °°° when $p < 0.0001$. (**C**) Percentages of ZIKV E protein+ cells (FACS) after the sEVs stimuli. The levels of the ZIKV E protein were compared (by an unpaired Student's t-test) between all conditions' values. Statistical significance was recognized as *, ¡,/, or ~ when $p < 0.05$, **, ¡¡,//, or ~~ when $p < 0.01$, and ***, ¡¡¡,///, or ~~~ when $p < 0.0001$.

Likewise, in parallel assays, all samples were evaluated for the presence of ZIKV E protein by fluorescence microscopy and by the cytopathic effects observation using light field microscopy (Figure

S12). The cytopathic effects in the light fields were indicated with black arrows (20×). An increased cytopathic effect with the formation of syncytium structures was present in the ZIKV-infected cultures but also in cultures stimulated with small and large ZIKV C6/36 EVs (RNase A + UV-treated and untreated) isolates. The ZIKV E protein (red) was detected in naïve C6/36 cells stimulated with the small and large ZIKV C6/36 EVs at similar levels and patterns of ZIKV-infected mosquito cells (fluorescence microscopy, 60×) and were undetected in cultures of naïve C6/36 cells in the presence of mock C6/36 EVs (small and large). The same was found for the non-EVs ZIKV supernatant or mock C6/36 cultures.

3.5. ZIKV-Infected C6/36 EVs Participate during Infection of Naïve Human Monocytes

Recently, it was reported that dengue virus (DENV) uses small EVs of C6/36 mosquito cells for its transmission from the vector to mammalian host cells, including human skin keratinocytes and ECs [16,30]. However, to date, it has not been determined if EVs from ZIKV-infected C6/36 cells participate during monocyte infection. As monocytes are the main target cells during ZIKV human host infection [17,47], we evaluated the possible participation of large and small EVs in the potential. Initially, we evaluated the viral infection of human monocytes (MOI 1) via the detection of the viral E protein at the surfaces of the membrane monocytes at 24, 48, 72, 96, and 120 h PI using a FACS assay (Figure 6).

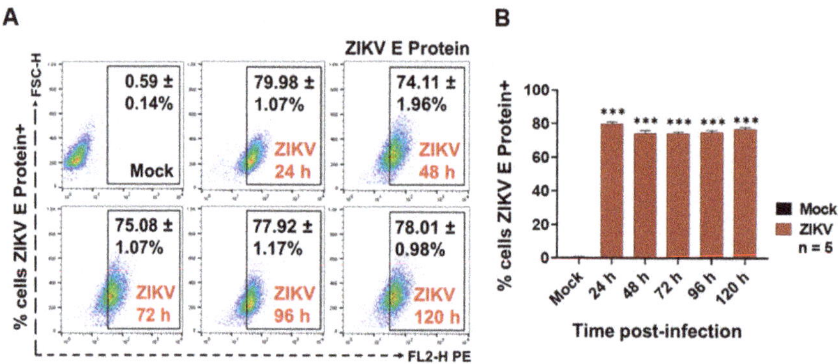

Figure 6. ZIKV (MOI 1) infects human monocytes. (**A**) ZIKV E protein detection at 24, 48, 72, 96, and 120 h PI by FACS assay. Dot plots are the representative mean ± SD of the positive cells from five independent experiments. (**B**) ZIKV-infected cells percentages obtained by FACS assay. The ZIKV E protein levels were compared (by an unpaired Student's t-test) with mock C6/36 (*) values. Statistical significance was recognized as *** when $p < 0.0001$.

ZIKV was able to establish a productive infection of human monocytes, since the viral E protein was detected at high levels on the cell membrane's surface in all post-infection time points of the assay (Figure 6A). However, the greater percentages of ZIKV-positive cells were present at 24 h PI (79.98 ± 1.07%), at 96 h PI (77.92 ± 1.17%), and at 120 h (78.01 ± 0.98%), which were statistically significant ($p < 0.0001$) when compared to the mock cells. Likewise, we found that the ZIKV infection of THP-1 human monocytes favor progressive activation and cell differentiation effects (Figure S13A). The cytopathic effect observation by light field microscopy supports the presence of higher amounts of adherent cells between 96 and 120 h PI. Consequently, we decided to evaluate the ZIKV infection of naive monocytes after the EVs stimulation at 96 h PI.

Different EVs stimulation assays were performed using naïve THP-1 cells in the presence of ZIKV (MOI 1), mock C6/36 EVs, or ZIKV C6/36 EVs (small and large isolates as the same for the RNase A + UV-treated and untreated samples) (Figure S5), which were evaluated for possible infection by means of ZIKV E protein detection by FACS (see Materials and Methods). We also evaluated the iZIKV infection capability on THP-1 cells (Figure S11C,D) and found that iZIKV did not infect naïve monocytes after 96 h of incubation, because the ZIKV E protein was undetectable.

As was expected, the viral E protein was detected in 73.41 ± 0.59% of ZIKV-infected monocytes and was statistically significant ($p < 0.0001$) when compared to the mock cells (*) (Figure 7A–C). The ZIKV E protein was also detected in higher amounts on naïve THP-1 cells that were stimulated with ZIKV C6/36 lEVs (54.89 ± 0.56% in RNase A + UV-treated (!) isolates and 70.23 ± 1.53% in untreated (#) isolates) and ZIKV C6/36 sEVs (38.29 ± 0.79% in RNase A + UV-treated (~) isolates and 41.34 ± 0.39% in untreated (/) isolates). The ZIKV E protein levels were statistically significant ($p < 0.0001$) when compared to the ZIKV-C6/36 EVs stimulated cultures against the mock cells, mock C6/36 EVs, and non-EV ZIKV SNT (°) cultures (Figure 7A–C).

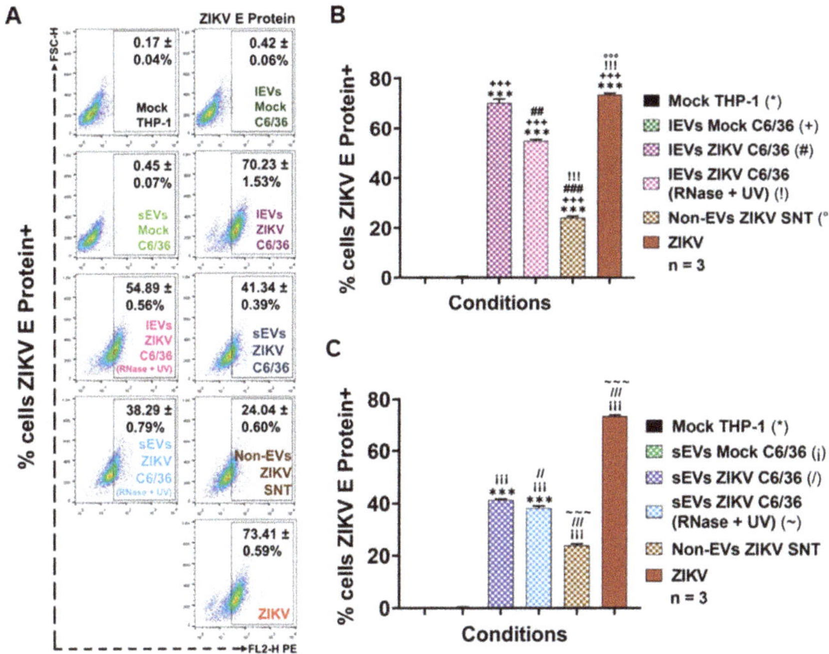

Figure 7. ZIKV E protein is present on the membrane's surface of naïve monocytes (THP-1 cells) after the stimulus with ZIKV-infected C6/36 EVs. (**A**) ZIKV E protein detection at different EV stimuli conditions (FACS assay). Dot plots are the representative mean ± SD of the positive cells from three independent experiments. (**B**) Percentages of ZIKV E protein+ cells (FACS) after the lEVs stimuli. The levels of the ZIKV E protein were compared (by an unpaired Student's t-test) between all conditions' values. Statistical significance was recognized as *, +, #, !, or ° when $p < 0.05$, **, ++, ##, !!, or °° when $p < 0.01$, and ***, +++, ###, !!!, or °°° when $p < 0.0001$. (**C**) Percentages of ZIKV E protein+ cells (FACS) after the sEVs stimuli. The levels of the ZIKV E protein were compared (by an unpaired Student's t-test) between all conditions' values. Statistical significance was recognized as *, ¡,/, or ~ when $p < 0.05$, **, ¡¡,//, or ~~ when $p < 0.01$, and ***, ¡¡¡,///, or ~~~ when $p < 0.0001$.

The present data suggest that ZIKV-infected C6/36 EVs not only support the infection of naïve mosquito cells but also participate during infection of mammalian host cells, as in the case of human monocytes, which are important immune effector cells during host–pathogen interplay.

3.6. ZIKV-Infected C6/36 EVs Promote Change in Monocyte Phenotype (CD14, CD16, and CD11b)

It is known that EVs may have different functions when interacting with other cells, modifying their naïve cellular behavior [48]. Therefore, if EVs released by ZIKV-infected mosquito cells were able to infect human monocytes, they could also favor monocyte activation and/or differentiation. To assess this objective, naïve human monocytes were stimulated by the presence of ZIKV (MOI 1), mock C6/36 EVs, or ZIKV C6/36 EVs (small and large isolates as the same for the RNase A + UV), which were evaluated for monocyte activation or differentiation (to adherent phenotype cells) by means of the monocytes' phenotypic shift from a classical (CD14++ CD16−) to an intermediate (CD14++ CD16+) or non-classical (CD14+ CD16++) phenotype, which seems to be the main producer of inflammatory mediators in response to viral infection [49].

By the stimulation assays in the presence of the different EVs samples, we observed (Figure 8A,B) that the classical monocyte phenotype changes to CD14++ CD16+ intermediate monocytes with respect to the mock THP-1 cells and those stimulated with the mock C6/36 EVs ($p < 0.0001$ for CD14 and $p < 0.01$ for CD16). Moreover, in a parallel assay, we observed that naïve monocytes were differentiated and expressed CD11b+ at the membrane's surface (Figure 8C) with elevated levels in ZIKV-infected monocytes and those stimulated with ZIKV C6/36 EVs ($p < 0.0001$) compared with the mock THP-1 cells and those stimulated with mock C6/36 EVs. On the other hand, naïve monocyte activation by the ZIKV C6/36 EVs was also observed by light-field microscopy (Figure S13B), showing the transformation of naïve cells to the adherent phenotype (black arrows), as similar levels occur in ZIKV-infected monocytes.

It has been suggested that monocyte subsets (intermediate and non-classical) play an essential role in immunopathology during *Flavivirus* infection [49], since non-classical monocytes seem to be the main producers of pro-inflammatory mediators in response to viral infection. The present data show that ZIKV C6/36 (small/large) EVs activate and differentiate naïve monocytes, changing cells to a pro-inflammatory state, in a similar mode to ZIKV (MOI 1) infection.

Next, to evaluate the possible expression of a pro-inflammatory response induced by EVs released from ZIKV-infected C6/36 cells, we performed a detection of the tumor necrosis factor-alpha (TNF-α) mRNA expression in naïve monocytes infected by ZIKV (MOI 1), and those stimulated with the mock C6/36 EVs or ZIKV C6/36 EVs (RNase A + UV-treated and untreated isolates). We found that, like ZIKV-infected monocytes, ZIKV C6/36 sEVs were able to induce TNF-α mRNA expression in human monocytes (Figure 8D). TNF-α is a pro-inflammatory cytokine that was recently determined to be an important host factor involved in neurological disorders and central nervous system inflammation during ZIKV human infection [50].

A growing number of evidence indicates that sEVs are involved in inflammatory processes or immune responses that play an important role in a large number of pathologic states, including infectious diseases. sEVs can modulate gene expression and the functions of the cells with which they interact, and their content depends on the cells from which they are released [23,27]. We found that sEVs from ZIKV-infected C6/36 cells induce immunophenotype changing (to intermediate/non-classical) in monocytes. This proinflammatory phenotype could be directly implicated in TNF-α mRNA expression.

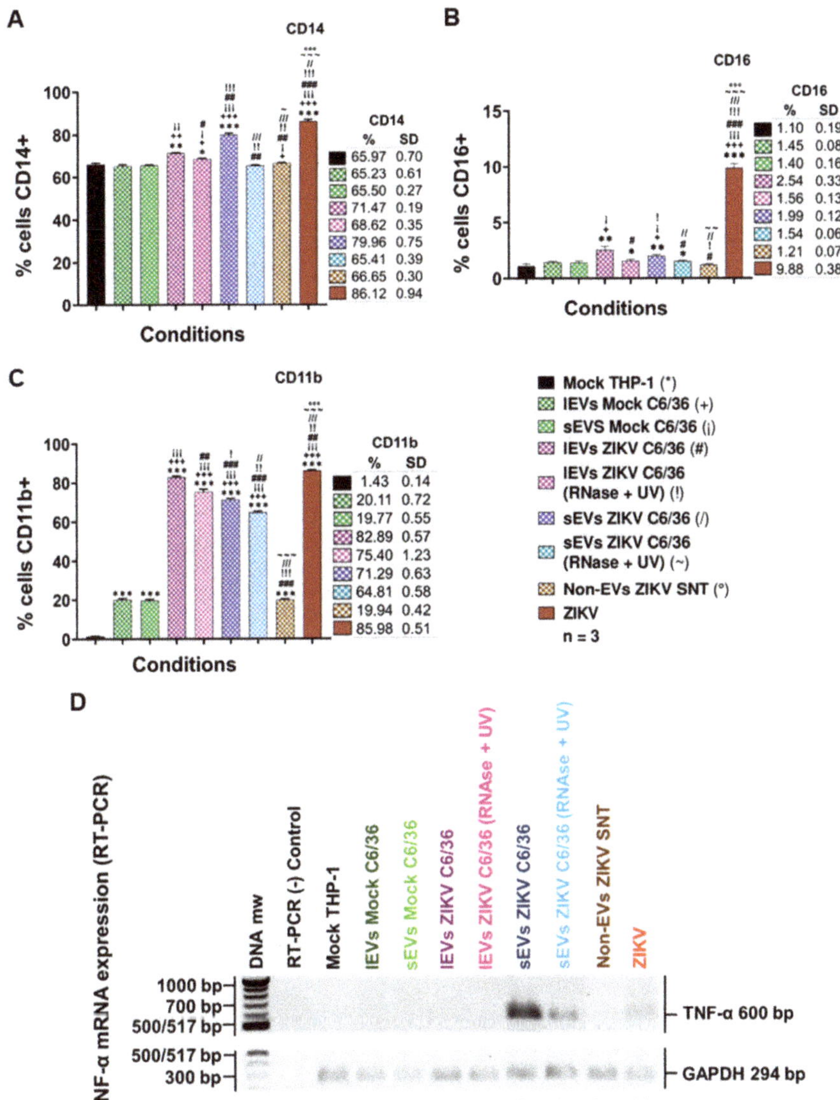

Figure 8. EVs from ZIKV-infected C6/36 favor the pro-inflammatory phenotype change in naïve human monocytes. (**A**) Monocytes CD14+ percentages (FACS) at different EV stimuli conditions from three independent experiments. (**B**) Monocytes CD16+ percentages (FACS) at different EVs stimuli conditions from three independent experiments. (**C**) Monocytes CD11b+ percentages (FACS) at different EVs stimuli conditions from three independent experiments. The CD14, CD16, or CD11b levels were compared (by an unpaired Student's t-test) between all conditions' values. Statistical significance was recognized as *, +, ¡, #, !,/, ~, or ° when $p < 0.05$, **, ++, ¡¡, ##, !!,//, ~~, or °° when $p < 0.01$, and ***, +++, ¡¡¡, ###, !!!,///, ~~~, or °°° when $p < 0.0001$. (**D**) Tumor necrosis factor-alpha (TNF-α) mRNA expression (RT-PCR) in naïve monocytes at different EVs stimuli conditions. The TNF-α genome conserved region (amplicon of 600 bp) was visualized on 2% ethidium bromide-stained 1.2% agarose gel.

3.7. ZIKV C6/36 EVs Participate during Infection of Naïve Endothelial Vascular Cells

The ZIKV mosquito EVs participation during infection of vascular ECs is unknown. We first determined whether ZIKV (MOI 1) was able to infect human endothelial vascular cells, by means of the viral E protein detection on cells membrane surface at 24, 48, 72, 96, and 120 h PI by the FACS assay. First, the optimal conditions for ZIKV vascular endothelial cell infection were evaluated (Figure 9).

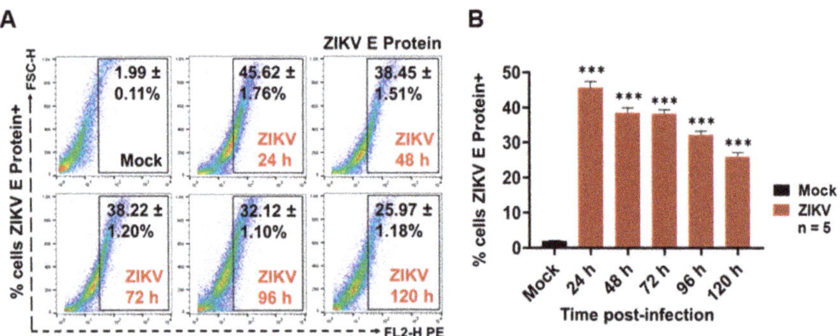

Figure 9. ZIKV (MOI 1) infects human endothelial (HMEC-1) cells. (**A**) ZIKV envelope (E) protein detection at 24, 48, 72, 96, and 120 h PI by the FACS assay. Dot plots are the representative mean ± SD of the positive cells from five independent experiments. (**B**) ZIKV-infected cells percentages obtained by FACS. The ZIKV E protein levels were compared (by an unpaired Student's t-test) with the mock C6/36 (*) value. Statistical significance was recognized as * when $p < 0.05$, ** when $p < 0.01$, and *** when $p < 0.0001$.

ZIKV was able to establish a productive infection in microvascular endothelial (HMEC-1) cells (Figure 9A,B), since viral E protein was detected at high levels on the cell membrane surface mainly at 24 (45.62 ± 1.76%) and 72 h (38.22 ± 1.20%) PI time points, with a significance level of $p < 0.0001$ when compared to the mock HMEC-1 cells (22.9- and 19.2-fold higher, respectively). Likewise, when these samples were observed by light-field microscopy, the ZIKV-infected ECs showed important cytopathic effects (black arrows), with a formation of syncytium structures at 48 and 72 h PI, but without monolayer detachment (Figure S14A). We also evaluated the iZIKV infection capability on HMEC-1 cells (Figure S11E,F) and found that iZIKV did not infect naïve ECs after 72 h of incubation, because the ZIKV E protein was undetectable.

Afterward, the possible participation of small/large C6/36 EVs in the potential infection of vascular ECs was evaluated. Next, different stimulation assays were performed using HMEC-1 naïve cells in the presence of ZIKV (MOI 1), the mock C6/36 EVs, and the ZIKV small/large C6/36 EVs (the same for the RNase A + UV-treated and untreated). All stimuli were evaluated by measuring the ZIKV E protein presence by FACS (Figure 10).

As shown in Figure 10A, viral E protein was present at high levels in ZIKV (MOI 1) infected HMEC-1 cells (30.22 ± 0.58%), but also at high levels in naïve ECs stimulated by the ZIKV C6/36 sEVs (RNase A + UV-treated (19.28 ± 0.80%) and untreated (26.40 ± 0.78%)) and ZIKV C6/36 lEVs (RNase A + UV-treated (15.28 ± 0.49%) and untreated (26.32 ± 0.56%)). The percentage of viral E protein was compared between all conditions' values against the mock HMEC-1; these values were statistically significant ($p < 0.0001$) (Figure 10B,C). These data suggest that ZIKV infected-C6/36 EVs (large and small) support the infection of mammalian host cells, including vascular ECs.

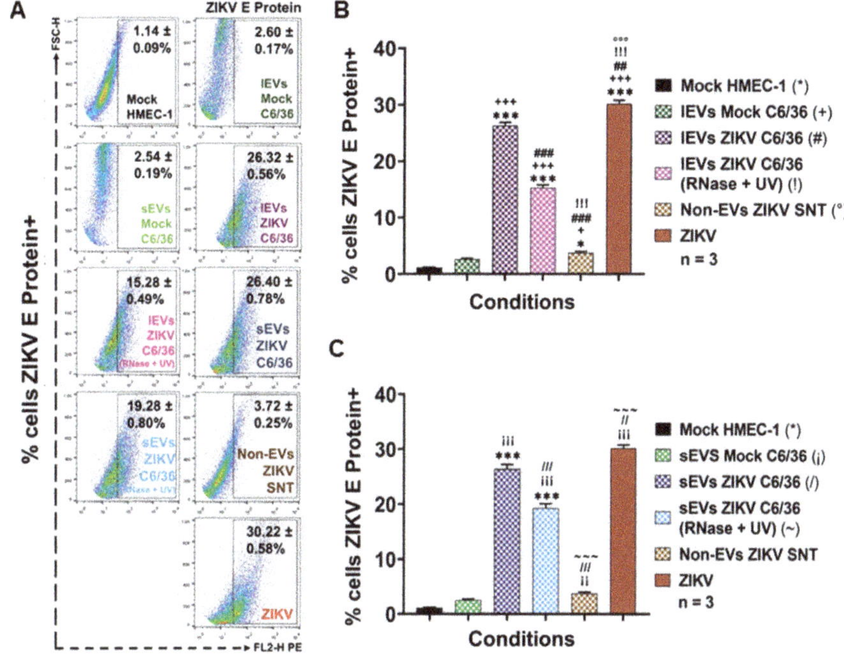

Figure 10. ZIKV E protein is present on the membrane's surface of naïve endothelial cells (HMEC-1) after the stimulus with ZIKV-infected C6/36 EVs. (**A**) ZIKV E protein detection at different EVs stimuli conditions (FACS assay). Dot plots are the representative mean ± SD of the positive cells from three independent experiments. (**B**) Percentages of ZIKV E protein+ cells (FACS) after the lEVs stimuli. The levels of the ZIKV E protein were compared (by an unpaired Student's t-test) between all conditions' values. Statistical significance was recognized as *, +, #, !, or ° when $p < 0.05$, **, ++, ##, !!, or °° when $p < 0.01$, and ***, +++, ###, !!!, or °°° when $p < 0.0001$. (**C**) Percentages of ZIKV E protein+ cells (FACS) after the sEVs stimuli. The levels of the ZIKV E protein were compared (by an unpaired Student's t-test) between all conditions' values. Statistical significance was recognized as *, ¡,/, or ~ when $p < 0.05$, **, ¡¡,//, or ~~ when $p < 0.01$, and ***, ¡¡¡,///, or ~~~ when $p < 0.0001$.

3.8. ZIKV C6/36 EVs Favor a Pro-Inflammatory and Pro-Coagulant State of Vascular Endothelial Cells and Promote the Endothelial Vascular Cells' Permeability

Recent observations in humans and animal models [51,52] suggest that, in severe Zika cases, different coagulation disorders occur. It has been shown that some viruses activate the coagulation system through tissue factor (TF) receptor expression [53]. We previously reported that the DENV upregulates the TF coagulation receptor in endothelial vascular cells, which triggers the generation of hemostatic proteases (thrombin) favoring the activation of protease-activated receptors or PARs, which, in turn, induce signaling inflammatory pathways (via phosphorylation of MAPKs p38 and ERK1/2, by transcription of NF-κB factor), thereby supporting the upregulation of VCAM-1 adhesion or pro-inflammatory molecules in ECs [54]. Next, we determined whether the ZIKV infection (MOI 1) of vascular endothelial cells (HMEC-1) or the stimulation by ZIKV C6/36 EVs favor a pro-inflammatory/pro-coagulant state of naïve endothelial vascular cells.

Therefore, we assessed the possible participation of small and large EVs issued from ZIKV-infected C6/36 cells for the induction of coagulation (TF) or inflammation (PAR-1) receptors at the membrane's surface of ECs. In a parallel assay, the adhesion ICAM-1 molecule was also evaluated. Different stimulation assays were performed using naïve HMEC-1 cells in the presence of ZIKV (MOI 1), the mock C6/36 EVs, or the ZIKV C6/36 EVs (small and large isolates, the same for the RNase A + UV), which were evaluated for the presence of TF, PAR-1, and ICAM-1 at their cell membranes (FACS) at 72 h post-stimulus (Figure 11A–C). Likewise, the cytopathic effect formation on these samples was evaluated by light-field microscopy (Figure S14B).

Elevated levels of TF (20.27 ± 0.51%) were detected in ZIKV-infected (MOI 1) vascular ECs (Figure 11A), but also in the presence of large ZIKV C6/36 EVs (16.75 ± 0.59%), which were both statistically significant ($p < 0.0001$) when compared with the mock HMEC-1 cultures (*). In small EV culture samples, the TF values were 4.82 ± 0.23%. The upregulation of the TF receptor may trigger the generation of hemostatic proteases (thrombin) favoring the activation of protease-activated receptors (PARs). Figure 11B shows the activation percentages of the PAR-1 in ECs infected with ZIKV (26.41 ± 0.56%), and the same is true for large ZIKV C6/36 EVs (23.17 ± 0.33% for untreated and 13.15 ± 0.38% for RNase A + UV-treated) and small ZIKV C6/36 EVs (21.55 ± 0.14% for untreated and 16.19 ± 0.79% for RNase A + UV-treated) culture samples. We found a significant difference ($p < 0.0001$) in all EVs stimuli compared with the mock HMEC-1 cultures. It is well known that PAR-1 favors signaling pathways for the expression of pro-adherent and pro-inflammatory molecules [55]. Therefore, we assessed ICAM-1 (Figure 11C) detection on EC surfaces in ZIKV-infected HMEC-1 cultures (12.87 ± 0.16%). The same was observed in cultures of naïve ECs in the presence of the mock C6/36 EVs or the ZIKV C6/36 EVs (small/large isolates, the same for the RNase A + UV) ($p < 0.001$). These data were corroborated by light-field microscopy (Figure S14B), where ZIKV-infected ECs showed an increased cytopathic effect (black arrows) with the formation of vacuolization and syncytial structures. Cytopathic effects were also observed in all naïve EC culture assays stimulated by the ZIKV C6/36 EVs.

Moreover, to evaluate the possible expression of the pro-inflammatory response by the stimulation of EVs released from ZIKV-infected C6/36 cells, we measured the TNF-α mRNA expression in naïve ECs infected by ZIKV (MOI 1) and those stimulated by the mock C6/36 EVs and the ZIKV C6/36 EVs (small/large isolates as the same for the RNase A + UV). We found that, similar to ZIKV infection, ZIKV C6/36 EVs were able to induce TNF-α mRNA expression in endothelial vascular cells (Figure 11D). Our data suggest the possible participation of the coagulation–inflammation process in the coagulation disorders present in severe cases of Zika. The endothelial vascular cell activation (damage) during ZIKV infection with an inflammatory response can cause EC dysfunction and weaken the endothelial barrier integrity. Thus, we evaluate the vascular endothelial barrier integrity in vitro using a Transwell assay (Figure 11E).

Our data indicate that endothelial vascular cells are susceptible to ZIKV infection and activation by both ZIKV (MOI 1) and ZIKV (small/large) C6/36 EVs with pro-inflammatory cytokine expression, which could increase endothelial monolayer permeability. Therefore, we determined whether ZIKV infection or ZIKV C6/36 EVs disturb the vascular endothelial barrier's integrity in vitro using a Transwell assay (see Materials and Methods; Figure S6), performed in a naïve EC culture in the presence of ZIKV (MOI 1), the mock C6/36 EVs, or the ZIKV C6/36 EVs.

For ZIKV-infected ECs, the permeability percentage (15.77 ± 1.23%) increased 2.3-fold compared to the mock HMEC-1 cultures ($p < 0.0001$). In the presence of the ZIKV C6/36 lEVs (RNase A + UV-treated (10.87 ± 1.29%) and untreated (10.95 ± 1.37%)), permeability increased, on average, 1.6-fold ($p < 0.05$); in cultures stimulated by ZIKV C6/36 sEVs (RNase A + UV-treated (10.29 ± 1.14%); when untreated (12.33 ± 1.19%)), the endothelial permeability increased 1.5- ($p < 0.05$) and 1.8-fold ($p < 0.01$), respectively (Figure 11E). These data suggest that ZIKV C6/36 EVs may participate in vascular endothelial damage with a weakening of the endothelial barrier integrity and support the mosquito EVs participation during the infection process, which could contribute to the pathogenesis of ZIKV infection in a human host.

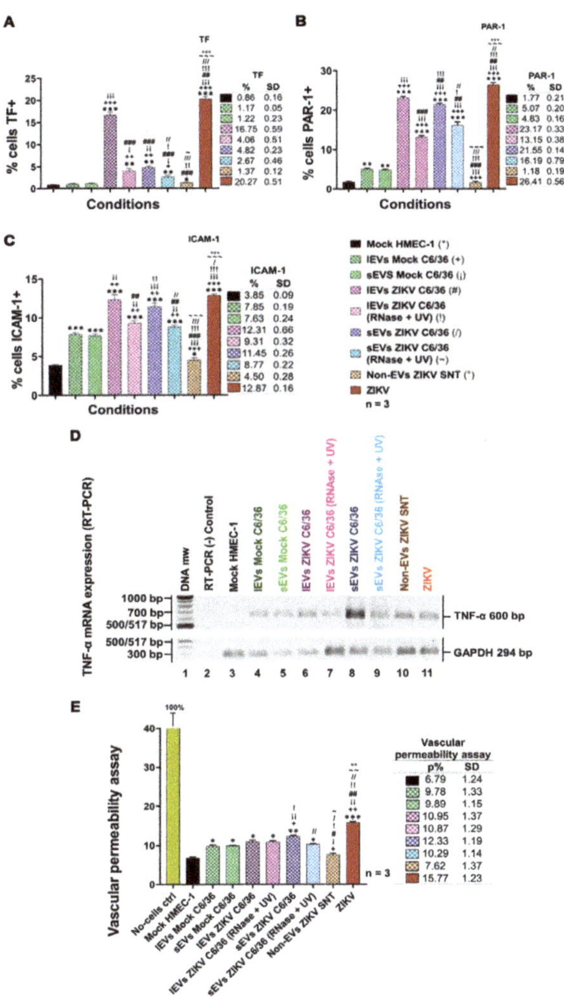

Figure 11. EVs from ZIKV-infected C6/36 modify towards a pro-coagulant, pro-inflammatory, and pro-adherent phenotype and favor permeability in naïve endothelial cells (ECs). (**A**) EC TF-1+ percentages (FACS) at different EVs stimuli conditions from three independent experiments. (**B**) EC Protease Activated Receptor+ (PAR-1) percentages at different EVs stimuli conditions from three independent experiments. (**C**) EC intercellular adhesion molecule-1+ (ICAM-1) percentages at different EVs stimuli conditions from three independent experiments. The TF, PAR-1, or ICAM-1 levels were compared (by an unpaired Student's t-test) between all conditions' values. (**D**) TNF-α mRNA expression (RT-PCR) in naïve ECs at different EVs stimuli conditions. The TNF-α genome conserved region (amplicon of 600 bp) was visualized on 2% ethidium bromide-stained 1.2% agarose gel. (**E**) Permeability percentages obtained by assessing the fluorescein isothiocyanate (FITC)-Dextran pass through the EC monolayers in the presence of different EV stimuli conditions. Three independent experiments were performed. For the 100% FITC-Dextran delivered control, a no-cell insert was used. The endothelial vascular permeability percentages were compared (by an unpaired Student's t-test) between all conditions' values. For all experiments, statistical significance was recognized as *, +, ¡, #, !, /, ~, or ° when $p < 0.05$, **, ++, ¡¡, ##, !!, //, ~~, or °° when $p < 0.01$, and ***, +++, ¡¡¡, ###, !!!, ///, ~~~, or °°° when $p < 0.0001$.

4. Discussion

Vector-borne diseases cause nearly one million deaths per year and represent 17% of all infectious illnesses worldwide [56]. This public health problem highlights the importance of understanding how arthropod vectors, microbes, and their mammalian hosts interact. At present, research efforts are focused on pathogen–host interactions, with a lack of attention on the significant contribution of vector-derived products in disease development. The molecular and cellular events occurring in vector–pathogen–host interactions are critical in determining the outcome of the vector-borne diseases. Recently, it was proposed that one strategy used by vectors, to promote a successful host infection, is the manipulation of EVs [41,57]. Infected vector cells secrete vesicles that may contain antigens, nucleic acids, and microbial cargos (or the whole pathogen), which exacerbate the pathogenesis and modulate the host responses [36]. EVs are involved in the exchange of bioactive molecules between cells. Although all EVs are vesicles constituted by lipid layers, their cargo reflects the state of the source cell, and their content can be altered in adverse conditions or manipulated by pathogens. Extracellular vesicles have an important role in the establishment of arboviral diseases [16,30,36]. EVs originating from arthropod vectors are an important strategy for immune evasion during viral transmission. For example, dengue virus (DENV) uses EVs derived from mosquitoes to infect mammalian cells. Mosquito-derived vesicles carry DENV proteins and a full-length viral genome. DENV transmission may occur through the interaction between the tetraspanin Tsp29Fb, a mosquito homolog of the human sEVs marker CD63, and the viral E protein [16]. Similarly, during the infection of the tick-cell line ISE6 with the Langat virus (LGTV), cells release sEVs that contain cargo from both the virus and the vector, which enable these EVs to transmit the virus to mammalian cells [36]. Ample evidence has been provided to show that sEVs carry and deliver viral genomes into recipient cells in vitro, as was reported for the hepatitis C virus (HCV), the hepatitis A virus (HAV), and human herpes virus 6 (HHV-6), among others [58–62]. Nevertheless, to date, it is unknown if, during mosquito ZIKV delivery to the vertebrate host cells, this arbovirus or the viral components such as viral proteins and the viral genome can be transferred by EVs from infected mosquito tissues to the host. We determined whether viral elements (viral RNA and envelope protein) could be transported by mosquito EVs, as was recently reported for other arthropod-borne flaviviruses.

In the present work, we found that, during ZIKV infection of C6/36 mosquito cells, small and large EVs were produced in high amounts. The isolated sEVs from ZIKV-infected cells were purified using paramagnetic nanobeads coated with anti-CD63 antibodies, thus demonstrating their endosomal origin. The separation of sEVs from virions by positive selection using magnetic beads coupled with an antibody against a tetraspanin enriched in sEVs is by far the best method [60,63]. The EVs characterization by nanoparticle tracking analysis (NTA) and transmission electron microscopy (TEM) from the mock C6/36 and the ZIKV-infected C6/36 EVs (Figures 2D and 3F) showed different EVs populations, which were heterogeneous in shape, with sizes up to 200 nm (1000 nm scale) in the case of lEVs, and less than 200 nm (500 nm scale) in diameter for sEVs. These populations showed well-defined bilipid membranes in a proper structural resolution. Our data agree with other reports demonstrating that cells secrete EVs as a heterogeneous population with different sizes and shapes [30,44,46]. Recently, it was proposed that the EVs size variation could be due to the internal content (proteins/RNAs) of vesicles produced from uninfected and infected cells [30]. By the NTA, we found that the lEVs from ZIKV-infected C6/36 presented an average size of 319.3 ± 11.5 nm (1.2-fold higher compared with the average size of lEVs from the mock cells) (Figure 2A). On the other hand, the sEVs from ZIKV-infected C6/36 presented an average size of 125.5 ± 1.6 nm (1.2-fold higher compared with the average size of sEVs from the mock cells) (Figure 3C). In this sense, the ZIKV C6/36 EVs RNase A + UV treatment assays (Figure 4 and Figure S4), performed to eliminate the possible presence of polluting free virus in samples, suggest a presence of viral RNA inside ZIKV-infected C6/36 EVs. The presence of the ZIKV E protein detected in ZIKV C6/36 lEVs and sEVs isolates was not observed in mock C6/36 EVs (Figure 2E,F and Figure 3G,H). These findings suggest that both sEVs and lEVs from ZIKV-infected mosquito cells carry the viral E protein and viral RNA. Different viruses manipulate EVs for their benefit in order to increase their persistence, pathogenesis, and transmission [58–62,64]. The hijacking of mosquito cell membranes by ZIKV could facilitate their escape from host immune responses, promoting the viral elements' spread. We first used the epithelial cells (Vero) from the monkey *Cercopithecus aethiops* (used as gold standard cells for infection assay) [32]. The presence of lytic plaques (Figure 4E) in cultures of Vero with the ZIKV C6/36 EVs (lEVs and sEVs) isolates were observed in more concentrated amounts, formed in high or undetermined quantities. However, plaque formation was not detected in the negative controls. These data suggest ZIKV-mosquito EVs participation during the infectious process of monkey epithelial cells. Additionally, the possibility of EVs participation during the infection process of naïve mosquito cells via cell-to-cell EVs transfer was evaluated. We demonstrated the presence of the ZIKV envelope protein at the mosquito cell membrane surface in cultures incubated with the ZIKV C6/36 EVs (large and small), implying the infection of naïve mosquito cells (Figure 5 and Figure S12). These results are consistent with those of Vora et al. (2018), showing that the full-length genome of DENV-2 detected in the EVs from DENV-infected mosquito cells was infectious in naïve mosquito and mammalian cells [16].

To date, it has not been determined whether EVs from ZIKV-infected mosquito cells are utilized as ZIKV viral element vehicles (genome/protein) to mammalian host cells. As monocytes and vascular ECs are the main targets during ZIKV human host infection, we evaluated the possible participation of EVs derived from the ZIKV-infected mosquito C6/36 cells in potential viral element carriers to human monocytes and vascular ECs. ZIKV was able to establish a productive infection in human monocytes and ECs, since the viral E protein was detected at high levels on the cell membrane surface at all PI time points of the assay (Figures 6 and 9). In stimulation assays of monocytes and ECs in the presence of ZIKV-infected C6/36 EVs, we found that ZIKV C6/36 EVs supported cell infection in naïve cells (Figures 7 and 10). Interestingly, the possible cellular mechanisms of the budding/trafficking of sEVs that could be used by HCV, HAV, HIV, Epstein-Barr virus (EBV) and Kaposi's sarcoma-associated herpesvirus (KSHV) to spread from cell-to-cell viral elements or viral particles were recently revised [60]. Zhou et al. (2019), by using primary cultures of murine cortical neurons, showed that ZIKV uses sEVs as a mediator of viral transmission between neurons. Neuronal sEVs contained both ZIKV and viral RNA/protein(s) that were highly infectious to naïve cells. RNase A and neutralizing antibody assays suggested the presence of viral RNA/proteins inside EVs [65].

It is known that EVs may have different functions when interacting with other cells, modifying their naïve cellular behavior [66]. For monocytes (Figure 8), in the presence of different EVs samples, it was observed that naïve cells were differentiated and that they expressed a CD11b+ differentiation marker at high levels compared to the mock THP-1 cells and those stimulated with the mock C6/36 EVs. Likewise, the classical monocyte phenotype was changed to the CD14++ CD16+ intermediate phenotype with respect to the mock THP-1 cells and the mock C6/36 EVs stimulation assays ($p < 0.0001$ for CD14 and $p < 0.01$ for CD16). Intermediate and non-classical monocytes seem to be the main producers of pro-inflammatory mediators in response to viral infection [17,49]. The activation of naïve monocytes via the ZIKV C6/36 sEVs favors TNF-α mRNA expression, which suggests that ZIKV C6/36 EVs–human monocyte interplay plays a role in establishing a pro-inflammatory state.

The possible participation of ZIKV C6/36 EVs in the infection and activation of vascular ECs was also evaluated. In cultures performed in HMEC-1 naïve cells in the presence of ZIKV (MOI 1), the mock C6/36 EVs, and the ZIKV small/large C6/36 EVs, the amount of ZIKV E protein was evaluated. Figure 10 shows the viral E protein presence at high levels in the ZIKV (MOI 1)-infected HMEC-1 cells as well as in the naïve ECs stimulated with ZIKV C6/36 EVs (RNase A + UV-treated and untreated). The E protein percentages were compared among values of all conditions against the mock HMEC-1 cells; these values were statistically significant ($p < 0.0001$) (Figure 10B,C). These data suggest that ZIKV-infected C6/36 EVs support the infection of mammalian host cells, including vascular ECs.

At the endothelial vascular cell level, our data indicate that ECs are susceptible to ZIKV infection and activation by ZIKV C6/36 EVs, and these EVs favor the induction of damage receptors, such as coagulation (TF) and inflammation (PAR-1) receptors, and adhesion molecule (ICAM-1) presence at the cell membrane's surface level (Figure 11). Recent observations in human and animal models [51,52,67] suggest that, in severe Zika cases, different coagulation disorders occur. It has been shown that several viruses activate the coagulation system especially through TF receptor expression [53]. Dengue virus has also been shown to cause coagulation disorders in ECs [54].

Anfasa et al. (2019) provided in vitro evidence that ZIKV infection of human umbilical vein endothelial cells (HUVECs) induces apoptosis and increases TF production, which triggers the activation of secondary hemostasis [68]. Additionally, to evaluate the possible expression of a pro-inflammatory response by EVs released from ZIKV-infected C6/36 cells, we measured the TNF-α mRNA expression in naïve ECs infected by ZIKV (MOI 1) and those stimulated with the mock C6/36 EVs or the ZIKV C6/36 EVs (small/large isolates the same for RNase A + UV-treated). We found that, like ZIKV infection, the ZIKV C6/36 EVs were able to induce TNF-α mRNA expression in endothelial vascular cells (Figure 11D). The ZIKV C6/36 EVs also participated in vascular endothelial damage, with a weakening of the endothelial barrier integrity, and this was demonstrated using a Transwell assay (Figure 11E). We previously reported that DENV infection of ECs upregulates the TF coagulation receptor in endothelial vascular cells, which triggers the generation of hemostatic proteases (thrombin) favoring the activation of protease-activated receptors or PARs, which, in turn, induces signaling inflammatory pathways (via phosphorylation of MAPKs p38 and ERK1/2, by transcription of the NF-κB factor), thereby supporting the upregulation of adhesion VCAM-1 or pro-inflammatory molecules in ECs, being part of the pathogenic mechanisms for the vascular endothelial injury present in severe Dengue cases [54]. At present, it is under evaluation whether these signaling pathways can support the participation of ZIKV-infected vector cells in the activation and damage of vascular ECs that can contribute to the pathogenesis of severe ZIKV cases. The present data suggest that ZIKV C6/36 EVs allow for ZIKV elements (viral genome/protein) to modulate host response and enhance viral fitness abilities.

In summary, during ZIKV infection of C6/36 mosquito cells, small and large EVs were produced (Figure 12A). The mosquito EVs released from ZIKV-infected cells carried viral E protein and viral RNA and were able to infect naïve mosquito and mammalian cells. The ZIKV-infected mosquito EVs, then, modified the naïve cellular behavior, since they promote the infection, activation, and differentiation of human monocytes that favor a pro-inflammatory monocyte state (Figure 12B). At the endothelial vascular cell level, our data indicate that vascular ECs are susceptible to ZIKV activation and infection by C6/36 EVs, which favor the induction of tissue damage receptors, such as coagulation (TF) and inflammation (PAR-1) receptors, and adhesion molecule presence (ICAM-1) at the membrane surface level with an increase in cell permeability (Figure 12C). Knowledge of the targets' cellular pathways that allow ZIKV to establish prolonged viral persistence could contribute to novel vaccines and therapies.

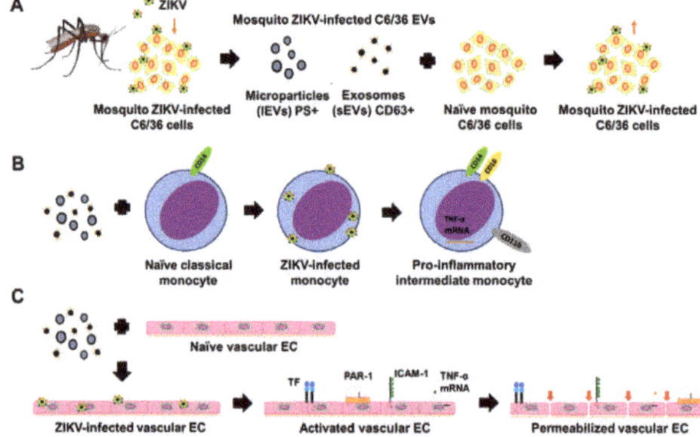

Figure 12. Extracellular vesicles (EVs) from ZIKV-infected mosquito (C6/36) cells participate in the modification of naïve cells' behavior by mediating cell-to-cell transmission of viral elements (Graphic description). (**A**) ZIKV C6/36 EVs favor naïve mosquito cell infection. Mosquito image created with BioRender.com. (**B**) ZIKV C6/36 EVs promote infection and shift to a pro-inflammatory phenotype in monocytes. (**C**) ZIKV C6/36 EVs participate in the vascular EC infection and activation.

Supplementary Materials: The supplementary materials are available online at http://www.mdpi.com/2073-4409/9/1/123/s1.

Author Contributions: Conceptualization: P.P.M.-R. and B.H.R.-O.; methodology: P.P.M.-R., E.Q.-G., V.M.-M., and B.H.R.-O.; validation: P.P.M.-R., E.Q.-G., and V.M.-M.; formal analysis: P.P.M.-R. and B.H.R.-O.; investigation: B.H.R.-O.; resources: B.H.R.-O, L.T.A.-M., and L.F.J.-G.; data curation: L.T.A.-M. and L.F.J.-G.; writing—original draft preparation: P.P.M.-R. and B.H.R.-O.; writing—review and editing: MDPI; visualization: P.P.M.-R. and B.H.R.-O.; supervision: B.H.R.-O.; project administration: V.M.-M.; funding acquisition: B.H.R.-O. All authors have read and agreed to the published version of the manuscript.

Funding: This research received no external funding.

Acknowledgments: Pedro Pablo Martínez-Rojas is a doctoral student from Programa de Doctorado en Ciencias Bioquímicas, Universidad Nacional Autónoma de México (UNAM), and was supported by Consejo Nacional de Ciencia y Tecnología (CONACYT), México. Elizabeth Quiroz-García is a doctoral student from Programa de Doctorado en Ciencias Biomédicas, UNAM and was supported by CONACYT. We thank the PAPIIT-UNAM Program for partially funding support. We also thank LabNalCit-UNAM (CONACYT) for the technical support in the acquisition of flow cytometry samples and Miguel Tapia Rodríguez, from Unidad de Microscopía, Instituto de Investigaciones Biomédicas UNAM, for his support in the use of the Olympus IX71 microscope.

Conflicts of Interest: The authors declare that there is no conflict of interest.

References

1. Plourde, A.R.; Bloch, E.M. A Literature Review of Zika Virus. *Emerg. Infect. Dis.* **2016**, *22*, 1185–1192. [CrossRef] [PubMed]

2. Abushouk, A.I.; Negida, A.; Ahmed, H. An updated review of Zika virus. *J. Clin. Virol.* **2016**, *84*, 53–58. [CrossRef] [PubMed]

3. Hills, S.L.; Fischer, M.; Petersen, L. Epidemiology of Zika Virus Infection. *J. Infect. Dis.* **2017**, *216*, S868–S874. [CrossRef] [PubMed]

4. Song, B.H.; Yun, S.I.; Woolley, M.; Lee, Y.M. Zika virus: History, epidemiology, transmission, and clinical presentation. *J. Neuroimmunol.* **2017**, *308*, 50–64. [CrossRef] [PubMed]

5. Petersen, L.R.; Jamieson, D.J.; Powers, A.M.; Honein, M.A. Zika Virus. *N. Engl. J. Med.* **2016**, *374*, 1552–1563. [CrossRef]

6. Platt, D.J.; Miner, J.J. Consequences of congenital Zika virus infection. *Curr. Opin. Virol.* **2017**, *27*, 1–7. [CrossRef]

7. Nugent, E.K.; Nugent, A.K.; Nugent, R.; Nugent, K. Zika Virus: Epidemiology, Pathogenesis and Human Disease. *Am. J. Med. Sci.* **2017**, *353*, 466–473. [CrossRef]

8. Karimi, O.; Goorhuis, A.; Schinkel, J.; Codrington, J.; Vreden, S.G.S.; Vermaat, J.S.; Stijnis, C.; Grobusch, M.P. Thrombocytopenia and subcutaneous bleedings in a patient with Zika virus infection. *Lancet* **2016**, *387*, 939–940. [CrossRef]

9. Karkhah, A.; Nouri, H.R.; Javanian, M.; Koppolu, V.; Masrour-Roudsari, J.; Kazemi, S.; Ebrahimpour, S. Zika virus: Epidemiology, clinical aspects, diagnosis, and control of infection. *Eur. J. Clin. Microbiol. Infect. Dis.* **2018**, *37*, 2035–2043. [CrossRef]

10. Wikan, N.; Smith, D. Zika virus: History of a newly emerging arbovirus. *Lancet Infect. Dis.* **2016**, *16*, 119–126. [CrossRef]

11. Imperato, P.J. The Convergence of a Virus, Mosquitoes, and Human Travel in Globalizing the Zika Epidemic. *J. Community Health* **2016**, *41*, 674–679. [CrossRef] [PubMed]

12. Sakkas, H.; Economou, V.; Papadopoulou, C. Zika virus infection: Past and present of another emerging vector-borne disease. *J. Vector Borne Dis.* **2016**, *53*, 305–311. [PubMed]

13. Rückert, C.; Weger-Lucarelli, J.; Garcia-Luna, S.M.; Young, M.C.; Byas, A.D.; Murrieta, R.A.; Fauver, J.R.; Ebel, G.D. Impact of simultaneous exposure to arboviruses on infection and transmission by *Aedes aegypti* mosquitoes. *Nat. Commun.* **2017**, *8*, 1–9. [CrossRef] [PubMed]

14. Rückert, C.; Ebel, G.D. How Do Virus-Mosquito Interactions Lead to Viral Emergence? *Trends Parasitol.* **2018**, *34*, 310–321. [CrossRef]

15. Cime-Castillo, J.; Delannoy, P.; Mendoza-Hernández, G.; Monroy-Martínez, V.; Harduin-Lepers, A.; Lanz-Mendoza, H.; Hernández-Hernández, F.d.L.; Zenteno, E.; Cabello-Gutiérrez, C.; Ruiz-Ordaz, B.H. Sialic acid expression in the mosquito *Aedes aegypti* and its possible role in dengue virus-vector interactions. *Biomed. Res. Int.* **2015**, *2015*, 1–16. [CrossRef]

16. Vora, A.; Zhou, W.; Londono-Renteria, B.; Woodson, M.; Sherman, M.B.; Colpitts, T.M.; Neelakanta, G.; Sultana, H. Arthropod EVs mediate dengue virus transmission through interaction with a tetraspanin domain containing glycoprotein Tsp29Fb. *Proc. Natl. Acad. Sci. USA* **2018**, *115*, E6604–E6613. [CrossRef]

17. Jurado, K.A.; Iwasaki, A. Zika virus targets blood monocytes. *Nat. Microbiol.* **2017**, *2*, 1460–1461. [CrossRef]

18. Miner, J.J.; Diamond, M.S. Zika Virus Pathogenesis and Tissue Tropism. *Cell Host Microbe* **2017**, *21*, 134–142. [CrossRef]

19. Routhu, N.K.; Byrareddy, S.N. Host-Virus Interaction of ZIKA Virus in Modulating Disease Pathogenesis. *J. Neuroimmune Pharmacol.* **2017**, *12*, 219–232. [CrossRef]

20. Quesenberry, P.J.; Aliotta, J.; Deregibus, M.C.; Camussi, G. Role of extracellular RNA-carrying vesicles in cell differentiation and reprogramming. *Stem Cell Res. Ther.* **2015**, *6*, 1–10. [CrossRef]

21. Pitt, J.M.; Kroemer, G.; Zitvogel, L. Extracellular vesicles: Masters of intercellular communication and potential clinical interventions. *J. Clin. Investig.* **2016**, *126*, 1139–1143. [CrossRef] [PubMed]

22. Théry, C.; Witwer, K.W.; Aikawa, E.; Alcaraz, M.J.; Anderson, J.D.; Andriantsitohaina, R.; Antoniou, A.; Arab, T.; Archer, F.; Atkin-Smith, G.K.; et al. Minimal information for studies of extracellular vesicles 2018 (MISEV2018): A position statement of the International Society for Extracellular Vesicles and update of the MISEV2014 guidelines. *J. Extracell. Vesicles* **2018**, *7*, 1–43. [CrossRef] [PubMed]

23. EL Andaloussi, S.; Mäger, I.; Breakefield, X.O.; Wood, M.J. Extracellular vesicles: Biology and emerging therapeutic opportunities. *Nat. Rev. Drug. Discov.* **2013**, *12*, 347–357. [CrossRef] [PubMed]

24. van Niel, G.; D'Angelo, G.; Raposo, G. Shedding light on the cell biology of extracellular vesicles. *Nat. Rev. Mol. Cell Biol.* **2018**, *19*, 213–228. [CrossRef] [PubMed]

25. Chahar, H.S.; Bao, X.; Casola, A. Exosomes and Their Role in the Life Cycle and Pathogenesis of RNA Viruses. *Viruses* **2015**, *7*, 3204–3225. [CrossRef] [PubMed]

26. van Dongen, H.M.; Masoumi, N.; Witwer, K.W.; Pegtel, D.M. Extracellular Vesicles Exploit Viral Entry Routes for Cargo Delivery. *Microbiol. Mol. Biol. Rev.* **2016**, *80*, 369–386. [CrossRef]

27. Schwab, A.; Meyering, S.S.; Lepene, B.; Iordanskiy, S.; van Hoek, M.L.; Hakami, R.M.; Kashanchi, F. Extracellular vesicles from infected cells: Potential for direct pathogenesis. *Front. Microbiol.* **2015**, *6*, 1–18. [CrossRef]

28. Burger, D.; Schock, S.; Thompson, C.; Montezano, A.C.; Hakim, A.M.; Touyz, R.M. Microparticles: Biomarkers and beyond. *Clin. Sci. (Lond.)* **2013**, *124*, 423–441. [CrossRef]

29. Meckes, D.G., Jr.; Raab-Traub, N. Microvesicles and viral infection. *J. Virol.* **2011**, *85*, 12844–12854. [CrossRef]

30. Reyes-Ruiz, J.M.; Osuna-Ramos, J.F.; De Jesús-González, L.A.; Hurtado-Monzón, A.M.; Farfan-Morales, C.N.; Cervantes-Salazar, M.; Bolaños, J.; Cigarroa-Mayorga, O.E.; Martín-Martínez, E.S.; Medina, F.; et al. Isolation and characterization of exosomes released from mosquito cells infected with dengue virus. *Virus Res.* **2019**, *266*, 1–14. [CrossRef]

31. Liu, S.; DeLalio, L.J.; Isakson, B.E.; Wang, T.T. AXL-Mediated Productive Infection of Human Endothelial Cells by Zika Virus. *Circ. Res.* **2016**, *119*, 1183–1189. [CrossRef] [PubMed]

32. Moser, L.A.; Boylan, B.T.; Moreira, F.R.; Myers, L.J.; Svenson, E.L.; Fedorova, N.B.; Pickett, B.E.; Bernard, K. Growth and adaptation of Zika virus in mammalian and mosquito cells. *PLoS Negl. Trop. Dis.* **2018**, *12*, e0006880. [CrossRef] [PubMed]

33. Zhang, L.; Shen, Z.L.; Feng, Y.; Li, D.Q.; Zhang, N.N.; Deng, Y.Q.; Qi, X.P.; Sun, X.M.; Dai, J.J.; Yang, C.G.; et al. Infectivity of Zika virus on primary cells support tree shrew as animal model. *Emerg. Microbes Infect.* **2019**, *8*, 232–241. [CrossRef] [PubMed]

34. Lässer, C.; Eldh, M.; Lötvall, J. Isolation and characterization of RNA-containing exosomes. *J. Vis. Exp.* **2012**, *59*, 1–6. [CrossRef]

35. Kay, J.; Grinstein, S. Sensing phosphatidylserine in cellular membranes. *Sensors* **2011**, *11*, 1744–1755. [CrossRef]

36. Zhou, W.; Woodson, M.; Neupane, B.; Bai, F.; Sherman, M.B.; Choi, K.H.; Neelakanta, G.; Sultana, H. Exosomes serve as novel modes of tick-borne flavivirus transmission from arthropod to human cells and facilitates dissemination of viral RNA and proteins to the vertebrate neuronal cells. *PLoS Pathog.* **2018**, *14*, e1006764. [CrossRef]

37. Patnaik, B.B.; Kang, S.M.; Seo, G.W.; Lee, H.J.; Patnaik, H.H.; Jo, Y.H.; Tindwa, H.; Lee, Y.S.; Lee, B.L.; Kim, N.J.; et al. Molecular cloning, sequence characterization and expression analysis of a CD63 homologue from the coleopteran beetle, *Tenebrio molitor*. *Int. J. Mol. Sci.* **2013**, *14*, 20744–20767. [CrossRef]

38. Yang, C.F.; Tu, C.H.; Lo, Y.P.; Cheng, C.C.; Chen, W.J. Involvement of Tetraspanin C189 in Cell-to-Cell Spreading of the Dengue Virus in C6/36 Cells. *PLoS Negl. Trop. Dis.* **2015**, *9*, e0003885. [CrossRef]

39. Gaunt, M.W.; Gould, E.A. Rapid subgroup identification of the flaviviruses using degenerate primer E-gene RT-PCR and site-specific restriction enzyme analysis. *J. Virol. Methods* **2005**, *128*, 113–127. [CrossRef]

40. Faye, O.; Faye, O.; Dupressoir, A.; Weidmann, M.; Ndiaye, M.; Alpha Sall, A. One-step RT-PCR for detection of Zika virus. *J. Clin. Virol.* **2008**, *43*, 96–101. [CrossRef]

41. Chávez, A.S.O.; O'Neal, A.J.; Santambrogio, L.; Kotsyfakis, M.; Pedra, J.H.F. Message in a vesicle-trans-kingdom intercommunication at the vector-host interface. *J. Cell Sci.* **2019**, *132*, 1–11. [CrossRef] [PubMed]

42. Chen, Y.H.; Du, W.; Hagemeijer, M.C.; Takvorian, P.M.; Pau, C.; Cali, A.; Brantner, C.A.; Stempinski, E.S.; Connelly, P.S.; Ma, H.C.; et al. Phosphatidylserine vesicles enable efficient en bloc transmission of enteroviruses. *Cell* **2015**, *160*, 619–630. [CrossRef] [PubMed]

43. Yáñez-Mó, M.; Siljander, P.R.; Andreu, Z.; Zavec, A.B.; Borràs, F.E.; Buzas, E.I.; Buzas, K.; Casal, E.; Cappello, F.; Carvalho, J.; et al. Biological properties of extracellular vesicles and their physiological functions. *J. Extracell. Vesicles* **2015**, *4*, 1–60. [CrossRef] [PubMed]

44. Bobrie, A.; Colombo, M.; Krumeich, S.; Raposo, G.; Théry, C. Diverse subpopulations of vesicles secreted by different intracellular mechanisms are present in exosome preparations obtained by differential ultracentrifugation. *J. Extracell. Vesicles* **2012**, *1*, 1–11. [CrossRef] [PubMed]

45. Dogrammatzis, C.; Deschamps, T.; Kalamvoki, M. Biogenesis of Extracellular Vesicles during Herpes Simplex Virus 1 Infection: Role of the CD63 Tetraspanin. *J. Virol.* **2019**, *93*, 1–15. [CrossRef] [PubMed]

46. Willms, E.; Johansson, H.J.; Mäger, I.; Lee, Y.; Blomberg, K.E.; Sadik, M.; Alaarg, A.; Smith, C.I.; Lehtiö, J.; El Andaloussi, S.; et al. Cells release subpopulations of exosomes with distinct molecular and biological properties. *Sci. Rep.* **2016**, *6*, 1–12. [CrossRef]

47. Michlmayr, D.; Andrade, P.; Gonzalez, K.; Balmaseda, A.; Harris, E. CD14+CD16+ monocytes are the main target of Zika virus infection in peripheral blood mononuclear cells in a paediatric study in Nicaragua. *Nat. Microbiol.* **2017**, *2*, 1462–1470. [CrossRef]

48. Yoshikawa, F.S.Y.; Teixeira, F.M.E.; Sato, M.N.; Oliveira, L.M.D.S. Delivery of microRNAs by Extracellular Vesicles in Viral Infections: Could the News be Packaged? *Cells* **2019**, *8*, 611. [CrossRef]

49. Naranjo-Gómez, J.S.; Castillo, J.A.; Rojas, M.; Restrepo, B.N.; Diaz, F.J.; Velilla, P.A.; Castaño, D. Different phenotypes of non-classical monocytes associated with systemic inflammation, endothelial alteration and hepatic compromise in patients with dengue. *Immunology* **2018**, *156*, 147–163. [CrossRef]

50. Delatorre, E.; Miranda, M.; Tschoeke, D.A.; Carvalho de Sequeira, P.; Alves Sampaio, S.; Barbosa-Lima, G.; Rangel Vieira, Y.; Leomil, L.; Bozza, F.A.; Cerbino-Neto, J.; et al. An observational clinical case of Zika virus-associated neurological disease is associated with primary IgG response and enhanced TNF levels. *J. Gen. Virol.* **2018**, *99*, 913–916. [CrossRef]

51. Sharp, T.M.; Muñoz-Jordán, J.; Perez-Padilla, J.; Bello-Pagán, M.I.; Rivera, A.; Pastula, D.M.; Salinas, J.L.; Martínez Mendez, J.H.; Méndez, M.; Powers, A.M.; et al. Zika Virus Infection Associated with Severe Thrombocytopenia. *Clin. Infect. Dis.* **2016**, *63*, 1198–1201. [CrossRef] [PubMed]

52. Van Dyne, E.A.; Neaterour, P.; Rivera, A.; Bello-Pagan, M.; Adams, L.; Munoz-Jordan, J.; Baez, P.; Garcia, M.; Waterman, S.H.; Reyes, N.; et al. Incidence and Outcome of Severe and Nonsevere Thrombocytopenia Associated With Zika Virus Infection-Puerto Rico, 2016. *Open Forum Infect. Dis.* **2018**, *6*, 1–9. [CrossRef] [PubMed]

53. Geisbert, T.W.; Hensley, L.E.; Jahrling, P.B.; Larsen, T.; Geisbert, J.B.; Paragas, J.; Young, H.A.; Fredeking, T.M.; Rote, W.E.; Vlasuk, G.P. Treatment of Ebola virus infection with a recombinant inhibitor of factor VIIa/tissue factor: A study in rhesus monkeys. *Lancet* **2003**, *362*, 1953–1958. [CrossRef]

54. Huerta-Zepeda, A.; Cabello-Gutiérrez, C.; Cime-Castillo, J.; Monroy-Martínez, V.; Manjarrez-Zavala, M.E.; Gutiérrez-Rodríguez, M.; Izaguirre, R.; Ruiz-Ordaz, B.H. Crosstalk between coagulation and inflammation during Dengue virus infection. *Thromb. Haemost.* **2008**, *99*, 936–943. [CrossRef] [PubMed]

55. Isermann, B. Homeostatic effects of coagulation protease-dependent signaling and protease activated receptors. *J. Thromb. Haemost.* **2017**, *15*, 1273–1284. [CrossRef]

56. World Health Organization. Vector-borne Diseases. Available online: https://www.who.int/news-room/factsheets/detail/vector-borne-diseases (accessed on 2 September 2019).

57. De Toro, J.; Herschlik, L.; Waldner, C.; Mongini, C. Emerging roles of exosomes in normal and pathological conditions: New insights for diagnosis and therapeutic applications. *Front. Immunol.* **2015**, *6*, 1–12. [CrossRef]

58. Ramakrishnaiah, V.; Thumann, C.; Fofana, I.; Habersetzer, F.; Pan, Q.; de Ruiter, P.E.; Willemsen, R.; Demmers, J.A.; Stalin Raj, V.; Jenster, G. Exosome-mediated transmission of hepatitis C virus between human hepatoma Huh7.5 cells. *Proc. Natl. Acad. Sci. USA* **2013**, *110*, 13109–13113. [CrossRef]

59. Cosset, F.L.; Dreux, M. HCV transmission by hepatic exosomes establishes a productive infection. *J. Hepatol.* **2014**, *60*, 674–675. [CrossRef]

60. Raab-Traub, N.; Dittmer, D.P. Viral effects on the content and function of extracellular vesicles. *Nat. Rev. Microbiol.* **2017**, *15*, 559–572. [CrossRef]

61. Mori, Y.; Koike, M.; Moriishi, E.; Kawabata, A.; Tang, H.; Oyaizu, H.; Uchiyama, Y.; Yamanishi, K. Human herpesvirus-6 induces MVB formation, and virus egress occurs by an exosomal release pathway. *Traffic* **2008**, *9*, 1728–1742. [CrossRef]

62. Bello-Morales, R.; López-Guerrero, J.A. Extracellular Vesicles in Herpes Viral Spread and Immune Evasion. *Front. Microbiol.* **2018**, *9*, 1–9. [CrossRef] [PubMed]

63. Oksvold, M.P.; Neurauter, A.; Pedersen, K.W. Magnetic bead-based isolation of exosomes. *Methods Mol. Biol.* **2015**, *1218*, 465–481. [PubMed]

64. Kaminski, V.L.; Ellwanger, J.H.; Chies, J.A.B. Extracellular vesicles in host-pathogen interactions and immune regulation-exosomes as emerging actors in the immunological theater of pregnancy. *Heliyon* **2019**, *5*. [CrossRef] [PubMed]

65. Zhou, W.; Woodson, M.; Sherman, M.B.; Neelakanta, G.; Sultana, H. Exosomes mediate Zika virus transmission through SMPD3 neutral Sphingomyelinase in cortical neurons. *Emerg. Microbes Infect.* **2019**, *8*, 307–326. [CrossRef] [PubMed]

66. Iraci, N.; Leonardi, T.; Gessler, F.; Vega, B.; Pluchino, S. Focus on Extracellular Vesicles: Physiological Role and Signalling Properties of Extracellular Membrane Vesicles. *Int. J. Mol. Sci.* **2016**, *17*, 171. [CrossRef] [PubMed]

67. Nguyen, S.M.; Antony, K.M.; Dudley, D.M.; Kohn, S.; Simmons, H.A.; Wolfe, B.; Salamat, M.S.; Teixeira, L.B.C.; Wiepz, G.J.; Thoong, T.H.; et al. Highly efficient maternal-fetal Zika virus transmission in pregnant rhesus macaques. *PLoS Pathog.* **2017**, *13*, e1006378. [CrossRef] [PubMed]

68. Anfasa, F.; Goeijenbier, M.; Widagdo, W.; Siegers, J.Y.; Mumtaz, N.; Okba, N.; van Riel, D.; Rockx, B.; Koopmans, M.P.G.; Meijers, J.C.M.; et al. Zika Virus Infection Induces Elevation of Tissue Factor Production and Apoptosis on Human Umbilical Vein Endothelial Cells. *Front. Microbiol.* **2019**, *10*, 1–11. [CrossRef]

Article

The Envelope Residues E152/156/158 of Zika Virus Influence the Early Stages of Virus Infection in Human Cells

Sandra Bos [1], Wildriss Viranaicken [1], Etienne Frumence [1], Ge Li [2], Philippe Desprès [1], Richard Y. Zhao [2,3,4,5,*] and Gilles Gadea [1,*]

[1] Université de la Réunion, INSERM U1187, CNRS UMR 9192, IRD UMR 249, Unité Mixte Processus Infectieux en Milieu Insulaire Tropical, Plateforme Technologique CYROI, 94791 Sainte Clotilde, La Réunion, France; sandrabos.lab@gmail.com (S.B.); wildriss.viranaicken@univ-reunion.fr (W.V.); etienne.frumence@univ-reunion.fr (E.F.); philippe.despres@univ-reunion.fr (P.D.)

[2] Department of Pathology, University of Maryland School of Medicine, Baltimore, MD 21201, USA; lige_cn@hotmail.com

[3] Department of Microbiology and Immunology, University of Maryland, Baltimore, MD 21201, USA

[4] Institute of Global Health, University of Maryland, Baltimore, MD 21201, USA

[5] Institute of Human Virology, University of Maryland, Baltimore, MD 21201, USA

* Correspondence: RZhao@som.umaryland.edu (R.Y.Z.); gilles.gadea@inserm.fr (G.G.); Tel.: +33-262-262-938-806 (G.G.)

Received: 6 September 2019; Accepted: 13 November 2019; Published: 15 November 2019

Abstract: Emerging infections of mosquito-borne Zika virus (ZIKV) pose an increasing threat to human health, as documented over the recent years in South Pacific islands and the Americas in recent years. To better understand molecular mechanisms underlying the increase in human cases with severe pathologies, we recently demonstrated the functional roles of structural proteins capsid (C), pre-membrane (prM), and envelop (E) of ZIKV epidemic strains with the initiation of viral infection in human cells. Specifically, we found that the C-prM region contributes to permissiveness of human host cells to ZIKV infection and ZIKV-induced cytopathic effects, whereas the E protein is associated with viral attachment and early infection. In the present study, we further characterize ZIKV E proteins by investigating the roles of residues isoleucine 152 (Ile152), threonine 156 (Thr156), and histidine 158 (His158) (i.e., the E-152/156/158 residues), which surround a unique *N*-glycosylation site (E-154), in permissiveness of human host cells to epidemic ZIKV infection. For comparison purpose, we generated mutant molecular clones of epidemic BeH819015 (BR15) and historical MR766-NIID (MR766) strains that carry each other's E-152/156/158 residues, respectively. We observed that the BR15 mutant containing the E-152/156/158 residues from MR766 was less infectious in A549-Dual™ cells than parental virus. In contrast, the MR766 mutant containing E-152/156/158 residues from BR15 displayed increased infectivity. The observed differences in infectivity were, however, not correlated with changes in viral binding onto host-cells or cellular responses to viral infection. Instead, the E-152/156/158 residues from BR15 were associated with an increased efficiency of viral membrane fusion inside infected cells due to conformational changes of E protein that enhance exposure of the fusion loop. Our data highlight an important contribution of E-152/156/158 residues to the early steps of ZIKV infection in human cells.

Keywords: flavivirus; Zika virus; envelope protein; glycosylation; fusion loop; viral fusion; cell entry

1. Introduction

Mosquito-borne Zika (ZIKV), dengue (DENV), Yellow fever (YFV) and Japanese encephalitis (JEV) viruses belonging to flavivirus genus (*Flaviviridae* family), are four enveloped RNA viruses of significant public health concern worldwide [1–4]. Recent ZIKV global outbreaks, with Brazil at the epicentre, highlighted how a previously neglected flavivirus can turn into a severe threat for human health. While human ZIKV infections remained only sporadic and with a limited impact for decades [5–8], recent outbreaks revealed that ZIKV caused clusters of severe congenital and neurological abnormalities in infants and peripheral nervous system impairments in adults [9–12]. Considering the dramatic increase of severe human cases, strategies to efficiently control this virus, either in terms of antiviral therapies or vaccines, are urgently needed and a granted requirement for more extensive studies.

Flaviviruses contain a genomic single-stranded positive RNA encoding a single large polyprotein that is subsequently cleaved by cellular and viral proteases into three structural proteins (C, prM/M and E) and seven nonstructural proteins (NS1 to NS5). The latter are responsible for virus replication, assembly and escape from host immune system, while structural proteins form the viral particle surrounding genomic viral RNA. Among structural proteins, the E protein is responsible for viral entry into host cells. Viral E protein first binds to cellular attachment factors and receptors, leading to virion internalisation primarily through a clathrin-mediated endocytic pathway [13]. In endosomes, fusion of viral and cellular membranes occurs after E protein conformational changes triggered by low pH [14]. The E protein peptide chain folds into three distinct domains: a central ß-barrel (domain EDI), an elongated dimerization region (domain EDII), which includes the fusion loop, and a C-terminal, immunoglobulin-like module (domain EDIII) [15]. Most flavivirus E proteins are post-translationally modified by addition of a single N-linked oligosaccharide on residue N-154 located within the EDI-loop [16]. Flavivirus E proteins represent one of the key determinants for viral pathogenesis. Flavivirus envelope supports virus tropism and single amino-acid changes can redirect virus tropism [17]. Flavivirus E proteins also represent a major target for neutralizing antibodies but, at the same time, can be involved in enhancement/cross-reactivity of reactive antibodies [18–21].

Recently, our studies on chimeric ZIKV clones between an epidemic Brazilian strain of ZIKV BeH819015 (hereafter called BR15) and a historical African strain MR766 highlighted an important role of two structural proteins prM/M, and E in ZIKV ability to infect human cells [16,22–24]. We further showed that they contribute to the initiation of viral infection. Analysis of chimeric viruses indicated that C-prM region plays a role in triggering cell death by ZIKV and E protein is associated with viral attachment to host cells during early infection [23,24]. Flavivirus E proteins usually contain two *N*-glycosylation sites at position E-56 and E-154. The first site is lacking in ZIKV E protein and contribution of the second site in ZIKV viral cycle, including in the mosquito vector, has been recently highlighted [25]. N-linked glycosylation of the E protein was shown to be an important determinant of ZIKV virulence in a mouse model of viral encephalitis [22,26]. In invertebrate vectors, ZIKV bearing an unglycosylated E protein was attenuated in its capacity to replicate in *Aedes aegypti* [27]. Although a *N*-glycosylation site is highly conserved among flaviviruses, which suggests of its biological importance, E proteins could remain unglycosylated as it has been observed in some ZIKV strains. To date, the exact mechanism by which the *N*-glycosylation motif region of the E protein contributes to ZIKV infectivity still remains elusive. In the present study, we further characterised structural protein contribution in ZIKV infectivity by focusing on the ZIKV E protein. Our goal was to determine whether three E residues—Ile152, Thr156, and His158 (hereafter called as E-152/156/158 residues)—which surround the Asn154 composing the *N*-glycosylation site NDT, may have any effect on ZIKV's ability to infect human cells.

2. Materials and Methods

2.1. Cells and Reagents

Vero cells (CCL-81, ATCC, Manassas, VA, USA), A549-Dual™ cells (a549d-nfis, InvivoGen, San Diego, CA, USA) and human embryonic kidney HEK-293 cells (CRL-1573, ATCC, Manassas, VA, USA) were cultured at 37 °C under a 5% CO_2 atmosphere in MEM medium, supplemented with 5% to

10% heat-inactivated foetal bovine serum (FBS). A549-Dual™ (A549^{DUAL}) cells were maintained in growth medium supplemented with nonessential amino acids, 10 µg·mL^{-1} blasticidin and 100 µg·mL^{-1} zeocin (InvivoGen, San Diego, CA, USA). Chloroquine phosphate was purchased from Sigma-Aldrich (Saint-Louis, MO, USA). Rat antibody specifically raised against ZIKV E protein Domain III was developed *in-house* and used in immunoblot with reducing conditions [28]. Mouse anti-pan flavivirus envelope E protein monoclonal antibody (mAb) 4G2 was purchased from RD Biotech (Besancon, France) and used in immunoblot with nonreducing conditions. Horseradish peroxidase-conjugated anti-rabbit and anti-mouse antibodies were purchased from Vector Laboratories (Burlingame, CA, USA).

2.2. Design of ZIKV Molecular Clones

ZIKV molecular clones (MR766, GenBank accession number LC002520, and BR15, GenBank accession number KU365778) were designed and produced according to the Infectious Subgenomic Amplicon method as previously described [23,29,30]. To introduce BR15 E-152/156/158 residues into MR766 (MR766$^{E-152I/156T/158H}$), we used mutagenesis primers (forward primer: 5′-ggctcccagcacagtgggatgatcgttaatgacacaggacatgaaactg-3′ and reverse primer: 5′-cagtttcatgtcctgtgtcattaacgatcatcccactgtgctgggagcc-3′) to generate two overlapping fragments Z1$^{MR766-E-MUT1}$ and Z1$^{MR766-E-MUT2}$ from the Z1^{MR766} fragment encoding the MR766 structural proteins in which encoding region of the E protein received the IVNDTGH motif (amino acids 152 to 158) from BR15. To generate BR15$^{E-152T/156I/158Y}$, a new Z1$^{BR15-E-I152T/T156I/H158Y}$ fragment was synthesised in which the sequence was modified so that encoding region of the E protein received the TVNDIGY motif (amino acids 152 to 158) from MR766. Synthetic genes were cloned into plasmid pUC57 by GeneCust (Boynes, France). Fragments were amplified by PCR from their respective plasmids using a set of primer pairs that was designed so that fragments overlapped with each other of about 30 to 50 nucleotides.

2.3. Recovering of Molecular Clones BR15$^{E-152T/156I/158Y}$ and MR766$^{E-152I/156T/158H}$

Molecular clones were produced as previously described [23,29]. Briefly, purified PCR fragments were electroporated into Vero cells. After 5 days, cell supernatants were recovered usually in absence of cytopathic effect and used to infect fresh Vero cells in a first round of amplification (P1). Viral clones were recovered at the onset of cytopathic effect and amplified for another round on Vero cells to produce a second round of amplification (P2), which was used for described studies. To produce MR766$^{E-152I/156T/158H}$ and BR15$^{E-152T/156I/158Y}$ mutant viral clones, Vero cells were respectively electroporated with PCR fragments Z1$^{MR766-E-MUT1}$, Z1$^{MR766-E-MUT2}$, Z2^{MR766}, Z3^{MR766}, and Z4^{MR766} and with Z1$^{BR15-E-MUT}$, Z2^{BR15}, Z3^{BR15}, and Z4^{BR15}. Recovered mutant viruses MR766$^{E-152I/156T/158H}$ and BR15$^{E-152T/156I/158Y}$ respectively consist of viral sequence of MR766 in which E-152/156/158 residues of BR15 ZIKV strain were introduced and viral sequence of BR15 in which E-152/156/158 residues were replaced with its counterpart from MR766 ZIKV strain.

2.4. Plaque-Forming Assay

Viral titres were determined by a standard plaque-forming assay as previously described with minor modifications [31]. Briefly, Vero cells grown in a 48-well culture plate were infected with serial tenfold dilutions of virus samples for 2 h at 37 °C, and then incubated with 0.8% carboxymethylcellulose (CMC) for 4 days. Cells were fixed with 3.7% formaldehyde (FA) in PBS and stained with 0.5% crystal violet in 20% ethanol. Viral titres were expressed as plaque-forming units (PFU) per mL (PFU·mL^{-1}).

2.5. Quantification of Viral Stocks

Zika virus samples were analysed by titration on Vero cells while genomic viral RNA was quantified by RT-qPCR, as previously described [23]. Briefly, viral RNA was extracted from virus particles using QIAmp Kit (Qiagen, Hilden, Germany). PCR standard curve used for quantification of ZIKV copy numbers was obtained with a pUC57/ZIKV-E amplicon plasmid containing a synthetic cDNA encompassing nucleotides 961 to 1301 of genomic RNA (MR766). The pair of ZIKV E primers was used to equally amplify pUC57/ZIKV-E amplicon and cDNA encompassing nucleotides 1046 to 1213 from genomic RNA of ZIKV molecular clones used in this study.

2.6. Immunoblot Assay

Cell lysates were performed in RIPA lysis buffer or buffer A (cell fractionation). All subsequent steps of immunoblotting were performed as previously described [32,33]. Primary antibodies were used at 1:500 dilutions. Anti-mouse immunoglobulin-horseradish peroxidase and anti-rat immunoglobulin-horseradish peroxidase conjugates were used as secondary antibodies (dilution 1:2000). Blots were revealed with ECL detection reagents (Amersham, Little Chalfont, United Kingdom).

2.7. Flow Cytometry Assay

A549-Dual™ cells were grown on six-well plates at 5×10^5 cells per well and infected at a multiplicity of infection (MOI) of 1. Infected cells were harvested and fixed with 3.7% formaldehyde in PBS for 20 min, permeabilized with 0.1% Triton X-100 in PBS for 4 min and then blocked with PBS-BSA for 10 min. Cells were stained with anti-E mAb 4G2 (1:1000) for 1 h. Antigen staining was visualized with goat anti-mouse Alexa Fluor 488 IgG (1:1000) for 20 min. Cells were subjected to a flow cytometric analysis using a CytoFLEX flow cytometer (Beckman Coulter, Brea, CA, USA). The percentage of positive cells was determined using FlowJo software (version 10, Tree Star, Inc., Ashland, OR, USA).

2.8. RT-qPCR

Total RNA including genomic viral RNA was extracted from cells (Qiagen, Hilden, Germany) and reverse transcription was performed using 500 ng of total RNA, random hexamer primers (intracellular viral RNA) or E reverse primer (virus particles) and moloney mouse leukemia virus reverse transcriptase (Life Technologies, Carlsbad, CA, USA) at 42 °C for 50 min. Quantitative PCR was performed on a CFX96 qPCR System (Bio-Rad, Hercules, CA, USA). Briefly, 10 ng of cDNA was amplified using 0.2 µM of each primer and 1X GoTaq Master Mix (Promega, Madison, WI, USA). When appropriate, data were normalised to the internal standard GAPDH. For each single-well amplification reaction, a threshold cycle (Ct) was calculated using the CFX96 qPCR program (Bio Rad, Hercules, CA, USA) in the exponential phase of amplification. Relative changes in gene expression were determined using the 2ΔΔCt method and reported relative to the control. Primers used in this study are listed in [31]. ZIKV E primers were designed to match both MR766-NIID and BeH819015 sequences (forward 5-gtcttggaacatggagg-3′ and reverse 5′-ttcaccttgtgttgggc-3′).

2.9. Virus Binding Assay

Cells were cultured at subconfluent density in 24-well plates. Cell monolayers were washed in cold PBS and cooled at 4 °C for at least 20 min in presence of cold MEM supplemented with 2% FBS. Pre-chilled cells were incubated at 4 °C with ZIKV at MOI of 1 in 1.5 mL of cold MEM supplemented with 2% FBS. After 1 h of incubation, virus inputs were removed and cells were washed with cold MEM supplemented with 2% FBS. Total cellular RNA was extracted using the RNeasy kit (Qiagen, Hilden, Germany) and RT-qPCR analysis on viral RNA was performed using primers for ZIKV E gene as described above.

2.10. Fusion Assay

Cells were cultured at subconfluent density in 12-well plates. Cell monolayers were cooled at 4 °C for at least 20 min in presence of cold MEM supplemented with 10% FBS. Pre-chilled cells were incubated at 4 °C with ZIKV at MOI of 1 in 1 mL of cold MEM supplemented with 10% FBS. After 1-h incubation, cells were shifted to 37 °C for another hour. Chloroquine was added to culture medium at 100 µM for a 2-h period. Next, culture medium was then replaced to avoid drug cytotoxic effects. Cells were harvested 30 h post temperature shifting for RNA extraction. Total RNA was subjected to RT-qPCR analysis as described.

2.11. Cytotoxicity Assay

Cell damages were evaluated by measuring lactate dehydrogenase (LDH) release. Supernatants of infected cells were recovered and subjected to CytoTox 96® nonradioactive cytotoxicity assay (Promega, Madison, WI, USA) according to manufacturer instructions. Absorbance of converted dye was measured at 490 nm using a microplate reader (Tecan, Mannedorf, Switzerland). Results of LDH activity in cell supernatants are presented with subtraction of values from mock-infected cells.

2.12. Measurement of the IFN-β Pathway Activation

Activation of the Interferon regulatory factors (IRF) pathway was monitored by measuring *Lucia* luciferase activity in A549-Dual™ cells. It was evaluated using QUANTI-Luc substrate (InvivoGen, San Diego, CA, USA) according to manufacturer's recommendations. IRF-induced luciferase levels were quantified using a FLUOstar Omega Microplate Reader (BMG LABTECH, Offenburg, Germany). Results are presented with subtraction of values from mock-infected cells.

2.13. TMD2-M/E Expression

To express recombinant E proteins from ZIKV in mammalian cells, TMD2-prM (TransMembrane Domain II) and E genes from MR766 and BR15, as well as a mutant BR15 bearing residues E-152 to E-158 from MR766, were synthesised by GeneCust (Boynes, France). Recombinant proteins comprised aa 275 to aa 775 of polyproteins, which correspond to the very end of prM protein (TMD2, used as signal peptide for E proteins) and the entire E protein. As Flavivirus prM protein plays a role of chaperone to ensure the proper folding of the E protein, we expected that viral chaperone activity eviction would have exacerbate differences in E protein folding [34]. Modifications to optimize viral E protein expression in human cells were done on original protein sequences. Then, mammalian codon-optimised sequences coding for TMD2-prM and E proteins were cloned into the *NheI* and *NotI* restriction sites of the pcDNA3.1(-) plasmid to generate pMR766, pBR15 and pBR15^{E-MUT}, respectively. Each plasmid was transfected in human HEK-293T cells using lipofectamine 3000 according to manufacturer's instructions.

2.14. Cell Fractionation

Cells were washed with PBS and lysed at a concentration of 1×10^4 cells per µl in protein separation buffer A (0.2% Triton X-100, 50 mM Tris-HCl pH 7.5, 150 mM NaCl, 2.5 mM EDTA) [32]. As misfolded proteins aggregate and become resistant to Triton X-100 solubilisation [32,35], Triton X-100-insoluble fraction was separated by centrifugation at 3400 g for 10 min. Pellets were enriched in misfolded proteins. Samples were analysed by immunoblot. Loading was normalised by the number of lysed cells. Band intensities were determined with ImageJ software (version 1.50i, NIH, Washington, WA, USA, 2016) and soluble/insoluble ratios calculated.

2.15. Statistical Analysis

All values are expressed as mean ± SD of at least two independent experiments. Comparisons between different treatments were analysed by a one-way or two-way ANOVA tests as deemed appropriate. Values of $p < 0.05$ were considered statistically significant for a post-hoc Tukey's test. All statistical tests were done using the software GraphPad Prism (version 8.01, GraphPad Software, La Jolla, CA, USA).

3. Results

3.1. Characterization of Mutant ZIKV Molecular Clones

To determine the contribution of E-152/156/158 residues in ZIKV E protein functions, we generated two mutant molecular clones: MR766$^{E-152I/156T/158H}$, hereafter called MR766^{E-MUT}, in which

E-152/156/158 residues of BR15 epidemic strain were introduced, and BR15$^{E-152T/156I/158Y}$, hereafter called BR15^{E-MUT}, in which E-152/156/158 residues were replaced with their counterparts from the MR766 historical African strain (Figure 1a). Genomes were assembled using the infectious subgenomic amplicon method [29]. Briefly, Vero cells were electroporated with overlapping fragments, in which appropriate mutations have been previously introduced. The two recovered clones were viable and twice amplified on Vero cells. Titres of P2 working viral stocks were determined in Vero cells and were ranging from 5×10^5 to 1×10^8 PFU·mL^{-1} (Figure 1b). MR766^{E-MUT} and BR15^{E-MUT} gave plaque morphologies that resembled those of respective MR766 and BR15 parental clones (Figure 1c), which is in agreement with previously published data [27]. In addition, we confirmed that the introduced mutations affected the electrophoretic mobility of ZIKV E proteins, suggesting that E-152/156/158 residues from BR15 E protein might enable its glycosylation (Figure 1d + Figure S1).

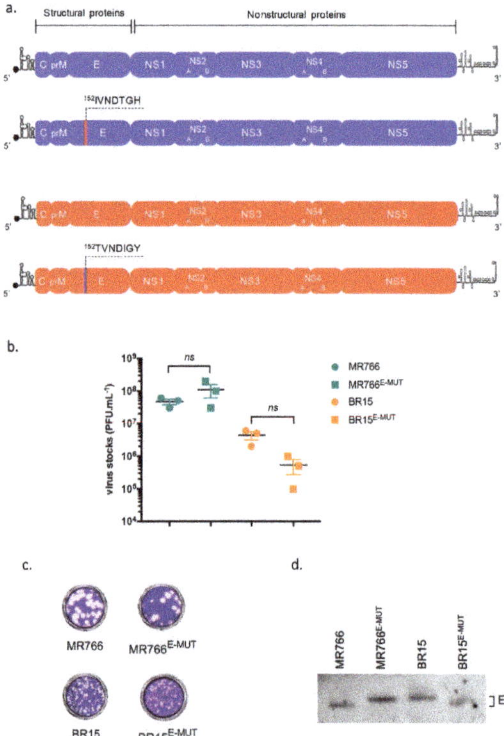

Figure 1. ZIKV mutant molecular clones. In (**a**), schematic representation of mutant viral clones BR15^{E-MUT} and MR766^{E-MUT} and their respective parental clones. In (**b**), histograms showing viral titres. Values represent means and standard errors of three independent experiments. In (**c**), examples of infectious plaques developed for BR15^{E-MUT} and MR766^{E-MUT}, and parental clones, after plaque-forming assay on Vero cells. In (**d**), Vero cells were infected with parental and mutant molecular clones (MR766 and BR15) at a MOI of 1. 24 h post-infection (hpi), cells were then lysed and subjected to an immunoblot, in non-reducing conditions. Anti-ZIKV EDIII immunoblot shows differences of electrophoretic mobility associated with residue mutations.

3.2. Residues E-152/156/158 from BR15 Potentiate Viral Infectivity

We first analysed infectivity of P2 virus stocks as described above. Particle-to-PFU ratios obtained from parental clones were around 900–1000 (Table 1), which is consistent with our previous observations [23]. We then analysed particle-to-PFU ratios of the two E mutant clones. Addition of residues E-152/156/158 from BR15 to MR766 resulted in a 2.3-fold decrease in the particle-to-PFU ratio. In contrast, when residues E-152/156/158 from MR766 were introduced to BR15, the particle-to-PFU ratio was markedly increased with more than ten-folds. These results suggest that residues E-152/156/158 from BR15 potentiate virion infectivity.

Table 1. Table showing particle-to-PFU ratios. Viral RNA extracted from viral stock P2 were subjected to quantification by RT-qPCR using E primers. Obtained Ct values were plotted in a standard curve (serial dilutions of plasmid copies) in order to get the number of viral RNA molecules per mL. These results were compared to viral stock quantifications by standard plaque-forming assay, which then gave the particle-to-PFU ratios, also named vRNA-to-PFU ratios. Values represent means and standard errors of two to four independent experiments.

VIRUS STOCK.	PARTICLE-TO-PFU RATIO	*p* VALUES (E-MUT vs. WT)
MR766	997 ± 36	
MR766$^{\text{E-MUT}}$	419 ± 87	<0.05
BR15	918 ± 91	
BR15$^{\text{E-MUT}}$	$11\,237 \pm 720$	<0.001

3.3. Alteration of Residues E-152 to E-158 of ZIKV E Protein Does Not Affect Virus Binding to Host Cells but May Affect Virus Progeny Production

We previously showed that historical and epidemic ZIKV strains display differences in their abilities to bind host cells, leading to differences in cell susceptibility to infection (18, 19). Here, we further investigated the ability of the described mutant clones to bind onto A549-Dual™ cells. Virus binding assays were performed and analysed by RT-qPCR to determine virus particle binding onto cell surface after an incubation period of 1 h. Panels A and B show no difference between mutant clones and their respective parental clones (Figure 2). These results contrast with other studies in mosquito cells [27], suggesting that viral receptors may vary between vertebrate and invertebrate cells. These data suggest that alteration of residues E-152 to E-158 of ZIKV E protein does not affect virus bindings to A549-Dual™ cells.

Instead, these results suggest that E-152/156/158 residues might influence ZIKV progeny production. Indeed, the progeny production of MR766$^{\text{E-MUT}}$ was modestly but reproducibly increased in comparison to that of MR766 (3×10^7 PFU·mL^{-1} vs. 1×10^7 PFU·mL^{-1}) at 72 hpi [23]. Conversely, kinetics of the BR15$^{\text{E-MUT}}$ progeny production were strongly altered compared to BR15 (2×10^6 PFU·mL^{-1} vs. 4×10^7 PFU·mL^{-1}) at 72 hpi, respectively [23]. Differences in progeny production were observed at as early as 24 hpi. Similar differences were also seen on the percentages of the infected cell at 48 hpi. Taken together, these results indicate that E-152/156/158 residues from BR15 potentiate viral infectivity, independently of the virus binding to host A549-Dual™ cells.

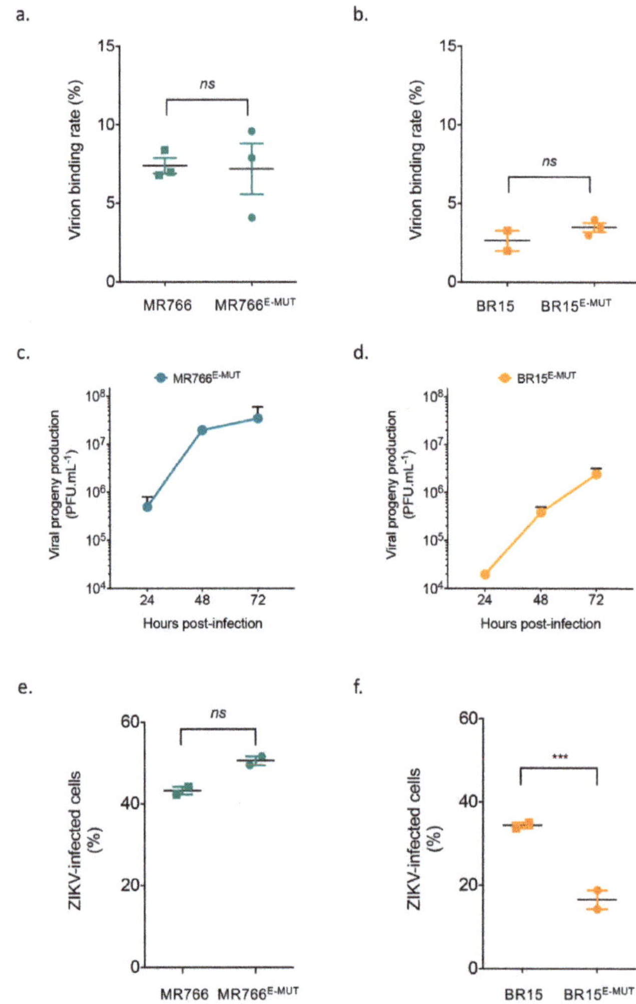

Figure 2. Analysis of virus binding and viral growth in A549-Dual™. In (**a**) and (**b**), for virus binding assays, cells were incubated with viral clones at the MOI of 1 for 1 h at 4 °C. The number of virus particles bound to cell surface was measured by RT-qPCR. Values represent means and standard errors of three independent experiments. In (**c**) and (**d**), A549-Dual™ were infected with BR15[E-MUT] and MR766[E-MUT] at MOI of 1. Infectious virus released into the supernatants of infected A549-Dual™ cells were quantified at 24, 48 and 72 hpi. Error bars represent standard errors of at least two independent experiments. In (**e**) and (**f**), A549-Dual™ were infected with BR15[E-MUT] and MR766[E-MUT] and parental clones at MOI of 1. Percentages of ZIKV-infected cells were determined at 48 h by flow cytometry using anti-E mAb 4G2 as primary antibody. Error bars represent standard errors of two independent experiments in duplicates. *ns*: not significant, ***: *p* value < 0.001

3.4. Mutations at E-152/156/158 Residues Have No Effect on ZIKV-Induced Cell Death or Interferon Pathways

To determine whether differences described with the mutant viruses were associated with specific host-cell responses, we first analysed virus-induced cell death at 48 h and 72 h post-infection. No difference in cytotoxicity measured by LDH release was observed between wild-type and mutant viruses (MR766 and BR15) (Figure 3, panels a and b). We then took advantage of the properties of A549-Dual™ cells to test whether mutant viruses can trigger different host cell innate immunity. A549-Dual™ cells were derived from A549 cells by stable integration of two reporter genes: *SEAP* gene (Secreted Embryonic Alkaline Phosphatase) and *Lucia* luciferase gene under the respective transcriptional control of an IFN-β minimal promoter, which is fused to NF-κB binding sites or an ISG54 minimal promoter in conjunction with interferon-sensitive response elements. We examined possible activation of the IRF pathway by monitoring production of *Lucia* luciferase at 48 dpi and 72 hpi. Similar responses were observed in both wild-type and mutant clones (Figure 3, panels c and d). The NF-κB pathway was not investigated, as we showed previously that this pathway is not activated upon ZIKV infection [23]. These results indicate that differences in the mutant virus properties could not be explained by specific host-cell responses, which are consistent with our previous observations suggesting a link between host-cell responses and ZIKV nonstructural proteins [23].

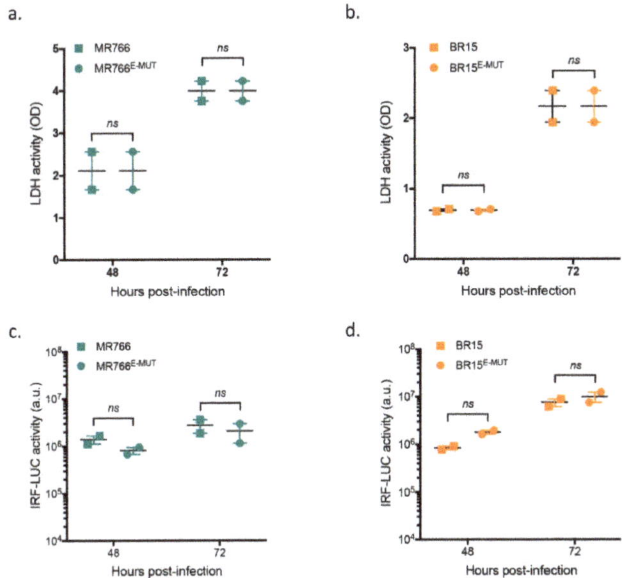

Figure 3. Analysis of infection-induced cell death and immune responses. A549-Dual™ were infected with BR15^{E-MUT} and MR766^{E-MUT} and parental clones at MOI of 1. In (**a**) and (**b**), LDH activity was measured at 48 and 72 hpi respectively. Values represents mean and standard errors of two independent experiments in triplicates. In (**c**) and (**d**), analysis of IRF pathway activation in response to viral infection. Activity of secreted *Lucia* luciferase was measured using QUANTI-Luc substrate at 48 and 72 hpi. Results are expressed as raw data of luminescence arbitrary units. Error bars represent standard errors of two independent experiments in triplicates. *ns*: not significant.

3.5. E-152/156/158 Residues from BR15 Facilitate Viral Fusion

We showed earlier that E-152/156/158 residues from BR15 have a growth advantage without apparent association with cellular attachment (Figure 2) or specific host-cell responses (Figure 3). We then studied viral fusion to test whether it could explain the observed growth advantage. Viral fusion of flaviviruses is commonly triggered from endosomes upon low-pH by a series of molecular changes within the E protein, resulting in the release of the nucleocapsid into cell cytoplasm. Chloroquine, a 4-aminoquinoline, is a weak base that inhibits endosome acidification and consequently restricts viral replication of many viruses through inhibition of pH-dependent steps. Recently, chloroquine was shown to inhibit Zika virus infection in different cellular models [36,37]. As BR15 and BR15^{E-MUT}

showed significant differences in the percentage of infected cells (Figure 2f), we decided to focus on these two molecular clones for the following virus fusion experiments. We treated A549-Dual™ cells infected with BR15 or BR15$^{E\text{-}MUT}$ with 100 µM of chloroquine 1 hpi for 2 h and then cells were moved back to regular medium. Intracellular viral RNA was quantified 30 hpi by RT-qPCR. BR15$^{E\text{-}MUT}$ fusion was significantly restricted by chloroquine treatment compared to that of BR15 (Figure 4). These data suggest that E-152/156/158 residues from BR15 favour viral fusion with host-cell membranes.

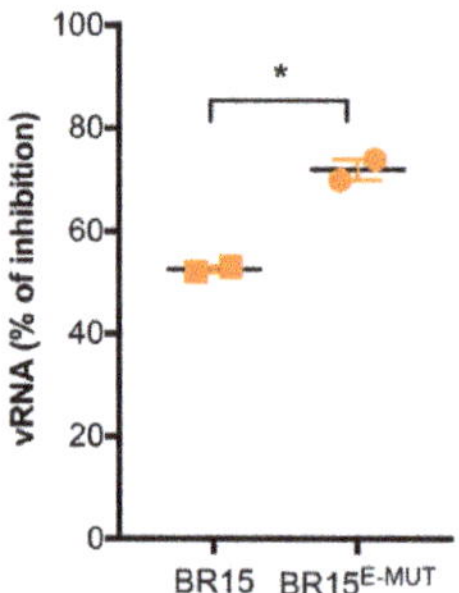

Figure 4. Viral fusion in A549-Dual™ cells. Pre-chilled cells were incubated at 4 °C with ZIKV at MOI of 1. After 1-h incubation, cells were shifted to 37 °C. Chloroquine was then added to the culture medium. Viral RNA was measured by RT-qPCR 30 h at 37 °C. Error bars represent standard errors of two independent experiments. *: *p* value < 0.1

3.6. E-152/156/158 Residues from BR15 Favour Conformational Changes within the Fusion Loop

As virus fusion with host cells is highly dependent on conformational changes of the E protein triggered at low-pH, we hypothesize that the reduced fusion we observed with the mutant molecular clone BR15$^{E\text{-}MUT}$ bearing E-152/156/158 residues from MR766 could be the consequence of conformational differences between E proteins of the two molecular clones. In order to test this hypothesis, BR15, BR15$^{E\text{-}MUT}$ and MR766 sequences coding for TMD2-prM/E were codon-optimised for expression in mammalian cells and cloned into a pcDNA3.1 vector. HEK-293T cells were transfected with different plasmids and positive cells were selected with antibiotics. The resulting stable cell lines were fractionated to evaluate the capacity of recombinant E proteins to fold properly, insolubility been a hallmark of misfolded proteins [32,35]. Resulting fractions were subjected to an immunoblot analysis. We first used a rat antibody developed in-house, specifically raised against E protein domain EDIII [28]. Figure 5a revealed that BR15$^{E\text{-}MUT}$ and MR766 TMD2-prM/E overexpression resulted in a greater E protein propensity to accumulate in insoluble fractions than that of BR15 overexpression, as shown by inversion of soluble/insoluble ratios. Interestingly, differences observed between BR15 and mutant BR15 TMD2-prM/E suggested that ZIKV E proteins bear different conformations that only depend on E-152/156/158 residues. To verify these observations, the same samples were immunoblotted using a 4G2 monoclonal antibody, which recognises a highly conserved fusion loop sequence of most flaviviruses. As shown in Figure 5b, 4G2 monoclonal antibody strongly reacts against E protein from BR15, whereas we could barely detect any signal with the two E proteins bearing E-152/156/158 residues from MR766. Preliminary in silico modelling of ZIKV E proteins suggest that changes in the glycosylation motif could affect structure of the glycosylation and fusion loops as well as interactions with surrounding residues (Figure S2) and surface hydrophobicity (not shown). These data confirm that E-152/156/158 residues in the EDI domain support conformational changes on the ZIKV E protein, which could be detected in the fusion loop of EDII domain. Finally, these results suggest that conformational changes occur in BR15 E protein upon mutation of E-152/156/158 residues.

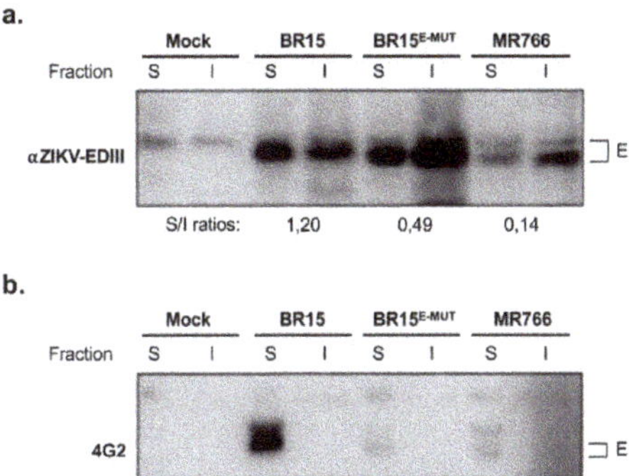

Figure 5. Conformational changes induced by residues E-152/156/158 of ZIKV E protein. In (**a**) and (**b**), HEK-293T cells were transfected with TMD2-prM/E constructs and antibiotics that were selected to raise stable cell lines. Cells were harvested and proteins extracts subjected to a fractionation. Protein fractions were immunoblotted with anti-ZIKV EDIII (**a**) or anti-E 4G2 (**b**) antibodies. S, soluble proteins; I, insoluble proteins. Band intensities were determined with ImageJ software and S/I ratios were calculated. Apparent discrepancies with cytometry experiments (Figure 2) regarding antibody reactivity are explained by experimental and recombinant protein overexpression versus viral infection conditions.

4. Discussion

The role of structural proteins in determining infectivity of human cells to ZIKV infection has been previously reported [23]. Zika viruses of historical or epidemic strains display differences in their abilities to bind host cells leading to differences in cell susceptibility to infection. To further characterise biological properties of contemporary ZIKV strains, which have been associated with recent epidemics and severe forms of human disease, we investigated the role of E-152/156/158 residues of the envelope protein, residues located around a unique *N*-glycosylation site (E-154) in viral entry into host cells. Our mutagenesis data showed that E-152/156/158 residues are responsible for the differences we observed in the infectivity of the virus and progeny production kinetics, without affecting viral attachment and host-cell responses. Further characterisations identified the E protein conformational changes in the fusion loop, supported by E-152/156/158 residues, as a major event in virus fusion and release of viral RNA into cell cytoplasm.

The first step in viral entry pathway involves nonspecific viral binding to cellular attachment factors. Negatively charged glycosaminoglycans, which are abundantly expressed on numerous cell types, are considered as low-affinity attachment factors by flaviviruses. These interactions serve to concentrate viruses on the cell surface and are mediated by the EDIII domain of E proteins [38]. Our data demonstrate that E-152/156/158 residues of ZIKV E protein do not influence virus binding, which suggests that domain EDIII conformation is not strongly affected by E-152/156/158 residues. We conclude that this initial step of ZIKV entry into cells does not depend on E-152/156/158 residues of the E protein.

Viral particle internalisation could occur through distinct routes, including clathrin-mediated endocytosis or non-classical clathrin-independent endocytic pathways. These distinct entry modes depend not only on host cells but also on viral serotype or strain [39]. Despite these differences in the internalisation process, genome release into the cytoplasm always occurs through E protein-mediated membrane fusion [40]. The low-pH environment within endosomes triggers a series of molecular changes within flavivirus E protein resulting in fusion of viral membrane with endosomal membrane and subsequent release of the nucleocapsid into cell cytoplasm [13,14]. The initial step in membrane fusion involves protonation-dependent disruption of E protein rafts at viral surface, resulting in conformational changes and formation of a fusion pore from which the nucleocapsid is released into the cytosol. Evidences suggested that the glycan loop modulates the overall Flavivirus E protein conformation and, specifically, fusion loop exposure [41,42]. Sevvana and colleagues proposed that, given the close proximity of ZIKV glycan and fusion loops, any interaction of the glycan loop with a receptor on a host-cell surface might promote exposure of the fusion loop and facilitate formation of fusogenic trimers leading to membrane fusion [43]. Another study based on antibody neutralizing activities demonstrated that residues surrounding ZIKV E protein glycan regulate virus antigenicity [44]. In a previous study, we also evaluated immunogenicity of a chimeric viral clone ZIKBeHMR-2, in which the region encoding envelope proteins of MR766 African strain backbone was replaced with its counterpart from BeH819015 epidemic strain [45,46]. Amino-acid substitutions I152T, T156I, and H158Y were introduced in the glycan loop of the E protein, making chimeric ZIKBeHMR-2 a nonglycosylated virus. Those results suggest that, rather than just determining the glycosylation, amino-acid residues at position 152, 156 and 158 play a pivotal role on accessibility of neutralizing antibody epitopes on mature virus particles. In this study, we demonstrated that changes in the glycan loop can modulate accessibility to the fusion loop. Although these observations were made independently of the presence of a glycan, it is conceivable that E protein N-glycosylation could provide another level of regulation on the access of the fusion loop. The role of N-glycosylation on ZIKV E protein has also been investigated using pseudoviral particles, showing that reduced infectivity was observed with mutant viral particles lacking the N-glycan [47]. Altogether, these studies suggest that conformational changes induced at the glycan loop most probably modulate fusion loop exposure and subsequent fusion of viral and cellular membranes, which strongly supports our observations.

One interesting finding from this study is that chloroquine treatment results in less BR15$^{E\text{-MUT}}$ entries than it does for BR15. This finding suggests that E-152/156/158 residues of BR15 increase pH sensitivity of E protein. To generate BR15$^{E\text{-MUT}}$, its sequence was modified so that the coding region of the E protein IVNDTGH (amino acids 152 to 158) in BR15 was replaced with TVNDIGY motif from MR766, meaning that not only N-glycosylation motif was abrogated but also that histidine E-158 was changed into a tyrosine. Histidine residues have been described as pH sensors in flavivirus membrane fusion [48]. Fusion initiation is crucially dependent on protonation of conserved histidine residues at the interface between domains EDI and EDIII of E protein, leading to dissolution of domain interactions and to fusion peptide exposure. Given the fusion differences we observed between wild-type and mutant BR15 molecular clones, further investigations on histidine E-158 protonation are required to determine its exact contribution to membrane fusion.

Finally, our analysis of virus inocula generated in Vero cells showed differences in particle-to-PFU ratios indicating that E-152/156/158 residues of BR15 E protein facilitate release of more infectious particles. In addition, we demonstrated with recombinant proteins that E-152/156/158 residues of epidemic ZIKV E protein also facilitate production of more soluble proteins. These results are supported by works of Mossenta and colleagues [47]. Whether these observations are due to conformational changes occurring during virion production remains undetermined. However, in a study on flavivirus cross-reactive epitopes, Crill and Chang suggested that close packing of fusion peptide against its subunit partner and glycan on the upper surface protects the fusion loop from irreversible pH-induced conformational changes during maturation and secretion [49]. All these observations suggest that E-152/156/158 residues of epidemic ZIKV E protein could also be an advantage during virion maturation process.

Altogether, our data indicate that the envelope residues E152/156/158 of Zika virus influence early stages of virus infection in human cells. This study highlights the importance of E-152/156/158 residues in ZIKV biology and specifically in their roles in supporting viral fusion. These new findings could potentially help to design innovative strategies for future ZIKV infection control.

Supplementary Materials: The following are available online at http://www.mdpi.com/2073-4409/8/11/1444/s1, Figure S1: Tunicamycin treatment of Vero cells infected with ZIKV mutant molecular clones, Figure S2: structures of ZIKV E wild-type and mutant proteins.

Author Contributions: Conceptualization, P.D. and G.G.; Funding acquisition, P.D.; Investigation, S.B., W.V., E.F. and G.G.; Visualization, S.B., W.V., E.F. and G.L.; Writing—original draft, G.G.; Writing—review & editing, S.B., W.V., E.F., P.D., R.Y.Z. and G.G.

Funding: This work was supported by ZIKAlert project (POE FEDER 2014–2020 Ile de la Réunion action 1.05 under grant agreement n SYNERGY:RE0001902). G.L. and R.Y.Z. were supported in part by funding from the National Institute of Health (NIH R21 AI129369) and an intramural fund from the University of Maryland Medical Center (to R.Y.Z.). E.F. received funding from FEDER Région (ZIKALIVax 2.0 program, N SYNERGIE RE0012406) and La Reunion University. S.B. has a PhD degree scholarship from La Réunion Island University (École Doctorale STS), funded by the French ministry, MESRI.

Acknowledgments: We gratefully acknowledge Pascale Krejbich-Trotot and Chaker El Kalamouni for helpful discussions. We also thank Hervé Pascalis for his kind help on structure comparison.

Conflicts of Interest: The authors declare no conflict of interest.

References

1. Lee, I.; Bos, S.; Li, G.; Wang, S.; Gadea, G.; Desprès, P.; Zhao, R.Y. Probing Molecular Insights into Zika Virus–Host Interactions. *Viruses* **2018**, *10*, 233. [CrossRef] [PubMed]
2. Bos, S.; Gadea, G.; Despres, P. Dengue: A growing threat requiring vaccine development for disease prevention. *Pathog. Glob. Heal.* **2018**, *112*, 294–305. [CrossRef] [PubMed]
3. Douam, F.; Ploss, A. Yellow Fever Virus: Knowledge Gaps Impeding the Fight Against an Old Foe. *Trends Microbiol.* **2018**, *26*, 913–928. [CrossRef] [PubMed]
4. Lindquist, L. Recent and historical trends in the epidemiology of Japanese encephalitis and its implication for risk assessment in travellers. *J. Travel Med.* **2018**, *25*, S3–S9. [CrossRef]
5. MacNamara, F.N. Zika virus: A report on three cases of human infection during an epidemic of jaundice in Nigeria. *Trans. R. Soc. Trop. Med. Hyg.* **1954**, *48*, 139–145. [CrossRef]
6. Fagbami, A.H.; Monath, T.P.; Fabiyi, A. Dengue virus infections in Nigeria: A survey for antibodies in monkeys and humans. *Trans. R. Soc. Trop. Med. Hyg.* **1977**, *71*, 60–65. [CrossRef]
7. Jan, C.; Languillat, G.; Renaudet, J.; Robin, Y. A serological survey of arboviruses in Gabon. *Bull. Soc. Pathol. Exot. Filiales* **1978**, *71*, 140–146.
8. Renaudet, J.; Jan, C.; Ridet, J.; Adam, C.; Robin, Y. A serological survey of arboviruses in the human population of Senegal. *Bull. Soc. Pathol. Exot. Filiales* **1978**, *71*, 131–140.
9. Cao-Lormeau, V.-M.; Blake, A.; Mons, S.; Lastère, S.; Roche, C.; Vanhomwegen, J.; Dub, T.; Baudouin, L.; Teissier, A.; Larre, P.; et al. Guillain-Barré Syndrome outbreak associated with Zika virus infection in French Polynesia: A case-control study. *Lancet* **2016**, *387*, 1531–1539. [CrossRef]
10. Cugola, F.R.; Fernandes, I.R.; Russo, F.B.; Freitas, B.C.; Dias, J.L.M.; Guimarães, K.P.; Benazzato, C.; Almeida, N.; Pignatari, G.C.; Romero, S.; et al. The Brazilian Zika virus strain causes birth defects in experimental models. *Nature* **2016**, *534*, 267–271. [CrossRef]
11. Merfeld, E.; Ben-Avi, L.; Kennon, M.; Cerveny, K.L. Potential mechanisms of Zika-linked microcephaly. *Wiley Interdiscip. Rev. Dev. Biol.* **2017**, e273. [CrossRef] [PubMed]
12. Parra, B.; Lizarazo, J.; Jiménez-Arango, J.A.; Zea-Vera, A.F.; González-Manrique, G.; Vargas, J.; Angarita, J.A.; Zuñiga, G.; Lopez-Gonzalez, R.; Beltran, C.L.; et al. Guillain-Barré Syndrome Associated with Zika Virus Infection in Colombia. *N. Engl. J. Med.* **2016**, *375*, 1513–1523. [CrossRef] [PubMed]
13. Smit, J.; Moesker, B.; Rodenhuis-Zybert, I.; Wilschut, J. Flavivirus Cell Entry and Membrane Fusion. *Viruses* **2011**, *3*, 160–171. [CrossRef] [PubMed]
14. Stiasny, K.; Fritz, R.; Pangerl, K.; Heinz, F.X. Molecular mechanisms of flavivirus membrane fusion. *Amino Acids* **2011**, *41*, 1159–1163. [CrossRef]
15. Rey, F.A.; Heinz, F.X.; Mandl, C.; Kunz, C.; Harrison, S.C. The envelope glycoprotein from tick-borne encephalitis virus at 2 Å resolution. *Nature* **1995**, *375*, 291–298. [CrossRef]
16. Pierson, T.C.; Diamond, M.S. Degrees of maturity: The complex structure and biology of flaviviruses. *Curr. Opin. Virol.* **2012**, *2*, 168–175. [CrossRef]
17. Monath, T.P.; Arroyo, J.; Levenbook, I.; Zhang, Z.-X.; Catalan, J.; Draper, K.; Guirakhoo, F. Single Mutation in the Flavivirus Envelope Protein Hinge Region Increases Neurovirulence for Mice and Monkeys but Decreases Viscerotropism for Monkeys: Relevance to Development and Safety Testing of Live, Attenuated Vaccines. *J. Virol.* **2002**, *76*, 1932–1943. [CrossRef]

18. Bradt, V.; Malafa, S.; von Braun, A.; Jarmer, J.; Tsouchnikas, G.; Medits, I.; Wanke, K.; Karrer, U.; Stiasny, K.; Heinz, F.X. Pre-existing yellow fever immunity impairs and modulates the antibody response to tick-borne encephalitis vaccination. *npj Vaccines* **2019**, *4*, 38. [CrossRef]

19. Zaidi, M.B.; Cedillo-Barron, L.; González y Almeida, M.E.; Garcia-Cordero, J.; Campos, F.D.; Namorado-Tonix, K.; Perez, F. Serological tests reveal significant cross-reactive human antibody responses to Zika and Dengue viruses in the Mexican population. *Acta Trop.* **2019**, *201*, 105201. [CrossRef]

20. Montecillo-Aguado, M.R.; Montes-Gómez, A.E.; García-Cordero, J.; Corzo-Gómez, J.; Vivanco-Cid, H.; Mellado-Sánchez, G.; Muñoz-Medina, J.E.; Gutiérrez-Castañeda, B.; Santos-Argumedo, L.; González-Bonilla, C.; et al. Cross-Reaction, Enhancement, and Neutralization Activity of Dengue Virus Antibodies against Zika Virus: A Study in the Mexican Population. *J. Immunol. Res.* **2019**, *2019*, 7239347. [CrossRef]

21. Dai, L.; Song, J.; Lu, X.; Deng, Y.-Q.; Musyoki, A.M.; Cheng, H.; Zhang, Y.; Yuan, Y.; Song, H.; Haywood, J.; et al. Structures of the Zika Virus Envelope Protein and Its Complex with a Flavivirus Broadly Protective Antibody. *Cell Host Microbe* **2016**, *19*, 696–704. [CrossRef] [PubMed]

22. Annamalai, A.S.; Pattnaik, A.; Sahoo, B.R.; Muthukrishnan, E.; Natarajan, S.K.; Steffen, D.; Vu, H.L.X.; Delhon, G.; Osorio, F.A.; Petro, T.M.; et al. Zika Virus Encoding Nonglycosylated Envelope Protein Is Attenuated and Defective in Neuroinvasion. *J. Virol.* **2017**, *91*, e01348-17. [CrossRef] [PubMed]

23. Bos, S.; Viranaicken, W.; Turpin, J.; El-Kalamouni, C.; Roche, M.; Krejbich-Trotot, P.; Desprès, P.; Gadea, G. The structural proteins of epidemic and historical strains of Zika virus differ in their ability to initiate viral infection in human host cells. *Virology* **2018**, *516*, 265–273. [CrossRef] [PubMed]

24. Li, G.; Bos, S.; Tsetsarkin, K.A.; Pletnev, A.G.; Desprès, P.; Gadea, G.; Zhao, R.Y. The Roles of prM-E Proteins in Historical and Epidemic Zika Virus-mediated Infection and Neurocytotoxicity. *Viruses* **2019**, *11*, 157. [CrossRef]

25. Gong, D.; Zhang, T.-H.; Zhao, D.; Du, Y.; Chapa, T.J.; Shi, Y.; Wang, L.; Contreras, D.; Zeng, G.; Shi, P.-Y.; et al. High-Throughput Fitness Profiling of Zika Virus E Protein Reveals Different Roles for Glycosylation during Infection of Mammalian and Mosquito Cells. *iScience* **2018**, *1*, 97–111. [CrossRef]

26. Carbaugh, D.L.; Baric, R.S.; Lazear, H.M. Envelope Protein Glycosylation Mediates Zika Virus Pathogenesis. *J. Virol.* **2019**, *93*, e00113-19. [CrossRef]

27. Fontes-Garfias, C.R.; Shan, C.; Luo, H.; Muruato, A.E.; Medeiros, D.B.A.; Mays, E.; Xie, X.; Zou, J.; Roundy, C.M.; Wakamiya, M.; et al. Functional Analysis of Glycosylation of Zika Virus Envelope Protein. *Cell Rep.* **2017**, *21*, 1180–1190. [CrossRef]

28. Viranaicken, W.; Nativel, B.; Krejbich-Trotot, P.; Harrabi, W.; Bos, S.; El Kalamouni, C.; Roche, M.; Gadea, G.; Desprès, P. ClearColi BL21(DE3)-based expression of Zika virus antigens illustrates a rapid method of antibody production against emerging pathogens. *Biochimie* **2017**, *142*, 179–182. [CrossRef]

29. Gadea, G.; Bos, S.; Krejbich-Trotot, P.; Clain, E.; Viranaicken, W.; El-Kalamouni, C.; Mavingui, P.; Desprès, P. A robust method for the rapid generation of recombinant Zika virus expressing the GFP reporter gene. *Virology* **2016**, *497*, 157–162. [CrossRef]

30. Aubry, F.; Nougairede, A.; de Fabritus, L.; Querat, G.; Gould, E.A.; de Lamballerie, X. Single-stranded positive-sense RNA viruses generated in days using infectious subgenomic amplicons. *J. Gen. Virol.* **2014**, *95*, 2462–2467. [CrossRef]

31. Frumence, E.; Roche, M.; Krejbich-Trotot, P.; El-Kalamouni, C.; Nativel, B.; Rondeau, P.; Missé, D.; Gadea, G.; Viranaicken, W.; Desprès, P. The South Pacific epidemic strain of Zika virus replicates efficiently in human epithelial A549 cells leading to IFN-β production and apoptosis induction. *Virology* **2016**, *493*, 217–226. [CrossRef] [PubMed]

32. Mattioli, L. ER storage diseases: A role for ERGIC-53 in controlling the formation and shape of Russell bodies. *J. Cell Sci.* **2006**, *119*, 2532–2541. [CrossRef] [PubMed]

33. Nativel, B.; Marimoutou, M.; Thon-Hon, V.G.; Gunasekaran, M.K.; Andries, J.; Stanislas, G.; Planesse, C.; Da Silva, C.R.; Césari, M.; Iwema, T.; et al. Soluble HMGB1 is a novel adipokine stimulating IL-6 secretion through RAGE receptor in SW872 preadipocyte cell line: Contribution to chronic inflammation in fat tissue. *PLoS ONE* **2013**, *8*, e76039. [CrossRef] [PubMed]

34. Lorenz, I.C.; Allison, S.L.; Heinz, F.X.; Helenius, A. Folding and Dimerization of Tick-Borne Encephalitis Virus Envelope Proteins prM and E in the Endoplasmic Reticulum. *J. Virol.* **2002**, *76*, 5480–5491. [CrossRef]

35. Wang, Y.; Bruce, A.T.; Tu, C.; Ma, K.; Zeng, L.; Zheng, P.; Liu, Y.; Liu, Y. Protein aggregation of SERCA2 mutants associated with Darier disease elicits ER stress and apoptosis in keratinocytes. *J. Cell Sci.* **2011**, *124*, 3568–3580. [CrossRef] [PubMed]

36. Delvecchio, R.; Higa, L.; Pezzuto, P.; Valadão, A.; Garcez, P.; Monteiro, F.; Loiola, E.; Dias, A.; Silva, F.; Aliota, M.; et al. Chloroquine, an Endocytosis Blocking Agent, Inhibits Zika Virus Infection in Different Cell Models. *Viruses* **2016**, *8*, 322. [CrossRef]

37. Zhang, S.; Yi, C.; Li, C.; Zhang, F.; Peng, J.; Wang, Q.; Liu, X.; Ye, X.; Li, P.; Wu, M.; et al. Chloroquine inhibits endosomal viral RNA release and autophagy-dependent viral replication and effectively prevents maternal to fetal transmission of Zika virus. *Antivir. Res.* **2019**, *169*, 104547. [CrossRef]

38. Chen, Y.; Maguire, T.; Hileman, R.E.; Fromm, J.R.; Esko, J.D.; Linhardt, R.J.; Marks, R.M. Dengue virus infectivity depends on envelope protein binding to target cell heparan sulfate. *Nat. Med.* **1997**, *3*, 866–871. [CrossRef]

39. Cruz-Oliveira, C.; Freire, J.M.; Conceição, T.M.; Higa, L.M.; Castanho, M.A.R.B.; Da Poian, A.T. Receptors and routes of dengue virus entry into the host cells. *FEMS Microbiol. Rev.* **2015**, *39*, 155–170. [CrossRef]

40. Pierson, T.C.; Kielian, M. Flaviviruses: Braking the entering. *Curr. Opin. Virol.* **2013**, *3*, 3–12. [CrossRef]

41. Lee, E.; Weir, R.C.; Dalgarno, L. Changes in the Dengue Virus Major Envelope Protein on Passaging and Their Localization on the Three-Dimensional Structure of the Protein. *Virology* **1997**, *232*, 281–290. [CrossRef] [PubMed]

42. Yoshii, K.; Yanagihara, N.; Ishizuka, M.; Sakai, M.; Kariwa, H. N-linked glycan in tick-borne encephalitis virus envelope protein affects viral secretion in mammalian cells, but not in tick cells. *J. Gen. Virol.* **2013**, *94*, 2249–2258. [CrossRef] [PubMed]

43. Sevvana, M.; Long, F.; Miller, A.S.; Klose, T.; Buda, G.; Sun, L.; Kuhn, R.J.; Rossmann, M.G. Refinement and Analysis of the Mature Zika Virus Cryo-EM Structure at 3.1 Å Resolution. *Structure* **2018**, *26*, 1169–1177.e3. [CrossRef] [PubMed]

44. Goo, L.; DeMaso, C.R.; Pelc, R.S.; Ledgerwood, J.E.; Graham, B.S.; Kuhn, R.J.; Pierson, T.C. The Zika virus envelope protein glycan loop regulates virion antigenicity. *Virology* **2018**, *515*, 191–202. [CrossRef]

45. Frumence, E.; Viranaicken, W.; Gadea, G.; Desprès, P. A GFP Reporter MR766-Based Flow Cytometry Neutralization Test for Rapid Detection of Zika Virus-Neutralizing Antibodies in Serum Specimens. *Vaccines* **2019**, *7*, 66. [CrossRef] [PubMed]

46. Frumence, E.; Viranaicken, W.; Bos, S.; Alvarez-Martinez, M.-T.; Roche, M.; Arnaud, D.J.; Gadea, G.; Desprès, P. A Chimeric Zika Virus between Viral Strains MR766 and BeH819015 Highlights a Role for E-glycan Loop in Antibody-mediated Virus Neutralization. *Vaccines* **2019**, *7*, 55. [CrossRef]

47. Mossenta, M.; Marchese, S.; Poggianella, M.; Slon Campos, J.L.; Burrone, O.R. Role of N-glycosylation on Zika virus E protein secretion, viral assembly and infectivity. *Biochem. Biophys. Res. Commun.* **2017**, *492*, 579–586. [CrossRef]

48. Fritz, R.; Stiasny, K.; Heinz, F.X. Identification of specific histidines as pH sensors in flavivirus membrane fusion. *J. Cell Biol.* **2008**, *183*, 353–361. [CrossRef]

49. Crill, W.D.; Chang, G.-J.J. Localization and Characterization of Flavivirus Envelope Glycoprotein Cross-Reactive Epitopes. *J. Virol.* **2004**, *78*, 13975–13986. [CrossRef]

Article

Route of Infection Influences Zika Virus Shedding in a Guinea Pig Model

Ashley E. Saver [1], Stephanie A. Crawford [1], Jonathan D. Joyce [2] and Andrea S. Bertke [2,*]

[1] Virginia-Maryland College of Veterinary Medicine, Virginia Polytechnic Institute & State University, Blacksburg, VA 24061, USA; aes27@vt.edu (A.E.S.); stepc08@vt.edu (S.A.C.)
[2] Department of Population Health Sciences, Virginia-Maryland College of Veterinary Medicine, Virginia Polytechnic Institute & State University, Blacksburg, VA 24061, USA; jjoyce84@vt.edu
* Correspondence: asbertke@vt.edu; Tel.: +1-540-231-5733

Received: 1 October 2019; Accepted: 13 November 2019; Published: 14 November 2019

Abstract: Due to the recent epidemic of Zika virus (ZIKV) infection and resulting sequelae, as well as concerns about both the sexual and vertical transmission of the virus, renewed attention has been paid to the pathogenesis of this unique arbovirus. Numerous small animal models have been used in various ZIKV pathogenicity studies, however, they are often performed using immunodeficient or immunosuppressed animals, which may impact disease progression in a manner not relevant to immunocompetent humans. The use of immunocompetent animal models, such as macaques, is constrained by small sample sizes and the need for specialized equipment/staff. Here we report the establishment of ZIKV infection in an immunocompetent small animal model, the guinea pig, using both subcutaneous and vaginal routes of infection to mimic mosquito-borne and sexual transmission. Guinea pigs developed clinical signs consistent with mostly asymptomatic and mild disease observed in humans. We demonstrate that the route of infection does not significantly alter viral tissue tropism but does impact mucosal shedding mechanics. We also demonstrate persistent infection in sensory and autonomic ganglia, identifying a previously unrecognized niche of viral persistence that could contribute to viral shedding in secretions. We conclude that the guinea pig represents a useful and relevant model for ZIKV pathogenesis.

Keywords: Zika virus; ZIKV; virus host interactions; pathogenesis; MR766; guinea pig; subcutaneous; vaginal; sexual transmission; virus transmission

---

1. Introduction

Zika virus (ZIKV), discovered in the Ugandan Zika forest in 1947, is a single-stranded RNA arbovirus of the genus *Flavivirus* and the family Flaviviridae [1,2]. Prior to 2007, only 14 human cases of ZIKV infection had been reported. However, in 2007, the first major epidemic of ZIKV, with 185 confirmed cases, occurred in the Yap Islands of the Federated States of Micronesia [3,4]. Since then, ZIKV has spread to 30+ countries, with millions of suspected cases, and has gained international attention due to an association with microcephaly and Guillain-Barré Syndrome (GBS) [5–9]. Subsequently, ZIKV has been identified as a significant global health threat.

ZIKV is primarily transmitted by mosquitoes. However, it can also be transmitted sexually or by blood transfusion [10–12]. After inoculation from an infected mosquito, the virus replicates in tissues local to the bite, drains to local lymph nodes, and then spreads hematogenously to secondary replication sites [13]. In adults, most infections (~80%) are asymptomatic, with only about 20% of infections developing a self-limiting illness. Symptoms vary in severity, and may include fever, headache, maculopapular rash, arthralgia, myalgia, fatigue, and conjunctivitis [14]. Additionally, ZIKV infection during pregnancy can cross the placenta, where it targets neural stem and progenitor cells in the developing fetus, leading to microcephaly, lissencephaly, and cognitive deficits, as well as ocular impairments such as chorioretinal atrophy and optic nerve disorders [15–17].

ZIKV is the only arbovirus known to be transmitted sexually [18]. Sexual transmission has been reported from male to female, male to male, and female to male, indicating that infectious virus persists in both semen (up to four months [18]) and vaginal secretions (up to six months [19]) [18,20–28]. However, the site of ZIKV persistence, leading to viral shedding in the genital secretions of males and females, is not clear. Although ZIKV has been reported to persist in testes, evidence of viral shedding in semen of vasectomized males suggests an additional site of persistence [29–31]. In women, the site of persistence has not been determined. We recently showed that ZIKV persistently infects primary adult cultured sensory neurons of the lumbosacral dorsal root ganglia (LS-DRG), which innervate the genitourinary tract (GUT), suggesting a potential alternative reservoir for viral shedding in urine and genital secretions [32]. The pathogenesis of ZIKV after sexual transmission has not been studied extensively, but sexual transmission may result in different routes of spread within the host and potentially alter tissue tropism when compared to mosquito-borne transmission.

Efforts to understand the pathogenesis of ZIKV following mosquito-borne and sexual transmission have led to the development of various animal models. Several studies have shown that immunocompetent adult wild-type mice have minimal susceptibility to ZIKV infection and demonstrate different disease manifestations than humans [33]. Thus, more recent studies have primarily used immunocompromised animals, such as mice lacking interferon (IFN) or IFN receptors, or immunocompetent mice treated with IFN-blocking antibodies [34–40]. Neonatal wild-type mice are susceptible to ZIKV infection, but they are also immunocompromised since rodents do not develop a mature immune response until at least one month of age [36,41–43]. Non-human primate models have provided valuable information [44–48]. However, non-human primate studies are limited in statistical power since relatively few animals can be used in studies. Additionally, non-human primate studies are expensive to perform and are limited to facilities that have the necessary infrastructure to house these animals. More recently, several studies have explored the use of swine as a model of ZIKV infection; however, most infected swine do not exhibit clinical signs and have demonstrated only low levels of viremia [49,50]. Additionally, swine pose similar constraints as non-human primates, as they require more space and are more expensive than small animal models. Thus, an immunocompetent small animal model is needed to study ZIKV pathogenesis by different routes of infection.

Guinea pigs (*Cavia porcellus*) have served as reliable models of flavivirus infection, and due to their physiologic similarities to human immune responses and symptoms, are also used as a genital infection model for several viruses [51–53]. Furthermore, guinea pigs are used as a model for cytomegalovirus (CMV) congenital syndrome, which causes similar fetal anomalies as ZIKV [54,55]. To date, five studies have reported the outcome of ZIKV infection in guinea pigs, with varying results. Subcutaneous (SQ) inoculation of strain PRVABC59 (Puerto Rico) resulted in fever, lethargy, hunching, ruffled fur, and decreased mobility, correlating with viremia and viral replication in spleen and brain [56]. Similarly, SQ inoculation of male guinea pigs with strain GZ01 (Venezuela) or FSS13025 (Cambodia) resulted in viremia and robust viral secretion in saliva and tears, as well as transmission to naïve co-caged mates, but no overt signs of disease [57]. However, intracerebral infection with strain MR766 (Uganda) or intraperitoneal inoculation with strain ArD41525 (Senegal) or CPC-0740 (Philippines) failed to produce signs of infection or viremia [1,58]. An additional study assessed fetal impact of mid-gestation infection in guinea pigs, finding viremia and robust antibody response following SQ inoculation with strain H/PF/2013 (French Polynesia) without effects on pups [59]. Although these studies produced variable results, likely due to differences in ZIKV strain, inoculum size, and inoculation route, none of these studies assessed vaginal infection or directly compared different routes of infection. Therefore, we evaluated the use of guinea pigs as an immunocompetent small animal model of ZIKV infection by both subcutaneous (SQ) and vaginal (VAG) inoculation routes, simulating mosquito-borne and sexual transmission, to compare pathogenesis, tissues sites of viral persistence and viral shedding in bodily secretions following different routes of infection.

2. Materials and Methods

2.1. Ethics Statement

This study was carried out according to the Animal Welfare Act (US Department of Agriculture, Washington DC, USA) and the Public Health Service (PHS) Policy on Humane Care and Use of Laboratory Animals after approval by the Institutional Animal Care and Use Committee (IACUC) of Virginia Polytechnic Institute and State University (Protocol Number: 17-124 approved 7/21/2017).

2.2. Virus

African lineage ZIKV MR766 (GenBank accession number: AY632553), recovered from a sentinel rhesus macaque in Uganda in 1947, was used for inoculations (BEI Resources, Manassas, VA, USA). Viral stocks were produced by passage once in Vero E6 cells (CRL-1586, ATCC, Manassas, VA, USA) [60,61]. Viral titer was determined by standard plaque assay on Vero E6 cells [60,61].

2.3. Guinea Pig Subcutaneous and Vaginal Infection

After a three day acclimation period, three-week-old female Hartley guinea pigs (Charles River Laboratories, Wilmington, MA, USA) were inoculated under anesthesia with 1×10^6 plaque forming units (PFU) ZIKV MR766 diluted to a total volume of 50 uL in PBS, subcutaneously (SQ, $n = 6$) by injection in the nape of the neck using a tuberculin syringe, or intravaginally (VAG, $n = 6$) by pipette.

Guinea pigs were monitored daily for 24 days for temperature (tympanic), weight, and development of clinical signs of ZIKV infection (fever, lethargy, hunching, ruffled fur, and decreased mobility), using a numerical value (0–5) based on severity of clinical signs, with 0 representing normal parameters and 5 representing severe clinical signs. Vaginal, salivary, and ocular samples were collected daily for 21 days using pre-moistened FLOQSwabs (Copan Diagnostics, Murrieta, CA, USA). Swabs were maintained on ice for the duration of swabbing all animals each day, and then stored at −80 °C until analysis by plaque assay.

Seven days post infection (dpi), two guinea pigs from each group were euthanized to assess acute infection characteristics. Blood samples were collected and centrifuged to separate serum, which was stored at −80 °C until analysis via RT-qPCR. Necropsy was performed immediately following euthanasia and tissues were collected, including spleen, lymph nodes, genitourinary tract (GUT), ovaries, lumbosacral DRG (LS-DRG, innervates the GUT), cervical DRG (C-DRG, innervates the neck), trigeminal ganglia (TG, innervates the face), superior cervical ganglion (SCG, innervates head and neck), ciliary ganglion (CG, motor and sensory innervation to the eye), brain, and eyes. Half of the tissues were fixed in 4% paraformaldehyde (PFA), embedded in optimum cutting temperature (OCT) media, and sectioned for immunofluorescence. The other half of the tissues were frozen for RNA isolation for RT-qPCR. At 37 dpi, allowing five weeks for viral clearance to ensure we had reached convalescence, the remaining guinea pigs were euthanized, necropsied, and samples were collected as described above to assess convalescent characteristics and viral persistence.

2.4. ZIKV Shedding

Daily ocular, vaginal and oral swabs were collected from 1–21 dpi, using pre-moistened FLOQswabs. Viral shedding was assessed by standard plaque assay on Vero E6 cells using ocular and vaginal swab samples collected daily 1–21 dpi. Oral swabs were unable to be assessed by plaque assay due to excessive outgrowth of oral microflora.

2.5. ZIKV RNA Isolation and qPCR of Clinical Samples

Tissues were homogenized on ice with a Bel-Art hand held tissue homogenizer with sterile pestles (Cole-Parmer, Vernon Hills IL, USA). RNA was extracted using an RNEasy Kit (Qiagen, Germantown, MD, USA) following manufacturer's protocol. The iScript cDNA Synthesis Kit was used to reverse transcribe RNA into complementary DNA (cDNA) (Bio-Rad, Hercules, CA, USA). Briefly, 20 μL reactions were used for cDNA synthesis (4 μL 5× iScript reaction mix, 1 μL iScript reverse transcriptase, 10 μL nuclease-free water, and 5 μL RNA template). Thermocycler conditions were: 5 min 25 °C; reverse transcription for 30 min 42 °C; 5 min 85 °C. qPCR was conducted on a Viia 7 real-time PCR system (Applied Biosystems, v1.2.4, Foster City, CA, USA) using iTaq Universal SYBR Green Supermix (BioRad) and the previously synthesized cDNA. Ten μL reactions were used (5 μL iTaq supermix, 0.5 μL forward (5′ AAR TAC ACA TAC CAR AAC AAA GTG GT) /reverse (5′ TCC RCT CCC YCT YTG GTC TTG) primer mix (10 μM each), 0.5 μL rRNA primer mix, 1 μL nuclease-free water, 3 μL cDNA). Thermocycler conditions were on fast setting, 20 s 95 °C; 40 cycles of 1 sec 95 °C, 20 sec 60 °C; followed by melt-curve analysis. Resulting Ct values were used to determine number of ZIKV RNA copies/500 ng total RNA based on standards of known quantity.

2.6. Immunofluorescence of Clinical Samples

Tissues were fixed in 4% PFA, embedded in OCT media (ThermoFisher, Waltham, MA, USA), and stored at −80 °C until sectioned into 10 μm cryosections using a Leica CM3050-S cryostat (Leica Biosystems, Buffalo Grove, IL, USA), and sections were stored at −80 °C until immunostaining. Sections were immunostained for ZIKV using a mouse-anti-ZIKV Envelope (E) protein primary antibody (FL0006, Kerafast, Boston, MA, USA), and visualized with a donkey-anti-mouse-AlexaFluor 488 secondary antibody (ab150101, Abcam, Cambridge, MA, USA). Sensory neurons were immunostained with anti-PGP9.5 (NB300-675, Novus Biologicals, Centennial, CO, USA) and satellite glial cells with anti-glutamine synthetase (ab73593, Abcam, Cambridge MA, USA) antibodies, followed by species-specific secondary antibodies conjugated to AlexaFluor 488 (Fisher Scientific, Hamptom, NH, USA). CNS neurons were immunostained with anti-NeuN conjugated to AlexaFluor 488 (ABN78A4, EMD Millipore, Burlington, MA, USA). Slides were treated with SlowFade Gold Antifade Mountant (ThermoFisher, Waltham, MA, USA) before cover slipping. Slides were visualized and imaged using an IX71 inverted fluorescence microscope (Olympus Life Sciences, Waltham, MA, USA).

2.7. Statistical Analysis

Statistical analyses were performed in Excel (Microsoft Inc., Redmond, WA, USA) using t-tests to compare the two groups. *p*-values are summarized in figures as * < 0.01, ** < 0.001, *** < 0.0001.

3. Results

3.1. Clinical Observations

The guinea pigs were observed for 24 days post infection (dpi) for clinical signs of ZIKV infection. Although none of the guinea pigs' temperatures exceeded the normal range (37.2–39.5 °C), vaginally (VAG)-infected guinea pigs had higher temperatures than subcutaneously (SQ)-infected guinea pigs, which was statistically significant at 4 dpi (Figure 1A, *p* < 0.01). All guinea pigs gained weight at similar rates following infection (Figure 1B).

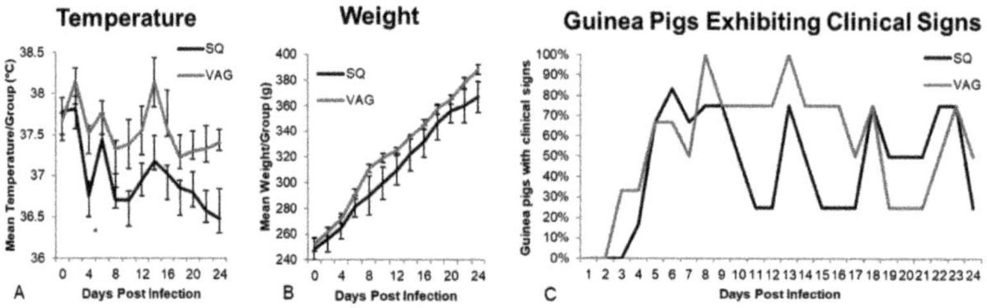

Figure 1. *Cont.*

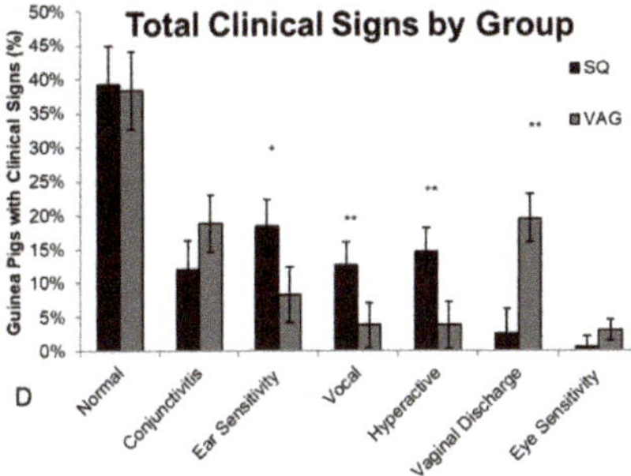
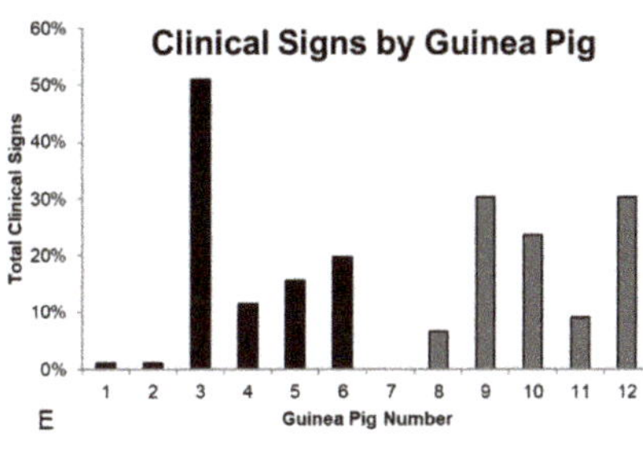

Figure 1. Clinical signs of infection following subcutaneous (SQ) and intravaginal (VAG) inoculation with 1×10^6 PFU ZIKV MR766. (**A**) Temperature (average/group, * $p < 0.01$); (**B**) Weight (average/group); (**C**) Percentage of guinea pigs exhibiting clinical signs, including conjunctivitis, ear sensitivity, vocalization, hyperactivity, vaginal discharge, or eye sensitivity; cyclical peaks began 8 dpi and repeated every 5 days; (**D**) 291 total clinical observations were recorded over 24 days. The number of times each clinical sign was observed was recorded and is shown as percentage of guinea pigs in each group exhibiting specific clinical signs (* $p < 0.05$, ** $p < 0.01$); (**E**) Total clinical signs (excluding "normal" observations) recorded for each guinea pig.

Based on clinical scoring, using a 5-point severity scale, ZIKV inoculation resulted in mild infections in both groups, although wide variability occurred among individual guinea pigs. Although mild, clinical signs were most apparent from 3–8 dpi and cyclical peaks of clinical signs occurred every five days thereafter (e.g., 8, 13, 18, 23 dpi, Figure 1C). While mild clinical signs of infection have been reported in previous guinea pig models, ours is the first to demonstrate a cyclical pattern to their nature. Over the 24-day observation period, 291 clinical observations were recorded; 39% of the 158 observations from the SQ group and 38% of the 133 observations from the VAG group were categorized as "no clinical signs" or "normal" (Figure 1D). Although we did not observe many classical overt signs of severe infection (e.g., ruffled fur, immobility, weight loss), some discrete signs of distress were observed in multiple animals. Conjunctivitis was observed in both SQ (12%) and VAG (19%) infected animals. However, guinea pigs infected SQ had a more frequent occurrence of ear sensitivity when tympanic temperatures were taken ($p = 0.014$), vocalization during handling ($p = 0.008$), and hyperactivity ($p = 0.002$) compared to the VAG infected group. In contrast, VAG infected guinea pigs had a significantly greater occurrence of vaginal discharge compared to SQ infected animals ($p < 0.001$ Figure 1D). Substantial variability was observed among all guinea pigs, regardless of inoculation route, with some animals showing minimal signs (Figure 1E). One guinea pig, GP#3, displayed the greatest number of observed signs of infection in the SQ group (GP#1-6), while GP#1 and GP#2 showed minimal signs of infection, which is consistent with ~20% of humans developing clinical disease from ZIKV infection. In the VAG group (GP#7-12), all but one guinea pig (GP#7) displayed signs of infection (Figure 1E). Determining the specific route of infection of humans in epidemic and endemic areas is rarely possible, thus the percentage of humans developing clinical symptoms following vaginal infection is not known. Therefore, the significance of these differences in relation to humans is challenging to interpret.

3.2. ZIKV Viral Shedding

Guinea pigs infected SQ shed infectious ZIKV in tears from 7-17 dpi, with peak shedding at 10 dpi (up to 10 PFU), although VAG-infected guinea pigs did not have detectable infectious ZIKV in tears (Figure 2A). Shedding of infectious ZIKV in tears after SQ injection (footpad) in immunocompromised or immunosuppressed mice has been demonstrated by Miner et al., who also noted the development of pan-uveitis [62]. Guinea pigs infected VAG shed infectious ZIKV in vaginal secretions throughout the study period, from 1–21 dpi, with peak shedding at 3 dpi (up to 425 PFU) (Figure 2B). The early peak of viral shedding in vaginal secretions at 3 dpi is consistent with local replication of the virus in vaginal or cervical tissues, prior to hematogenous spread to secondary tissues. Similar results have been documented in humans where infectious ZIKV has been recovered from vaginal secretions up to 10–14 days post symptom onset (pso) [28,63], although vaginal swabs have not been assessed the first few days following infection in humans to determine if very early viral shedding occurs. In contrast, no infectious virus was detected in vaginal secretions from guinea pigs inoculated SQ (Figure 2A). Overgrowth of the microflora of the oral cavity prevented successful plaque assay of oral swab samples.

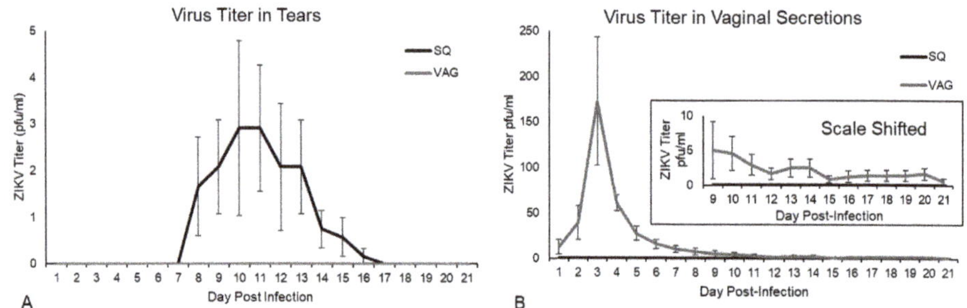

Figure 2. Viral shedding in guinea pigs inoculated subcutaneous (SQ) and intravaginal (VAG) with 1x10^6 PFU ZIKV MR766. Virus titer, determined by plaque assay on Vero E6 cells, detected in: (**A**) tears; (**B**) vaginal secretions. Inset shows the graph with the Y-axis shifted to show average titer in vaginal secretions through 21 dpi in VAG infected animals.

3.3. ZIKV in Guinea Pig Tissues

A low-level serum viremia (1.09 × 10^3–3.47 × 10^3 ZIKV RNA copies/500 ng total RNA) was detected in both groups of guinea pigs during acute infection (Figure 3). This low-level viremia was still present in both groups of guinea pigs during convalescence (3.25 × 10^1–1.36 × 10^3 ZIKV RNA copies/500 ng total RNA), although reduced in SQ-infected guinea pigs. Wide variability was detected among all animals; thus, no statistically significant differences in viremia were found between the guinea pig groups during acute or convalescent time points. ZIKV RNA copy numbers in all tissues tested, other than the cerebellum of SQ-infected guinea pigs during the acute phase, were higher than serum viremia at both time points (7 dpi and 37 dpi) in both groups (Figure 3).

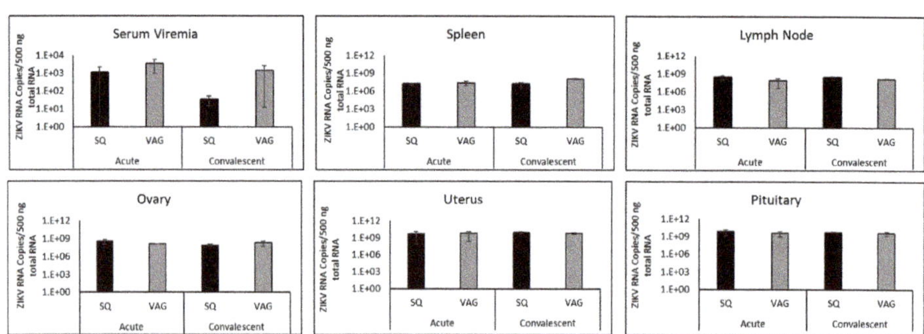

Figure 3. Viral load detected by RT-qPCR in serum and tissues of guinea pigs inoculated subcutaneously (SQ) or vaginally (VAG) with ZIKV at acute (7 dpi) and convalescent (37 dpi) time points.

Similar ZIKV RNA loads were found in tissues of guinea pigs following SQ and VAG infection (Figure 3). Viral load detected during acute infection (7 dpi) did not change substantially by the convalescent time point (37 dpi), suggesting an extended period of time is required for clearance of viral RNA even though no signs of disease were present at the end point of the study. However, an increase in ZIKV RNA copies was detected in the spleens of VAG-infected guinea pigs from acute (2.41×10^7) to convalescent (1.14×10^8) time points, and a one-log higher copy number of ZIKV was detected in the spleens of VAG compared to the SQ group ($p = 0.0141$) at the convalescent time point. ZIKV RNA detected in lymph nodes was similar to that found in spleens (Figure 3), although no statistically significant differences were found.

The ovaries and uteri of both guinea pig groups contained ZIKV during both time points (Figure 3). Surprisingly, there was no statistically significant difference between the guinea pig groups at either time point due to inoculation route. Greater ZIKV copy numbers were noted in convalescent uteri $(9.59 \times 10^9$ SQ, 5.75×10^9 VAG) than convalescent ovaries $(7.65 \times 10^7$ SQ, 1.87×10^8 VAG) in both groups (p = 0.025 SQ; p = 0.021 VAG), suggesting a higher level of ZIKV persistent replication in the uterus during infection. The pituitary glands in each group supported robust ZIKV replication during acute infection $(8.06 \times 10^9$ SQ and 4.07×10^9 VAG'), which continued into the convalescent period in both groups $(5.37 \times 10^9$ SQ and 3.76×10^9 VAG), suggesting that ZIKV infection may affect glandular function.

ZIKV antigen was detected by immunofluorescence (IF) staining in the spleen and cervical lymph nodes at both acute and convalescent time points in SQ- and VAG-infected guinea pigs, with minimal visible differences between the groups (Figure 4). At the convalescent time point, ZIKV antigen was detected in diffuse vibrant clusters of cells, as opposed to distinct individual cells at the acute time point, suggesting possible cell-to-cell spread within lymph nodes (Figure 4).

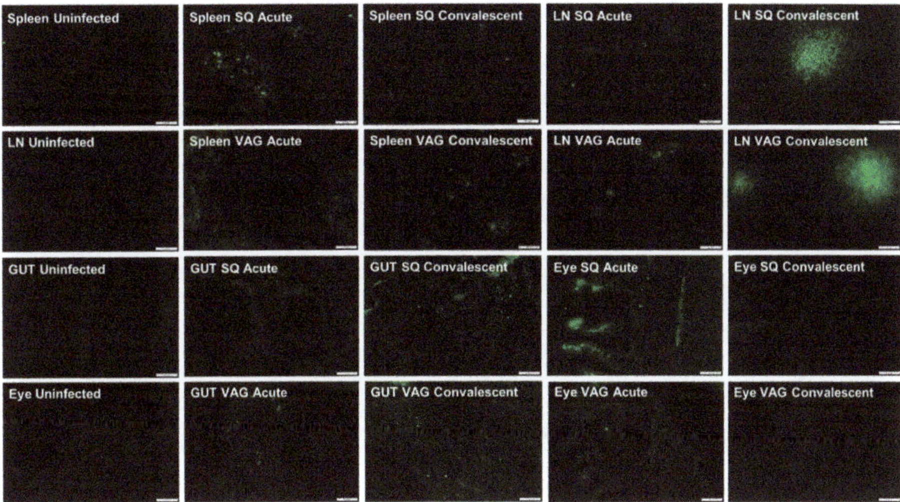

Figure 4. Representative immunofluorescence images of spleens, lymph nodes (LN), genitourinary tracts (GUT), and eyes from guinea pigs inoculated SQ or VAG with ZIKV at acute (7 dpi) and convalescent (37 dpi) time points. GUT images are from the uterus, which was the only part of the GUT we identified as positive for ZIKV antigen. Eye images show ciliary bodies, which were positive for ZIKV antigen.

The genitourinary tract (GUT) of both groups were IF stained for ZIKV to determine if ZIKV persisted in tissues through the convalescent time point. ZIKV antigen was detected in the GUT, specifically in the uterine wall, with virus persisting into the convalescent time point (Figure 4). At 7 dpi, ZIKV antigen was found localized to the uterus after VAG infection. ZIKV antigen was noted to be heaviest in VAG-infected guinea pigs at both time points compared to SQ-infected guinea pigs.

ZIKV antigen was observed in the eye of both SQ and VAG infected animals at both time points, but most prominently in the ciliary body of SQ infected guinea pigs during acute infection (Figure 4). Although present in VAG-infected animals, ZIKV antigen was noted to be heaviest in SQ infected guinea pigs compared to VAG-infected guinea pigs. This comports with our finding that guinea pigs infected SQ shed infectious ZIKV in ocular secretions. Similar immunofluorescent localization of ZIKV in the ciliary body has been reported in immunocompromised mouse models [64]. Although we did not detect substantial histopathology consistent with inflammation in the eyes of our adult guinea pigs, histopathological signs of inflammation have been localized to the ciliary bodies of fetal rhesus macaques (vertical transmission) and in a case series assessing congenital ZIKV syndrome in humans [65]. It is worth noting that no infectious virus was recovered from ocular secretions at acute or convalescent time points in VAG-infected guinea pigs, even though ZIKV viral RNA was detected in the eyes of this group, although less than after SQ infection. This suggests the possibility that a more robust viral replication is occurring in tissues that provide ocular secretions, such as the lacrimal glands.

3.4. ZIKV in Guinea Pig Nervous Systems

Sensory ganglia (LS-DRG, C-DRG, and TGs) of the peripheral nervous system of both groups supported stable and persistent ZIKV RNA at both time points, with no significant differences between routes of infection. A slightly lower ZIKV copy number was detected in TGs from VAG- infected guinea pigs at the convalescent time point, although the results did not reach statistical significance (Figure 5).

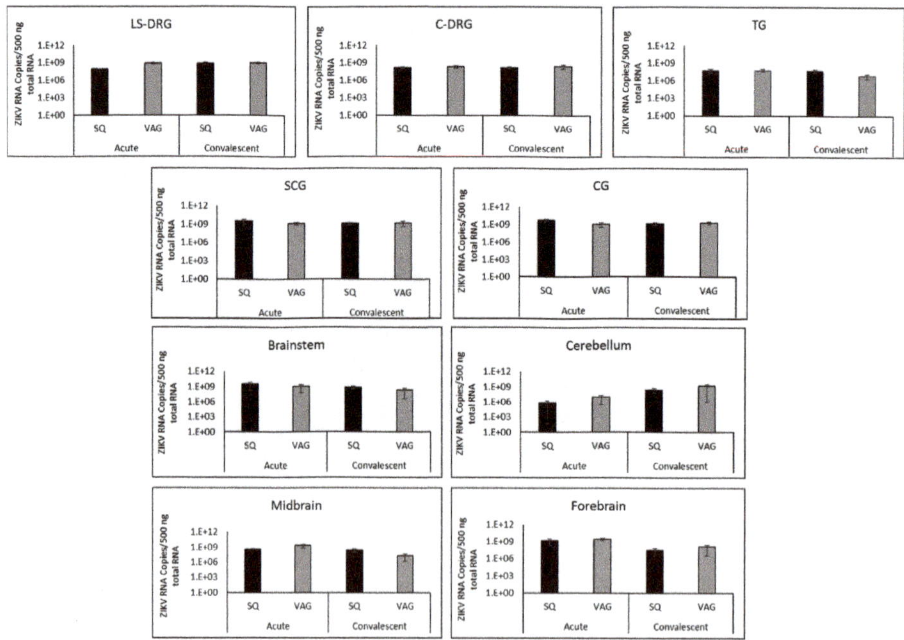

Figure 5. Viral load detected by RT-qPCR in nervous system of guinea pigs inoculated subcutaneously (SQ) or vaginally (VAG) with ZIKV at acute (7 dpi) and convalescent (37 dpi) time points. LS-DRG (lumbosacral dorsal root ganglia, sensory), C-DRG (cervical dorsal root ganglia, sensory), TG (trigeminal ganglia, sensory), SCG (superior cervical ganglia, sympathetic), CG (ciliary ganglia, parasympathetic).

Autonomic ganglia (SCG and CG) also contained similar ZIKV copy numbers following SQ and VAG infection, detectable during both acute and convalescent time points (Figure 3). These ganglia, in both groups, at both time points, supported ZIKV copy numbers slightly higher than those detected in sensory ganglia (1.24×10^9–5.59×10^9 RNA copies/500ng total RNA). There was no significant difference between ZIKV copy numbers in sympathetic (SCG) versus parasympathetic (CG) ganglia.

Structures throughout the brains of both SQ and VAG infected guinea pigs had high levels of detectable ZIKV RNA at acute and convalescent time points (Figure 5). ZIKV RNA copies decreased approximately 10-fold from acute to convalescent time points in brainstem (1.1–3.3×10^9 acute to 2.1-7.5×10^8 convalescent), midbrain (3.9×10^8–2.0×10^9 acute to 2.0×10^7–2.2×10^8 convalescent), and forebrain (1.6–2.6×10^9 acute to 23.2×10^7–1.3×10^8 convalescent). In contrast, the cerebellum had fewer ZIKV RNA copies during acute infection (7.2×10^5 SQ and 8.6×10^6 VAG) compared to the convalescent time point (2.4×10^8 SQ and 1.7×10^9 VAG). In fact, the cerebellum contained lower copy numbers of ZIKV RNA than any other region of the nervous system during acute infection.

We previously demonstrated that cultured primary adult mouse sensory DRGs became infected with ZIKV and persistently released infectious virus for at least five days without dying [32]. We also determined that satellite glial cells (SGCs) in those cultures became infected and were killed by the virus within 24 h post inoculation. Thus, we hypothesized that the LS-DRG may be an alternative reservoir of persistent virus that could be shed in genital secretions, since the LS-DRG innervates the GUT. Thus, we assessed the presence of ZIKV in the LS-DRG to determine if neurons or SGCs became persistently infected with ZIKV following SQ or VAG infection. We detected ZIKV antigen by IF in the LS-DRG in both groups (Figure 6). During acute infection (7 dpi), ZIKV was detected by immunofluorescence surrounding sensory neurons within the ganglia and co-localizing with satellite glial cell (SGC) marker glutamine synthetase (GS). However, ZIKV was not detected within the neurons themselves, which were visualized by immunofluorescence for sensory neuronal marker PGP9.5. ZIKV persisted within the ganglia through the convalescent time point, particularly in the VAG-infected animals. SGCs normally wrap around sensory neurons within the ganglion (see LS-DRG Uninfected GS in Figure 6) but in ZIKV infected animals, regardless of route of inoculation or time point, the morphology of SGCs was substantially altered compared to uninfected animals, suggesting viral destruction of the SGCs. In contrast to the findings of our in vitro study, the sensory neurons were not infected in LS-DRG. ZIKV antigen was also detected in the SGCs of the C-DRG following SQ infection, but not in neurons. ZIKV antigen was found in C-DRG in only one of the VAG-infected guinea pigs during the acute time point.

Within the central nervous system, ZIKV antigen was observed in the brains of both groups at both acute and convalescent time points. Infection was diffuse throughout the brains of both groups, with fluorescence most notably in cortical tissues in the frontal lobe and in the hippocampus (Figure 7). Within each of these regions of the brain, isolated neurons identified as positive for ZIKV showed an altered morphology, substantially larger than nearby uninfected neurons and surrounded by a "halo". This localization of ZIKV has been observed in a rhesus macaque model of infection (pregnant mother and fetus), as well as in human fetuses [17,66]. Histopathology, suggesting ZIKV localization to cortical tissues, has also been demonstrated in numerous murine models [36,38,62].

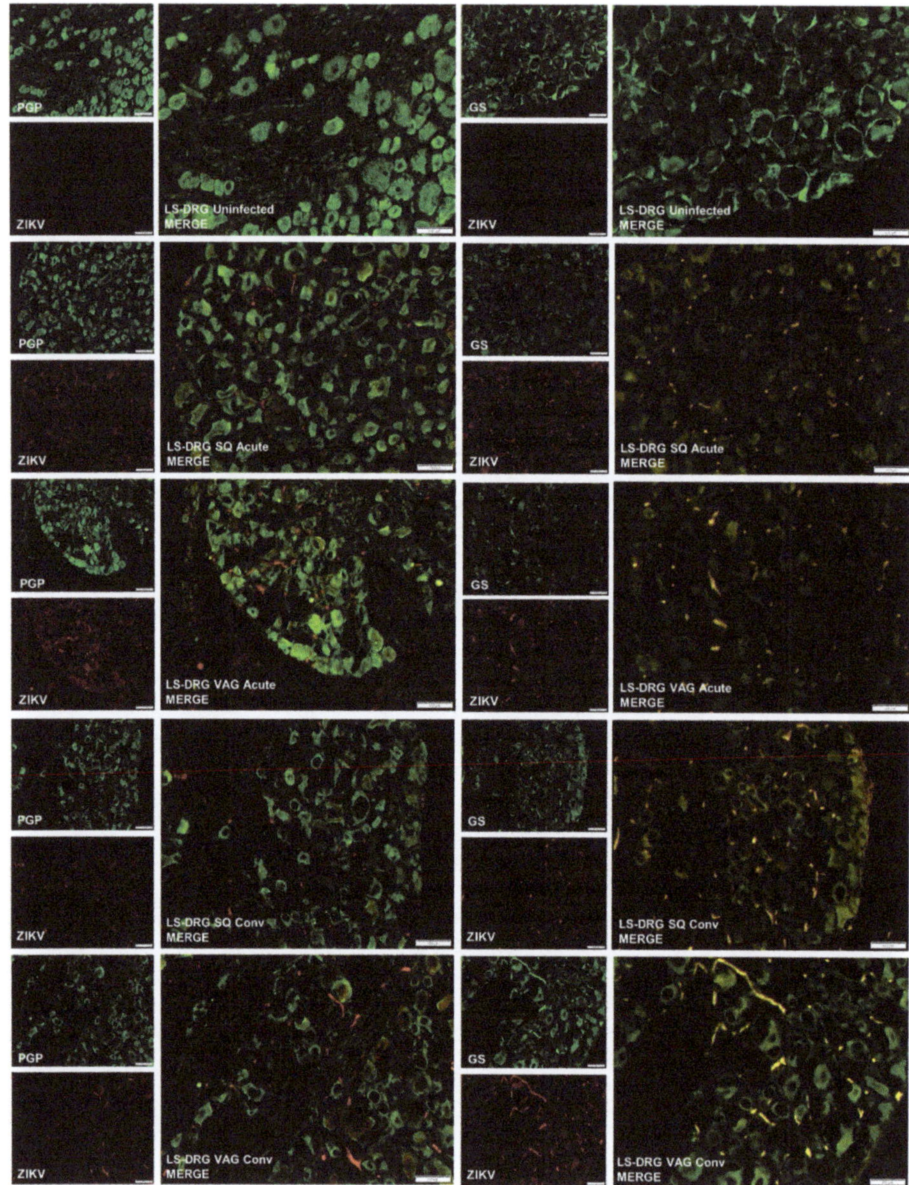

Figure 6. Representative immunofluorescence images of lumbosacral dorsal root ganglia (LS-DRG) from uninfected guinea pigs, and guinea pigs inoculated subcutaneously (SQ) or vaginally (VAG) with ZIKV at acute (7 dpi) and convalescent (37 dpi) time points. Neurons were immunostained for the sensory neuronal marker PGP9.5 (green) or the satellite glial cell marker glutamine synthetase (GS, green), and ZIKV (red). Merged images are shown enlarged to show co-localization of ZIKV with satellite glial cells, not sensory neurons. Note the morphological changes of satellite glial cells between uninfected (upper right) and ZIKV infected guinea pigs.

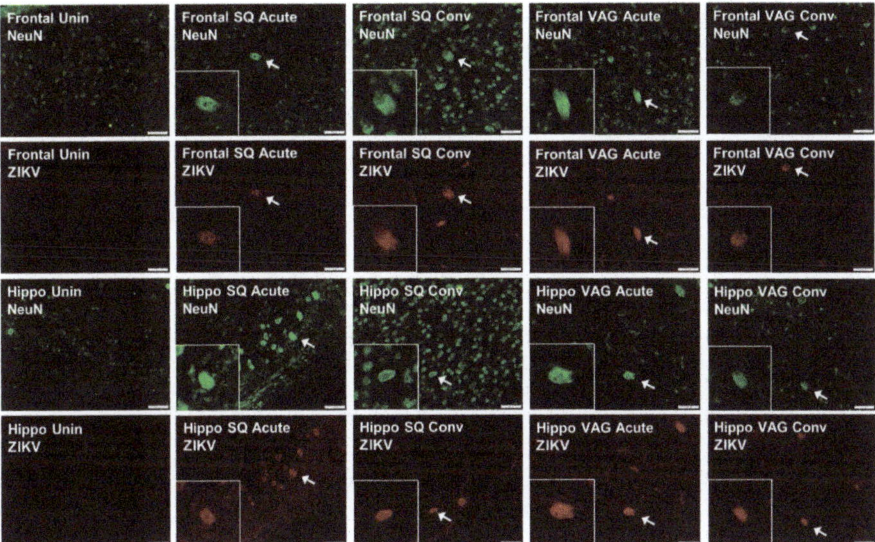

Figure 7. Representative immunofluorescence images of brain, including cortex in the frontal lobe (Frontal) and hippocampus (Hippo) from uninfected guinea pigs and guinea pigs inoculated subcutaneously (SQ) or vaginally (VAG) with ZIKV at acute (7 dpi) and convalescent (37 dpi) time points. Neurons were immunostained for the neuronal marker NeuN (green) and ZIKV (red). Insets are 200% enlargements of infected neurons, depicted by the white arrow, showing altered morphology.

4. Discussion

Numerous murine models exist for the study of ZIKV infection. However most of these models use immunodeficient or immunosuppressed mice, many lacking an intact IFN pathway, which may influence the pathogenesis of infection (e.g., severe viremia, disease, frequent death) in a manner not applicable to that found in immunocompetent humans (e.g., asymptomatic infection, self-limiting illness, rare death) [40,67–70]. Several immunocompetent non-human primate models have been used to study the pathogenesis of ZIKV infection, however these models are limited due to their prohibitive cost, resulting small sample sizes, reduced statistical power, and requirements for specialized facilities and staff [44–48]. Due to concerns of sexual and vertical transmission and neurological sequelae of ZIKV infection, an immunocompetent small animal model is warranted.

A guinea pig model of ZIKV infection presents an attractive alternative to the above models due to the physiologic similarities between humans and guinea pigs, which include reproductive physiology and estrous cycle, and homology between immune systems (major histocompatibility molecules (MHC), complement systems, IFNγ pathways, IL-8/12 receptors, and CD8 sequences) [51]. Other attractive characteristics of the guinea pig include the ability to establish infection in an immunocompetent host, general ease of handling and maintenance, utilization of larger sample sizes, and availability of immunological assays and techniques [51]. Additionally, guinea pigs have been used as reliable models of flavivirus infection (e.g., Japanese encephalitis virus), as well as for sexual transmission studies (herpes simplex virus) and congenital syndrome caused by vertical transmission (cytomegalovirus) [54,55,71–74]. In this study, we demonstrate successful infection and persistence of ZIKV in immunocompetent female Hartley guinea pigs after a physiologically relevant inoculation $(1 \times 10^6$ PFU) in clinically significant routes of transmission (SQ, VAG), which mimic mosquito-borne and sexual transmission.

ZIKV infection was established in all guinea pigs regardless of route of infection. Minimal clinical signs of infection were observed, although subtle signs were noted, such as ear sensitivity, vocalization, and hyperactivity. This is consistent with human infection, as the majority (~80%) of ZIKV infections in humans are asymptomatic [75–78]. The classic maculopapular rash observed in some humans with symptomatic ZIKV infection was not observed in our guinea pigs, and has not been reported in other guinea pig models. Dermatological manifestations have only been reported in non-human primates around injection sites and more recently in tree shrews, although the tree shrews did not demonstrate any other signs of ZIKV infection such as fever or weight loss [45]. The mechanism for the development of skin rash associated with ZIKV infection is not fully understood. More severe/overt signs of disease have been elicited in guinea pigs by inoculation with more contemporary ZIKV strains (e.g., PRVABC59) [56], suggesting that these animals may be useful for investigating pathogenicity differences between ZIKV strains. We are the first to observe a five-day cyclical/undulating nature of clinical signs of infection, although the clinical significance of this observation relative to humans is not clear. Also, our studies are the first to compare SQ and VAG routes of infection, including the observations of the increase in vaginal discharge in animals inoculated vaginally, which may be a factor in potential sexual transmission of ZIKV.

A low-level viremia was detected via RT-qPCR in the serum of both groups throughout the study. Most current models were unable to detect viremia beyond 5 dpi; however, one model reported detection of low-level viremia up to 14 dpi, and one reported no detection at any time point [56–59]. Our detection of a sustained and persistent low-level serum viremia in both SQ and VAG infected groups up to 37 dpi represents the longest detection of serum viremia in a guinea pig model. This finding is consistent with detection of ZIKV viremia in whole blood samples in humans from 14 to 100 days, for a median duration of 22 days, while another serosurvey showed viremia for up to 8 weeks in some patients [79,80]. Localization of ZIKV replication to secondary lymphoid organs (spleen, lymph nodes), genitourinary tract (uterus, ovary), brain (brainstem, cerebellum, midbrain, forebrain, pituitary gland), and eyes agree with results reported in murine models, non-human primate models, and human case reports [35,37,40,44,56,57,65,81]. However, the only statistically significant difference we identified in tissue tropism between SQ and VAG routes of infection, based on tissue viral loads, was in spleens at convalescence (VAG > SQ), suggesting that ZIKV may be cleared more quickly after mosquito-borne transmission than after sexual transmission. This is also supported by the higher levels of ZIKV RNA we detected in serum at the convalescent time point in VAG infected compared to SQ infected animals. It is interesting to note the lower ZIKV RNA copy number in the cerebellums in both groups compared to any other region of the brain during acute infection, indicating a delay in entry or replication in the cerebellum for an unknown reason.

Infectious ZIKV was recovered from vaginal secretions from 1-21 dpi and from tears 7-17 dpi, with peak recovery at 3 dpi and 10 dpi, respectively. Our recovery of ZIKV from secretions is consistent with the recovery of ZIKV from vaginal secretions in humans from 10-14 days pso and in tears up to 30 days pso [28,63,82]. Interestingly, ZIKV was recovered from vaginal secretions in only VAG-infected guinea pigs and from tears in only SQ-infected guinea pigs, even though we detected ZIKV antigens in similar tissues in both groups and found no statistically significant differences in ZIKV RNA copy number in ocular or genitourinary tissues between the groups. Additionally, vaginal secretions supported a longer period of viral shedding and higher viral titers during peak shedding than tears, which may indicate a more robust and prolonged viral replication locally in VAG-infected animals. The increased vaginal discharge we noted in VAG-infected guinea pigs may contribute to higher shedding rates of infectious ZIKV in this group compared to SQ-infected animals, which may increase the risk of sexual transmission.

With respect to ZIKV shedding in tears, Deng et al. also isolated infectious ZIKV from the tears of SQ- and intranasally-infected guinea pigs, demonstrating contact transmission of ZIKV between SQ-infected guinea pigs and their naïve cage mates, potentially mediated through viral shedding in tears [57]. Our recovery of infectious ZIKV in tears from 7-17 dpi (albeit at low titers), combined with the detection of ZIKV RNA in tears of both index and contact animals by Deng et al., provides additional evidence to suggest contact transmission can occur in animal models by viral shedding in tears. As an extension of these findings, it is worth noting a case report in which contact transmission of ZIKV is suspected to have occurred between an elderly patient with a fatal ZIKV infection with high serum viremia and an otherwise healthy family member participating in his care who came into contact with the patient's tears while not wearing personal protective equipment, subsequently developing a maculopapular facial rash and ZIKV antigenuria [83]. Taken together, these results suggest contact transmission of ZIKV via tears can occur between humans in certain rare instances.

We set out to determine if there were differences in tissue tropism and sites of ZIKV persistence that could contribute to extended periods of viral shedding in genital secretions that could contribute to sexual transmission, as well as vertical transmission. We had previously speculated that ZIKV persistence in LS-DRG, which innervate the GUT, may provide an alternative reservoir for viral shedding in genital secretions, particularly after sexual transmission [32]. We did not find significant differences in viral load or viral antigen in DRG following SQ or VAG infection, nor did we find significant differences in the GUT. However, we determined that ZIKV persists in the DRG and uterus following both SQ and VAG infection. To our knowledge this is the first in vivo animal model to investigate the role of sensory and autonomic ganglia in the maintenance of ZIKV infection. ZIKV antigen was localized to satellite glial cells (SGCs) surrounding the sensory neurons within LS-DRG in both groups during acute infection by 7 dpi, and the ganglia remained infected for at least 37 dpi. SGCs wrap around sensory neurons within the ganglia, providing support for the neurons as well as protection, forming a barrier between capillary endothelial cells and neurons within the ganglia and thus preventing access of blood-borne pathogens to the neurons. Based on our previous in vitro studies, in which SGCs were destroyed and naked neurons became persistently infected with ZIKV, we had anticipated that ZIKV would lytically infect SGCs, gaining access to sensory neurons within the DRG. DRGs from infected guinea pigs showed altered morphology and loss of SGCs surrounding sensory neurons within the ganglia, consistent with the destruction of SGCs we observed previously in primary DRG cultures [32]. However, the neurons themselves were not infected in either group. Detection of ZIKV antigen in SGCs, but not neurons of LS-DRG (or C-DRGs), suggests that ZIKV gains access to sensory ganglia through hematogenous dissemination but SGCs effectively prevent the virus from reaching and infecting the sensory neurons in vivo. Infection and destruction of SGCs surrounding neurons could disrupt synaptic transmission and potentially contribute to peripheral neuropathies. SGCs have been implicated in protection of neurons from blood-borne pathogens, as well as exacerbation of infection by mediating a robust inflammatory response within the ganglion. Although additional studies are needed to assess SGC survival, our results support a pathogenic model in which SGCs protect the sensory neurons from viremic ZIKV infection but contribute to viral spread and persistence in non-neuronal cells within the DRG. Additionally, our studies demonstrated ZIKV persistence in both sympathetic and parasympathetic autonomic ganglia. As autonomic ganglia innervate secretory glands and regulate release of secretions, persistence within autonomic ganglia also represents a previously undefined reservoir of persistent ZIKV that may contribute to viral shedding in secretions.

In summary, we sought to determine if route of infection influences pathogenesis of disease, tissue tropism and persistent reservoirs of ZIKV that may contribute to viral shedding. Although we did not identify differences in viral load or tissue tropism, route of infection contributed to substantial differences in viral shedding in secretions. Our studies support a pathogenic model in which ZIKV replicates locally at the site of infection, and then spreads hematogenously throughout the host. Following subcutaneous infection, simulating mosquito-borne transmission, ZIKV is more effective at shedding from ocular sections, although the site of persistence and mechanism are not completely clear. Following vaginal infection, local replication in the genitourinary tract induces increased vaginal secretions, which carry infectious virus that could contribute to sexual transmission. Since the majority of the ZIKV antigen that we found in the GUT after vaginal infection was localized to the uterine walls, the possibility exists that ZIKV sexual transmission may increase risk for the developing fetus during pregnancy. Further studies are needed to address pathogenic mechanisms of ZIKV by different routes of infection and the guinea pig model is well-suited for these endeavors.

Author Contributions: Conceptualization, A.E.S., A.S.B.; methodology, A.S.B.; validation, A.E.S., A.S.B.; formal analysis, A.E.S., J.D.J.; investigation, A.E.S., S.A.C., J.D.J., A.S.B.; data curation, A.S.B.; writing—original draft preparation, A.E.S.; writing—review and editing, A.E.S., S.A.C., J.D.J., A.S.B.; visualization, A.E.S., S.A.C., J.D.J., A.S.B.; supervision, A.S.B.; project administration, A.E.S., A.S.B.; funding acquisition, A.S.B.

Funding: This research received no external funding. AS and SC were participants in the Summer Veterinary Student Research Program and were funded by Virginia-Maryland College of Veterinary Medicine. The article processing charge is partially funded by the Virginia Tech Subvention Fund.

Acknowledgments: Guinea pigs for this study were generously provided by Charles River Laboratories. Special thanks to Hannah Ivester for staining uninfected tissues for Figure 4.

Conflicts of Interest: The authors declare no conflict of interest.

References

1. Dick, G.W. Zika virus. II. Pathogenicity and physical properties. *Trans. R. Soc. Trop Med. Hyg.* **1952**, *46*, 521–534. [CrossRef]

2. Musso, D.; Gubler, D.J. Zika Virus. *Clin. Microbiol Rev.* **2016**, *29*, 487–524. [CrossRef] [PubMed]

3. Duffy, M.R.; Chen, T.H.; Hancock, W.T.; Powers, A.M.; Kool, J.L.; Lanciotti, R.S.; Pretrick, M.; Marfel, M.; Holzbauer, S.; Dubray, C.; et al. Zika virus outbreak on Yap Island, Federated States of Micronesia. *N. Engl. J. Med.* **2009**, *360*, 2536–2543. [CrossRef] [PubMed]

4. Lanciotti, R.S.; Kosoy, O.L.; Laven, J.J.; Velez, J.O.; Lambert, A.J.; Johnson, A.J.; Stanfield, S.M.; Duffy, M.R. Genetic and serologic properties of Zika virus associated with an epidemic, Yap State, Micronesia, 2007. *Emerg. Infect. Dis.* **2008**, *14*, 1232–1239. [CrossRef] [PubMed]

5. Cauchemez, S.; Besnard, M.; Bompard, P.; Dub, T.; Guillemette-Artur, P.; Eyrolle-Guignot, D.; Salje, H.; Van Kerkhove, M.D.; Abadie, V.; Garel, C.; et al. Association between Zika virus and microcephaly in French Polynesia, 2013–2015: A retrospective study. *Lancet* **2016**, *387*, 2125–2132. [CrossRef]

6. Hennessey, M.; Fischer, M.; Staples, J.E. Zika Virus Spreads to New Areas - Region of the Americas, May 2015-January 2016. *Mmwr Morb Mortal Wkly. Rep.* **2016**, *65*, 55–58. [CrossRef] [PubMed]

7. Jouannic, J.M.; Friszer, S.; Leparc-Goffart, I.; Garel, C.; Eyrolle-Guignot, D. Zika virus infection in French Polynesia. *Lancet* **2016**, *387*, 1051–1052. [CrossRef]

8. Wang, Z.; Wang, P.; An, J. Zika virus and Zika fever. *Virol. Sin.* **2016**, *31*, 103–109. [CrossRef]

9. Weaver, S.C.; Costa, F.; Garcia-Blanco, M.A.; Ko, A.I.; Ribeiro, G.S.; Saade, G.; Shi, P.Y.; Vasilakis, N. Zika virus: History, emergence, biology, and prospects for control. *Antivir. Res.* **2016**, *130*, 69–80. [CrossRef]

10. Grischott, F.; Puhan, M.; Hatz, C.; Schlagenhauf, P. Non-vector-borne transmission of Zika virus: A systematic review. *Travel Med. Infect. Dis* **2016**, *14*, 313–330. [CrossRef]

11. Colt, S.; Garcia-Casal, M.N.; Pena-Rosas, J.P.; Finkelstein, J.L.; Rayco-Solon, P.; Weise Prinzo, Z.C.; Mehta, S. Transmission of Zika virus through breast milk and other breastfeeding-related bodily-fluids: A systematic review. *PLoS Negl. Trop Dis.* **2017**, *11*, e0005528. [CrossRef] [PubMed]

12. Gregory, C.J.; Oduyebo, T.; Brault, A.C.; Brooks, J.T.; Chung, K.W.; Hills, S.; Kuehnert, M.J.; Mead, P.; Meaney-Delman, D.; Rabe, I.; et al. Modes of Transmission of Zika Virus. *J. Infect. Dis.* **2017**, *216*, S875–S883. [CrossRef] [PubMed]

13. Turtle, L.; Griffiths, M.J.; Solomon, T. Encephalitis caused by flaviviruses. *QJM* **2012**, *105*, 219–223. [CrossRef] [PubMed]

14. Shuaib, W.; Stanazai, H.; Abazid, A.G.; Mattar, A.A. Re-Emergence of Zika Virus: A Review on Pathogenesis, Clinical Manifestations, Diagnosis, Treatment, and Prevention. *Am. J. Med.* **2016**, *129*, 879 e877–879 e812. [CrossRef]

15. Goodfellow, F.T.; Willard, K.A.; Wu, X.; Scoville, S.; Stice, S.L.; Brindley, M.A. Strain-Dependent Consequences of Zika Virus Infection and Differential Impact on Neural Development. *Viruses* **2018**, *10*. [CrossRef]

16. Peloggia, A.; Ali, M.; Nanda, K.; Bahamondes, L. Zika virus exposure in pregnancy and its association with newborn visual anomalies and hearing loss. *Int. J. Gynaecol. Obs.* **2018**, *143*, 277–281. [CrossRef]

17. Bhatnagar, J.; Rabeneck, D.B.; Martines, R.B.; Reagan-Steiner, S.; Ermias, Y.; Estetter, L.B.; Suzuki, T.; Ritter, J.; Keating, M.K.; Hale, G.; et al. Zika Virus RNA Replication and Persistence in Brain and Placental Tissue. *Emerg. Infect. Dis.* **2017**, *23*, 405–414. [CrossRef]

18. Moreira, J.; Peixoto, T.M.; Siqueira, A.M.; Lamas, C.C. Sexually acquired Zika virus: A systematic review. *Clin. Microbiol Infect.* **2017**, *23*, 296–305. [CrossRef]

19. Reyes, Y.; Bowman, N.M.; Becker-Dreps, S.; Centeno, E.; Collins, M.H.; Liou, G.A.; Bucardo, F. Prolonged Shedding of Zika Virus RNA in Vaginal Secretions, Nicaragua. *Emerg Infect. Dis.* **2019**, *25*, 808–810. [CrossRef]

20. Davidson, A.; Slavinski, S.; Komoto, K.; Rakeman, J.; Weiss, D. Suspected Female-to-Male Sexual Transmission of Zika Virus - New York City, 2016. *Mmwr. Morb. Mortal. Wkly. Rep.* **2016**, *65*, 716–717. [CrossRef]

21. Deckard, D.T.; Chung, W.M.; Brooks, J.T.; Smith, J.C.; Woldai, S.; Hennessey, M.; Kwit, N.; Mead, P. Male-to-Male Sexual Transmission of Zika Virus–Texas, January 2016. *Mmwr. Morb. Mortal. Wkly. Rep.* **2016**, *65*, 372–374. [CrossRef] [PubMed]

22. Hills, S.L.; Russell, K.; Hennessey, M.; Williams, C.; Oster, A.M.; Fischer, M.; Mead, P. Transmission of Zika Virus Through Sexual Contact with Travelers to Areas of Ongoing Transmission - Continental United States, 2016. *Mmw Mor Morta Wkly. Rep.* **2016**, *65*, 215–216. [CrossRef] [PubMed]

23. Matheron, S.; d'Ortenzio, E.; Leparc-Goffart, I.; Hubert, B.; de Lamballerie, X.; Yazdanpanah, Y. Long-Lasting Persistence of Zika Virus in Semen. *Clin. Infect. Di* **2016**, *63*, 1264. [CrossRef] [PubMed]

24. Turmel, J.M.; Abgueguen, P.; Hubert, B.; Vandamme, Y.M.; Maquart, M.; Le Guillou-Guillemette, H.; Leparc-Goffart, I. Late sexual transmission of Zika virus related to persistence in the semen. *Lancet* **2016**, *387*, 2501. [CrossRef]

25. Visseaux, B.; Mortier, E.; Houhou-Fidouh, N.; Brichler, S.; Collin, G.; Larrouy, L.; Charpentier, C.; Descamps, D. Zika virus in the female genital tract. *Lancet Infect. Di* **2016**, *16*, 1220. [CrossRef]

26. Prisant, N.; Breurec, S.; Moriniere, C.; Bujan, L.; Joguet, G. Zika Virus Genital Tract Shedding in Infected Women of Childbearing age. *Clin. Infect. Di* **2017**, *64*, 107–109. [CrossRef]

27. Counotte, M.J.; Kim, C.R.; Wang, J.; Bernstein, K.; Deal, C.D.; Broutet, N.J.N.; Low, N. Sexual transmission of Zika virus and other flaviviruses: A living systematic review. *PLoS Med.* **2018**, *15*, e1002611. [CrossRef]

28. Penot, P.; Brichler, S.; Guilleminot, J.; Lascoux-Combe, C.; Taulera, O.; Gordien, E.; Leparc-Goffart, I.; Molina, J.M. Infectious Zika virus in vaginal secretions from an HIV-infected woman, France, August 2016. *Euro. Surveill.* **2017**, *22*. [CrossRef]

29. Arsuaga, M.; Bujalance, S.G.; Diaz-Menendez, M.; Vazquez, A.; Arribas, J.R. Probable sexual transmission of Zika virus from a vasectomised man. *Lancet Infect. Dis.* **2016**, *16*, 1107. [CrossRef]

30. Duggal, N.K.; Ritter, J.M.; Pestorius, S.E.; Zaki, S.R.; Davis, B.S.; Chang, G.J.; Bowen, R.A.; Brault, A.C. Frequent Zika Virus Sexual Transmission and Prolonged Viral RNA Shedding in an Immunodeficient Mouse Model. *Cell Rep.* **2017**, *18*, 1751–1760. [CrossRef]

31. Froeschl, G.; Huber, K.; von Sonnenburg, F.; Nothdurft, H.D.; Bretzel, G.; Hoelscher, M.; Zoeller, L.; Trottmann, M.; Pan-Montojo, F.; Dobler, G.; et al. Long-term kinetics of Zika virus RNA and antibodies in body fluids of a vasectomized traveller returning from Martinique: A case report. *Bmc Infect. Dis.* **2017**, *17*, 55. [CrossRef] [PubMed]

32. Swartwout, B.K.; Zlotnick, M.G.; Saver, A.E.; McKenna, C.M.; Bertke, A.S. Zika Virus Persistently and Productively Infects Primary Adult Sensory Neurons In Vitro. *Pathogens* **2017**, *6*. [CrossRef] [PubMed]

33. Pena, L.J.; Miranda Guarines, K.; Duarte Silva, A.J.; Sales Leal, L.R.; Mendes Felix, D.; Silva, A.; de Oliveira, S.A.; Junqueira Ayres, C.F.; Junior, A.S.; de Freitas, A.C. In vitro and in vivo models for studying Zika virus biology. *J. Gen. Virol.* **2018**, *99*, 1529–1550. [CrossRef] [PubMed]

34. Lazear, H.M.; Govero, J.; Smith, A.M.; Platt, D.J.; Fernandez, E.; Miner, J.J.; Diamond, M.S. A Mouse Model of Zika Virus Pathogenesis. *Cell Host Microbe* **2016**, *19*, 720–730. [CrossRef] [PubMed]

35. Rossi, S.L.; Tesh, R.B.; Azar, S.R.; Muruato, A.E.; Hanley, K.A.; Auguste, A.J.; Langsjoen, R.M.; Paessler, S.; Vasilakis, N.; Weaver, S.C. Characterization of a Novel Murine Model to Study Zika Virus. *Am. J. Trop Med. Hyg.* **2016**, *94*, 1362–1369. [CrossRef] [PubMed]

36. Fernandes, N.C.; Nogueira, J.S.; Ressio, R.A.; Cirqueira, C.S.; Kimura, L.M.; Fernandes, K.R.; Cunha, M.S.; Souza, R.P.; Guerra, J.M. Experimental Zika virus infection induces spinal cord injury and encephalitis in newborn Swiss mice. *Exp. Toxicol. Pathol.* **2017**, *69*, 63–71. [CrossRef] [PubMed]

37. Dowall, S.D.; Graham, V.A.; Rayner, E.; Atkinson, B.; Hall, G.; Watson, R.J.; Bosworth, A.; Bonney, L.C.; Kitchen, S.; Hewson, R. A Susceptible Mouse Model for Zika Virus Infection. *PLoS Negl. Trop Dis.* **2016**, *10*, e0004658. [CrossRef]

38. Aliota, M.T.; Caine, E.A.; Walker, E.C.; Larkin, K.E.; Camacho, E.; Osorio, J.E. Characterization of Lethal Zika Virus Infection in AG129 Mice. *PLoS Negl. Trop Dis.* **2016**, *10*, e0004682. [CrossRef]

39. Smith, D.R.; Hollidge, B.; Daye, S.; Zeng, X.; Blancett, C.; Kuszpit, K.; Bocan, T.; Koehler, J.W.; Coyne, S.; Minogue, T.; et al. Neuropathogenesis of Zika Virus in a Highly Susceptible Immunocompetent Mouse Model after Antibody Blockade of Type I Interferon. *PLoS Negl. Trop Dis.* **2017**, *11*, e0005296. [CrossRef]

40. Tang, W.W.; Young, M.P.; Mamidi, A.; Regla-Nava, J.A.; Kim, K.; Shresta, S. A Mouse Model of Zika Virus Sexual Transmission and Vaginal Viral Replication. *Cell Rep.* **2016**, *17*, 3091–3098. [CrossRef]

41. Holsapple, M.P.; West, L.J.; Landreth, K.S. Species comparison of anatomical and functional immune system development. *Birth Defects Res. B Dev. Reprod Toxicol.* **2003**, *68*, 321–334. [CrossRef] [PubMed]

42. Li, S.; Armstrong, N.; Zhao, H.; Hou, W.; Liu, J.; Chen, C.; Wan, J.; Wang, W.; Zhong, C.; Liu, C.; et al. Zika Virus Fatally Infects Wild Type Neonatal Mice and Replicates in Central Nervous System. *Viruses* **2018**, *10*. [CrossRef] [PubMed]

43. Manangeeswaran, M.; Ireland, D.D.; Verthelyi, D. Zika (PRVABC59) Infection Is Associated with T cell Infiltration and Neurodegeneration in CNS of Immunocompetent Neonatal C57Bl/6 Mice. *Plos Pathog.* **2016**, *12*, e1006004. [CrossRef] [PubMed]

44. Osuna, C.E.; Lim, S.Y.; Deleage, C.; Griffin, B.D.; Stein, D.; Schroeder, L.T.; Omange, R.W.; Best, K.; Luo, M.; Hraber, P.T.; et al. Zika viral dynamics and shedding in rhesus and cynomolgus macaques. *Nat. Med.* **2016**, *22*, 1448–1455. [CrossRef] [PubMed]

45. Dudley, D.M.; Aliota, M.T.; Mohr, E.L.; Weiler, A.M.; Lehrer-Brey, G.; Weisgrau, K.L.; Mohns, M.S.; Breitbach, M.E.; Rasheed, M.N.; Newman, C.M.; et al. A rhesus macaque model of Asian-lineage Zika virus infection. *Nat. Commun.* **2016**, *7*, 12204. [CrossRef] [PubMed]

46. Dudley, D.M.; Newman, C.M.; Lalli, J.; Stewart, L.M.; Koenig, M.R.; Weiler, A.M.; Semler, M.R.; Barry, G.L.; Zarbock, K.R.; Mohns, M.S.; et al. Infection via mosquito bite alters Zika virus tissue tropism and replication kinetics in rhesus macaques. *Nat. Commun.* **2017**, *8*, 2096. [CrossRef]

47. Coffey, L.L.; Pesavento, P.A.; Keesler, R.I.; Singapuri, A.; Watanabe, J.; Watanabe, R.; Yee, J.; Bliss-Moreau, E.; Cruzen, C.; Christe, K.L.; et al. Zika Virus Tissue and Blood Compartmentalization in Acute Infection of Rhesus Macaques. *PLoS ONE* **2017**, *12*, e0171148. [CrossRef]

48. Aid, M.; Abbink, P.; Larocca, R.A.; Boyd, M.; Nityanandam, R.; Nanayakkara, O.; Martinot, A.J.; Moseley, E.T.; Blass, E.; Borducchi, E.N.; et al. Zika Virus Persistence in the Central Nervous System and Lymph Nodes of Rhesus Monkeys. *Cell* **2017**, *169*, 610–620 e614. [CrossRef]

49. Darbellay, J.; Lai, K.; Babiuk, S.; Berhane, Y.; Ambagala, A.; Wheler, C.; Wilson, D.; Walker, S.; Potter, A.; Gilmour, M.; et al. Neonatal pigs are susceptible to experimental Zika virus infection. *Emerg. Microbes Infect.* **2017**, *6*, e6. [CrossRef]

50. Ragan, I.K.; Blizzard, E.L.; Gordy, P.; Bowen, R.A. Investigating the Potential Role of North American Animals as Hosts for Zika Virus. *Vector Borne Zoonotic Dis.* **2017**, *17*, 161–164. [CrossRef]

51. Padilla-Carlin, D.J.; McMurray, D.N.; Hickey, A.J. The guinea pig as a model of infectious diseases. *Comp. Med.* **2008**, *58*, 324–340. [PubMed]

52. Bertke, A.S.; Patel, A.; Imai, Y.; Apakupakul, K.; Margolis, T.P.; Krause, P.R. Latency-associated transcript (LAT) exon 1 controls herpes simplex virus species-specific phenotypes: Reactivation in the guinea pig genital model and neuron subtype-specific latent expression of LAT. *J. Virol.* **2009**, *83*, 10007–10015. [CrossRef] [PubMed]

53. Holbrook, M.R.; Gowen, B.B. Animal models of highly pathogenic RNA viral infections: Encephalitis viruses. *Antivir. Res.* **2008**, *78*, 69–78. [CrossRef] [PubMed]

54. Griffith, B.P.; Aquino-de Jesus, M.J. Guinea pig model of congenital cytomegalovirus infection. *Transpl. Proc.* **1991**, *23*, 29–31, discussion 31.

55. Schleiss, M.R. Nonprimate models of congenital cytomegalovirus (CMV) infection: Gaining insight into pathogenesis and prevention of disease in newborns. *Ilar J.* **2006**, *47*, 65–72. [CrossRef]

56. Kumar, M.; Krause, K.K.; Azouz, F.; Nakano, E.; Nerurkar, V.R. A guinea pig model of Zika virus infection. *Virol J.* **2017**, *14*, 75. [CrossRef]

57. Deng, Y.Q.; Zhang, N.N.; Li, X.F.; Wang, Y.Q.; Tian, M.; Qiu, Y.F.; Fan, J.W.; Hao, J.N.; Huang, X.Y.; Dong, H.L.; et al. Intranasal infection and contact transmission of Zika virus in guinea pigs. *Nat. Commun.* **2017**, *8*, 1648. [CrossRef]

58. Miller, L.J.; Nasar, F.; Schellhase, C.W.; Norris, S.L.; Kimmel, A.E.; Valdez, S.M.; Wollen-Roberts, S.E.; Shamblin, J.D.; Sprague, T.R.; Lugo-Roman, L.A.; et al. Zika Virus Infection in Syrian Golden Hamsters and Strain 13 Guinea Pigs. *Am. J. Trop Med. Hyg* **2018**, *98*, 864–867. [CrossRef]

59. Bierle, C.J.; Fernandez-Alarcon, C.; Hernandez-Alvarado, N.; Zabeli, J.C.; Janus, B.C.; Putri, D.S.; Schleiss, M.R. Assessing Zika virus replication and the development of Zika-specific antibodies after a mid-gestation viral challenge in guinea pigs. *PLoS ONE* **2017**, *12*, e0187720. [CrossRef]

60. Vicenti, I.; Boccuto, A.; Giannini, A.; Dragoni, F.; Saladini, F.; Zazzi, M. Comparative analysis of different cell systems for Zika virus (ZIKV) propagation and evaluation of anti-ZIKV compounds in vitro. *Virus Res.* **2018**, *244*, 64–70. [CrossRef]

61. Freppel, W.; Mazeaud, C.; Chatel-Chaix, L. Production, Titration ad Imaging of Zika Virus in Mammalian Cells. *Bio-Protocl* **2018**, *8*. [CrossRef]

62. Miner, J.J.; Sene, A.; Richner, J.M.; Smith, A.M.; Santeford, A.; Ban, N.; Weger-Lucarelli, J.; Manzella, F.; Ruckert, C.; Govero, J.; et al. Zika Virus Infection in Mice Causes Panuveitis with Shedding of Virus in Tears. *Cell Rep.* **2016**, *16*, 3208–3218. [CrossRef] [PubMed]

63. Murray, K.O.; Gorchakov, R.; Carlson, A.R.; Berry, R.; Lai, L.; Natrajan, M.; Garcia, M.N.; Correa, A.; Patel, S.M.; Aagaard, K.; et al. Prolonged Detection of Zika Virus in Vaginal Secretions and Whole Blood. *Emerg Infect. Dis.* **2017**, *23*, 99–101. [CrossRef] [PubMed]

64. Zhao, Z.; Yang, M.; Azar, S.R.; Soong, L.; Weaver, S.C.; Sun, J.; Chen, Y.; Rossi, S.L.; Cai, J. Viral Retinopathy in Experimental Models of Zika Infection. *Invest. Ophthalmol Vis. Sci.* **2017**, *58*, 4355–4365. [CrossRef]

65. Fernandez, M.P.; Parra Saad, E.; Ospina Martinez, M.; Corchuelo, S.; Mercado Reyes, M.; Herrera, M.J.; Parra Saavedra, M.; Rico, A.; Fernandez, A.M.; Lee, R.K.; et al. Ocular Histopathologic Features of Congenital Zika Syndrome. *Jama Ophthalmol.* **2017**, *135*, 1163–1169. [CrossRef]

66. Hirsch, A.J.; Roberts, V.H.J.; Grigsby, P.L.; Haese, N.; Schabel, M.C.; Wang, X.; Lo, J.O.; Liu, Z.; Kroenke, C.D.; Smith, J.L.; et al. Zika virus infection in pregnant rhesus macaques causes placental dysfunction and immunopathology. *Nat. Commun.* **2018**, *9*, 263. [CrossRef]

67. Clancy, C.S.; Van Wettere, A.J.; Siddharthan, V.; Morrey, J.D.; Julander, J.G. Comparative Histopathologic Lesions of the Male Reproductive Tract during Acute Infection of Zika Virus in AG129 and Ifnar(-/-) Mice. *Am. J. Pathol.* **2018**, *188*, 904–915. [CrossRef]

68. Winkler, C.W.; Woods, T.A.; Rosenke, R.; Scott, D.P.; Best, S.M.; Peterson, K.E. Sexual and Vertical Transmission of Zika Virus in anti-interferon receptor-treated Rag1-deficient mice. *Sci. Rep.* **2017**, *7*, 7176. [CrossRef]

69. Uraki, R.; Hwang, J.; Jurado, K.A.; Householder, S.; Yockey, L.J.; Hastings, A.K.; Homer, R.J.; Iwasaki, A.; Fikrig, E. Zika virus causes testicular atrophy. *Sci. Adv.* **2017**, *3*, e1602899. [CrossRef]

70. Sheng, Z.Y.; Gao, N.; Wang, Z.Y.; Cui, X.Y.; Zhou, D.S.; Fan, D.Y.; Chen, H.; Wang, P.G.; An, J. Sertoli Cells Are Susceptible to ZIKV Infection in Mouse Testis. *Front. Cell Infect. Microbiol.* **2017**, *7*, 272. [CrossRef]

71. Lee, H.H.; Hong, S.K.; Yoon, S.H.; Jang, S.J.; Baik, Y.Y.; Song, M.D.; Park, PJ; Lee, K.H.; Kim, C.G.; Kim, B.; et al. Immunogenicity of Japanese encephalitis virus envelope protein by Hyphantria cunea nuclear polyhedrosis virus vector in guinea pig. *Appl. Biochem. Biotechnol.* **2012**, *167*, 259–269. [CrossRef] [PubMed]

72. Liu, X.Y.; Yu, Y.X.; Xu, H.S.; Liang, G.D.; Wang, H.Y.; Jia, L.L.; Dong, G.M. [Comparison of viremia formation between guinea-pigs infected with wild and attenuated (SA14-14-2) Japanese encephalitis viruses]. *Clin. Exp. Vaccine Res.* **2010**, *24*, 343–345.

73. Ryu, E. Attenuation of Japanese encephalitis virus by brain passage of guinea pigs. *Int. J. Zoonoses* **1976**, *3*, 145–150. [PubMed]

74. Stanberry, L.R. Evaluation of herpes simplex virus vaccines in animals: The guinea pig vaginal model. *Rev. Infect. Dis* **1991**, *13* (Suppl. 11), S920–923. [CrossRef]

75. Moghadas, S.M.; Shoukat, A.; Espindola, A.L.; Pereira, R.S.; Abdirizak, F.; Laskowski, M.; Viboud, C.; Chowell, G. Asymptomatic Transmission and the Dynamics of Zika Infection. *Sci. Rep.* **2017**, *7*, 5829. [CrossRef]

76. Haby, M.M.; Pinart, M.; Elias, V.; Reveiz, L. Prevalence of asymptomatic Zika virus infection: A systematic review. *Bull. World Health Organ.* **2018**, *96*, 402–413D. [CrossRef]

77. Flamand, C.; Fritzell, C.; Matheus, S.; Dueymes, M.; Carles, G.; Favre, A.; Enfissi, A.; Adde, A.; Demar, M.; Kazanji, M.; et al. The proportion of asymptomatic infections and spectrum of disease among pregnant women infected by Zika virus: Systematic monitoring in French Guiana, 2016. *Euro. Surveill* **2017**, *22*. [CrossRef]

78. Song, B.H.; Yun, S.I.; Woolley, M.; Lee, Y.M. Zika virus: History, epidemiology, transmission, and clinical presentation. *J. Neuroimmunol.* **2017**, *308*, 50–64. [CrossRef]

79. Mansuy, J.M.; Mengelle, C.; Pasquier, C.; Chapuy-Regaud, S.; Delobel, P.; Martin-Blondel, G.; Izopet, J. Zika Virus Infection and Prolonged Viremia in Whole-Blood Specimens. *Emerg Infect. Dis.* **2017**, *23*, 863–865. [CrossRef]

80. Paz-Bailey, G.; Rosenberg, E.S.; Doyle, K.; Munoz-Jordan, J.; Santiago, G.A.; Klein, L.; Perez-Padilla, J.; Medina, F.A.; Waterman, S.H.; Gubern, C.G.; et al. Persistence of Zika Virus in Body Fluids - Final Report. *N. Engl. J. Med.* **2018**, *379*, 1234–1243. [CrossRef]

81. Koide, F.; Goebel, S.; Snyder, B.; Walters, K.B.; Gast, A.; Hagelin, K.; Kalkeri, R.; Rayner, J. Development of a Zika Virus Infection Model in Cynomolgus Macaques. *Front. Microbiol.* **2016**, *7*, 2028. [CrossRef] [PubMed]

82. Tan, J.J.L.; Balne, P.K.; Leo, Y.S.; Tong, L.; Ng, L.F.P.; Agrawal, R. Persistence of Zika virus in conjunctival fluid of convalescence patients. *Sci. Rep.* **2017**, *7*, 11194. [CrossRef] [PubMed]

83. Swaminathan, S.; Schlaberg, R.; Lewis, J.; Hanson, K.E.; Couturier, M.R. Fatal Zika Virus Infection with Secondary Nonsexual Transmission. *N. Engl. J. Med.* **2016**, *375*, 1907–1909. [CrossRef] [PubMed]

 *cells*

Article

The ZIKA Virus Delays Cell Death Through the Anti-Apoptotic Bcl-2 Family Proteins

Jonathan Turpin, Etienne Frumence, Philippe Desprès, Wildriss Viranaicken * and Pascale Krejbich-Trotot *

PIMIT, Processus Infectieux en Milieu Insulaire Tropical, Université de La Réunion, INSERM UMR 1187, CNRS 9192, IRD 249, Plateforme CYROI, 97490 Sainte-Clotilde, Ile de La Réunion, France; jonathan.turpin@univ-reunion.fr (J.T.); etiennefrum@gmail.com (E.F.); philippe.despres@univ-reunion.fr (P.D.)
* Correspondence: wildriss.viranaicken@univ-reunion.fr (W.V.); pascale.krejbich@univ-reunion.fr (P.K-T.);
  Tel.: +33-262938829 (W.V.)

Received: 16 September 2019; Accepted: 26 October 2019; Published: 29 October 2019

Abstract: Zika virus (ZIKV) is an emerging human mosquito-transmitted pathogen of global concern, known to be associated with complications such as congenital defects and neurological disorders in adults. ZIKV infection is associated with induction of cell death. However, previous studies suggest that the virally induced apoptosis occurs at a slower rate compared to the course of viral production. In this present study, we investigated the capacity of ZIKV to delay host cell apoptosis. We provide evidence that ZIKV has the ability to interfere with apoptosis whether it is intrinsically or extrinsically induced. In cells expressing viral replicon-type constructions, we show that this control is achieved through replication. Finally, our work highlights an important role for anti-apoptotic Bcl-2 family protein in the ability of ZIKV to control apoptotic pathways, avoiding premature cell death and thereby promoting virus replication in the host-cell.

Keywords: Zika virus; apoptosis; viral replication; Bcl-2 protein family

---

1. Introduction

Zika virus (ZIKV), which is a flavivirus belonging to the Flaviviridae family, like Dengue virus (DENV), yellow fever virus (YFV) or West Nile virus (WNV), has become a major medical problem worldwide. The human disease known as Zika fever is characterized by mild flu-like symptoms including fever, maculopapular rash, headache and sometimes conjunctivitis, arthralgia and myalgia. Symptoms usually subside within a week [1]. However, during the latest outbreaks, serious pathological features of the disease have been reported. Complications such as microcephaly in newborns or Guillain–Barré syndrome (GBS) in adults were documented during the French Polynesia outbreak in 2013 and in Brazil in 2015 [2,3]. ZIKV is an arbovirus, mainly transmitted to humans through the bite of a mosquito vector from *Aedes* species [4]. Due to an increasingly global distribution of *Aedes*, ZIKV emergence is a threat in many areas that are no longer necessarily located in intertropical areas [5]. The ZIKV particle is composed of a single strand RNA molecule of about 11 kb, inside a nucleocapsid, surrounded by a host-derived membrane that contains two virus encoded proteins (E and M). Phylogenetic analysis of viral sequences has identified two main virus lineages, African and Asian [6], the latter being the main cause of large current epidemics with millions of cases of infection, in particular those that recently affected Brazil and the Americas [7]. Once ZIKV has entered the human body, it targets many types of cells such as epithelial cells, in order to replicate and produce a viral progeny. The life cycle of ZIKV, like other flaviviruses, leads to the release of its single-strand positive sense genomic RNA in the cell cytoplasm where it is translated into a single polyprotein, which is then cleaved by host and viral proteases into three structural proteins (C, prM/M and E), and seven nonstructural proteins (NS; NS1, NS2A, NS2B, NS3, NS4A, NS4B and NS5) [8]. The viral cycle continues with the replication process and the production and maturation of envelope proteins, encapsidation, budding and release of virions by exocytosis.

As with many other viruses, interactions between ZIKV and its host trigger a variety of host responses in the body's attempt to resolve the infection [9]. Among these responses, apoptosis plays an important role as a host defense mechanism [10]. Apoptosis can quickly remove intracellular niches of viral replication and thus bypass the virus as it multiplies and spreads. As a result, many viruses have developed strategies to evade, delay or divert the cell death responses, often to their advantage [11]. One example is the case of chikungunya virus (CHIKV), an arbovirus of the alphavirus family, which takes advantage of massive apoptosis to hide in disseminating blebs and thus optimizes its spread [12]. Typically, alphaviruses such as CHIKV (but also Sindbis virus, Ross River virus and Semliki Virus) replicate extremely quickly and the infected cells are characterized by rapid and concomitant apoptosis [13]. Unlike alphaviruses, flaviviruses replication is relatively slow. For several of them such as DENV, Japanese encephalitis virus (JEV) and WNV, apoptosis has been shown to be inhibited during the early stages of the viral cycle [14]. The role of viral proteins in the control of apoptosis has been extensively studied, with many observations in support of both pro-apoptotic activity and antiapoptotic effects. For WNV, a nuclear localization of the capsid was shown to induce a caspase-9-dependent apoptosis [15]. Whereas the WNV capsid protein was shown to suppress the activation of caspases 3 and 8 via Akt through a phosphatidylinositol 3-kinase-dependent mechanism (PI3K) [16]. Concerning ZIKV, in cellulo models have shown that infection can lead to cytopathic effects that are typical of apoptosis, and in previous work we observed late-onset apoptosis 48 h after infection of A549 cells with ZIKV isolate PF13 [17]. In some other cell types (human fetal astrocytes), moderate apoptosis can occur even later and possibly contribute to persistent infection [18]. In the particular case of Zika pathology, homeostasis disorder; involving a lack of apoptosis control, a persistent inflammatory response and even viral persistence in the brain has been reported to explain the microcephaly observed in infected newborns [19].

In this study we examined the time course of cellular death associated with ZIKV infection. We confirm that, in A549 cells infected with the epidemic strain from Asian lineage (BeH819015, BR15^{MC}), apoptosis is quantitatively moderate and occurs late, after the maximum production of viral progeny. We investigated whether this delay was due to a protective effect of the virus itself. When intrinsic or extrinsic apoptosis was induced within 2 h after infection, we could observe a significant decrease in cell death. As this protection was also obtained in cells expressing ZIKV "replicons", we deduced that viral replication was efficient at inhibiting apoptosis. ABT-737, an inhibitor of the anti-apoptotic Bcl-2 family proteins, abrogates the protective effects provided by ZIKV. This implies that, with a subversion mechanism that remains to be elucidated, ZIKV is able to maintain an anti-apoptotic status in infected cells while it completes its viral cycle.

2. Materials and Methods

2.1. Viruses, Cell Lines, Antibodies and Reagents

The clinical isolate PF-25013-18 (PF13) and the molecular clones of ZIKV (BR15^{MC} and MR766^{MC}) have been previously described [20,21]. Vero cells (ATCC, CCL-81) and HEK 293T (CRL-3216) were cultured at 37 °C under a 5% CO_2 atmosphere in MEM medium (PAN Biotech, Aidenbach, Germany), supplemented with 5% or 10% heat-inactivated fetal bovine serum (FBS) (PAN Biotech, Aidenbach, Germany). A549-Dual™ cells (InvivoGen, Toulouse, France, a549d-nfis) designated hereafter as A549 cells were maintained in MEM medium supplemented with 10% heat-inactivated FBS. A549 cells were maintained in growth medium supplemented with 10 µg.mL^{-1} blasticidin and 100 µg.mL^{-1} zeocin (InvivoGen, Toulouse, France).

To detect the ZIKV envelope E protein (ZIKV-E), we used the mouse anti-pan flavivirus envelope E protein mAb 4G2 produced by RD Biotech or the rabbit anti-ZIKV-E-DIII that was kindly provided by Dr. Valerie Choumet (Institut Pasteur). The rabbit anti-BAX (#2772) and anti-active caspase 3

(#9664) antibodies and the mouse anti-cytc 6H2.B4 (#12963) were purchased from Cell Signalling Technology (Ozyme, Saint-Cyr-l'École, France). The mouse anti-BAX clone 2D2 was from Invitrogen (Thermofisher, Les Ulis, France). The anti-mitochondria antibody MTC02-3298 was from Abcam (Cambridge, UK). Donkey anti-mouse Alexa Fluor 488 and anti-rabbit Alexa Fluor 594 IgG antibodies were from Invitrogen (Thermofisher, Les Ulis, France). Horseradish peroxidase-conjugated anti-rabbit (ab97051) and anti-mouse (ab6789) antibodies were purchased from Abcam (Cambridge, UK).

All reagents were from Sigma Aldrich (Humeau, La Chapelle-Sur-Erdre, France) except when indicated. ABT-737 (ab141336) was purchased from Abcam (Cambridge, UK).

2.2. Plaque Forming Assay

Viral titers were determined by a standard plaque-forming assay as previously described with minor modifications [17]. Briefly, Vero cells grown in 48-well culture plates were infected with tenfold dilutions of virus samples for 2 h at 37 °C and then incubated with 0.8% carboxymethylcellulose (CMC) for 4 days. The cells were fixed with 3.7% formaldehyde in PBS and stained with 0.5% crystal violet in 20% ethanol. Viral titers were expressed as plaque-forming units per mL (PFU.mL^{-1}).

2.3. Western Blotting (WB)

Cell lysates were performed in Radioimmunoprecipitation assay buffer (RIPA buffer). All subsequent steps of immunoblotting followed previous descriptions [22,23]. Primary antibodies were used at 1:1000 dilutions. Anti-rabbit immunoglobulin-horseradish peroxidase and anti-mouse immunoglobulin-horseradish peroxidase conjugates were used as secondary antibody (dilution 1:2000). Blots were revealed with enhanced chemiluminescence (ECL) detection reagents using an Amersham Imager 680 (GE, Buc, France).

2.4. Induction of Apoptosis

Apoptosis inducers were added 2 h post-infection (hpi) or 2 h before infection (hbi) for ZIKV infected cells. For the replicons, cells were treated 24 h after transfection or passage of stable cells.

For extrinsic apoptosis, cells were treated 2 hpi or 2 hbi with TNFα (10 ng.mL^{-1}) and cycloheximide (10 µg.mL^{-1}) (TNFα/CHX). The addition of cycloheximide prevents the activation of NFkB and the inhibition of apoptosis [24]. Drugs were added as indicated in the figure legends, before quantification of cell death.

For intrinsic apoptosis, etoposide at 10 µM or blasticidin at 5 µg.mL^{-1} (InvivoGen, Toulouse, France) were added 16 h to 18 h as indicated in the legends. Alternatively, cells were treated with a dose of 400 mJ of UV (Uvitec, Cambridge, UK) and collected for death measurements 16 h after treatment.

For inhibition of the anti-apoptotic Bcl-2 family proteins, ABT-737 (1 µM) was added at the same time as TNF/CHX and the cells were treated as above. Previous experiments were set-up to show that ABT-737 alone under these conditions did not induce cell death (Figure 8B).

2.5. Immunofluorescence Assay

A549 and HEK cells grown, infected and treated on glass coverslips were fixed with 3.7% formaldehyde at room temperature for 10 min. Fixed cells were permeabilized with 0.1% Triton X-100 in PBS for 4 min. Coverslips were incubated with primary antibodies (1:1000 dilution) in 1% PBS BSA. Antigen staining was visualized with Alexa Fluor-conjugated secondary antibodies (1:1000, Invitrogen). Nucleus morphology was revealed by DAPI staining. The coverslips were mounted with VECTASHIELD®(Clinisciences, Nanterre, France) and fluorescence was observed using a Nikon Eclipse E2000-U microscope. Images were captured and processed using a Hamamatsu ORCA2 ER camera and the imaging software NIS-Element AR (Nikon, Tokyo, Japan). From the immunofluorescence imaging, percentage of ZIKV-E-positive cells with a stained active mitochondrial BAX or cytosolic cytochrome-c or cleaved caspase 3 was estimated after counting a minimum of six microscopic fields i.e., about 1000 cells examined for each condition.

2.6. Cytotoxicity Assay

Necrotic cell damage was evaluated measuring lactate dehydrogenase (LDH) release resulting from a plasma membrane rupture. The supernatant of infected cells was recovered and subjected to a cytotoxicity assay, performed using the CytoTox 96®non-radioactive cytotoxicity assay (Promega, Madison, USA) according to manufacturer's instructions. Absorbance of converted dye was measured at 490 nm using a microplate reader (Tecan, Grödig, Austria).

2.7. Cell Viability Assay (MTT)

MTT (3-[4,5-dimethylthiazol-2-yl]-2,5-diphenyltetrazolium bromide) at 5 mg/mL^{-1} was added on A549 cells cultured in 96-well plate at a density of 5000 cells per well. Following 1 h incubation, MTT medium was removed and the insoluble formazan was solubilized with 100 μL of DMSO. Absorbance of converted dye was measured at 570 nm with a background subtraction at 690 nm.

2.8. Caspase-3/7 Activity

Cells were cultured in 96-well plate at a density of 5×10^3 cells per well. Caspase 3/7 activity in crude cell lysates was measured using the Caspase Glo® 3/7 Assay Kit (Promega, Madison, USA) according to the manufacturer's protocols. Caspase activity was quantified by luminescence using a FLUOstar Omega Microplate Reader (BMG LABTECH, Orthenberg, Germany).

2.9. Generation of ZIKV Replicon by the ISA Method

The production of HEK 293T expressing a stable ZIKV RNA replicon with GFP as a reporter protein, named Rep ZIKV-GFP in the study, was based on the sequence of ZIKV strain MR766 Uganda 47-NIID (Genbank access # LC002520) and the ISA (infectious subgenomic amplicons) method as described previously [25]. As a negative control, HEK 293T cells were transfected with pSilencer-puro 2.6 (Ambion, Thermofisher, Les Ulis, France) and pEGFP-C1 (Clontech, Ozyme, Saint-Cyr-l'École, France) plasmids with a ratio of 1 to 10 using lipofectamine 3000 according to supplier's instructions (Thermofisher, Les Ulis, France) and selected for 5 days in puromycin at 1 μg.mL^{-1}.

The production of A549 cells transiently expressing a ZIKV RNA replicon was based on the ISA method and four amplicons overlapping the sequences of BeH819015 isolated in Brazil in 2015 [25] (Figure 7A). Cells were electroporated in the presence of the viral amplicons (Z1 to Z4 fragments) using the Gene pulser II apparatus according to supplier's instructions (Biorad, Marnes-la-Coquette, France) and treated within 2 days with apoptotic inducers. Cell controls in these experiments were A549DUALtransfected with a plasmid encoding GFP (pEGFP-N1) and A549 transfected with the same amplicons as above but lacking the first segment (the Z1 amplicon) and named REP NEG (Figure 7A).

2.10. Statistical Analysis

All values are expressed as mean ± SD of at least three independent experiments, as indicated in figure legends. After normality tests, comparisons between different treatments were analyzed by a one-way ANOVA tests. Values of $p < 0.05$ were considered statistically significant for a post-hoc Tukey's test. All statistical tests were done using the software Graph-Pad Prism version 7.01. Degrees of significance are indicated in the figure captions as follow: * $p < 0.05$; ** $p < 0.01$; *** $p < 0.001$, **** $p < 0.0001$, ns = not significant.

3. Results

3.1. ZIKV Does Not Trigger Apoptosis Until the Release of Most of its Progeny

Our research team had previously demonstrated that a South Pacific epidemic clinical isolate of ZIKV (PF13-25013-18) was able to infect A549 epithelial cells. These cells are particularly permissive to the virus and therefore constitute a suitable model for studying in cellulo host-virus interactions [17].

In order to characterize the cellular death profile that accompanies ZIKV infection more precisely, we conducted a study of the cytopathic effects induced with the viral molecular clone of the epidemic strain from Asian lineage, BeH819015 isolated in Brazil in 2015 (BR15^{MC}) [21]. We infected A549 cells with BR15^{MC} at a multiplicity of infection (MOI) of 1 and followed for 3 days, the characteristics of the viro-induced cell death (Figure 1). We further monitored the induction and execution of apoptosis specifically in infected cells to compare them with the results of viral production (Figure 2).

The measurement of LDH activity in infected cell culture supernatants, which results from a loss of cell integrity mainly reflecting secondary necrosis or estimation of cell viability by measurement of mitochondrial activity by MTT assay, revealed that cell mortality was detected at 48 h post infection (hpi) to reach a high level 72 hpi (Figure 1A,B). At 24 hpi there was no detectable sign of cell death by apoptosis when we looked at the activity or presence of cleaved caspase 3 (Figures 1C and 2C). Relocalization of the pro-apoptotic factor BAX to mitochondria, an early marker of apoptosis (Supplemental Figure S1A), was only observed at 48 hpi (Figure 2A,B) and only occurred in approximately 10% of the cells that were immunolabeled with an antibody directed against the viral envelope protein E (ZIKV-E; Figure 2B). Examination of another signal of engagement in apoptosis, namely the presence of cytosolic cytochrome-c (Figure 2A, magnified image with (b) arrow on the cyt-c immunodetection panel), led to the same observation (Figure 2C). It should be noted that only some of the cells immunolabeled for ZIKV-E had apoptotic characteristics (Figure 2A). The percentage of uninfected cells with signs of death by apoptosis was always equivalent to that observed in the controlled cell cultures over time (Figure S1B). Analysis of apoptosis execution, such as the measurement of caspase 3/7 activity (Figure 1C), immuno-detection in infected cells of cleaved activated caspase 3 (Figure 2A,D) or by Western blotting (WB; Figure 3C) supported the delay in cell death with respect to the course of viral multiplication. Moreover, significant signs of engagement in apoptosis occurred when the released viral progeny have already reached their maximum (Figure 2E). A late and low proportion of ZIKV-infected cells engaged in apoptosis were also retrieved during the infection of Vero cells (Figure S2). This apoptosis kinetics differed significantly from the one induced during the Ross River alphavirus (RRV) infection. Thus during RRV infection the maximum of caspase 3 activity was observed at 24 hpi and was at least 10 fold higher than the one measured in ZIKV-infected cells. These observations suggest a ZIKV-infection specific feature of the viro-induced apoptosis, together with a regulatory mechanism implemented by the ZIKV to delay apoptosis.

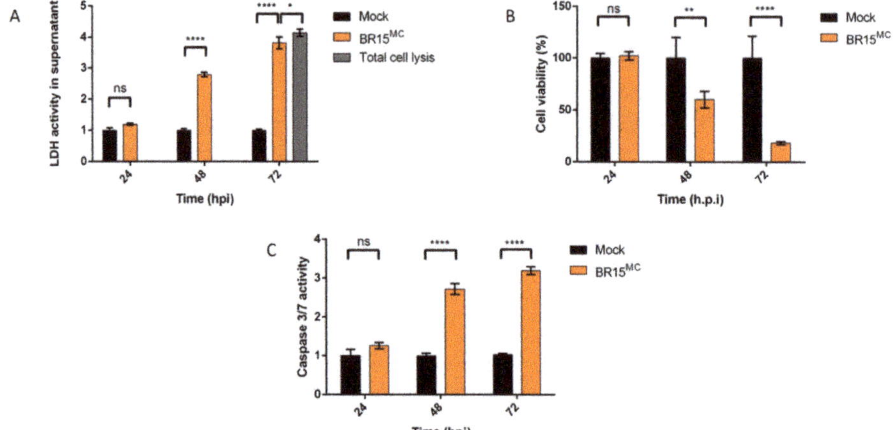

Figure 1. Cell death during a Zika virus (ZIKV) infection of A549 cells. A549 cells were infected with BR15^{MC} at a multiplicity of infection (MOI) of 1. LDH activity was measured in cell supernatant of mock infected cells, BR15 infected cells and in cells treated with triton X-100 as a positive control of total cell lysis value (grey bar) and was normalized to mock infected cells value (**A**), cell viability (MTT assay) (**B**) and caspase 3/7 activity (**C**) were measured at 24, 48 and 72 h post infection (hpi) and normalized to mock infected cells values. Values represent the mean and standard deviation of three independent experiments. Data were analyzed by a one-way ANOVA test with post-hoc Tukey's test (* $p < 0.05$; ** $p < 0.01$; **** $p < 0.0001$; ns = not significant).

To rule out that delayed apoptosis in infected cells was not a feature of the epithelial cell line A549 or Vero cells, we verified the respective courses of infection and cell death in other cell models. In the U251MG line of human brain glioblastoma-astrocytoma cells, infection kinetics was accompanied by a complete absence of apoptosis within the first 48 h of infection before the maximum of viral progeny production (Figure S3). Thus we confirm that death by apoptosis induced by our BR15^{MC} viral molecular clone as for the Asian epidemic clinical isolate PF13 occurs late and is relatively moderate compared to the kinetics of induced viral death that can be observed in the case of infection by other arboviruses like alphaviruses (Figure S2 and [12]).

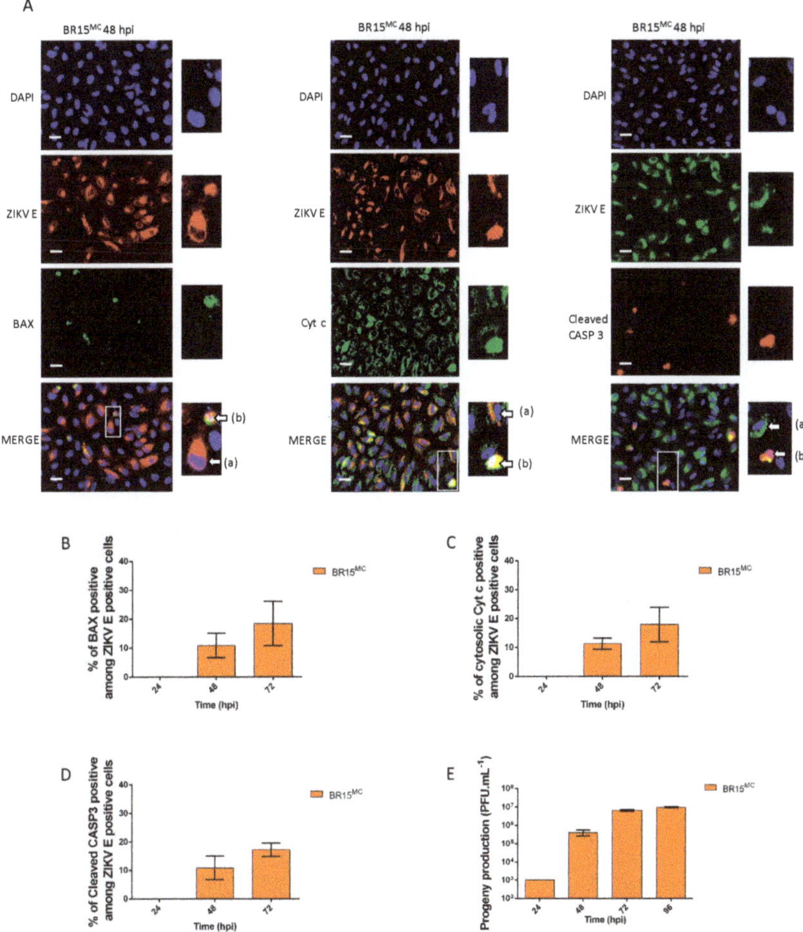

Figure 2. $BR15^{MC}$ does not cause significant activation of apoptosis until late in infection. A549 cells were infected with $BR15^{MC}$ at MOI of 1. (**A**) Cells were immunostained for active mitochondrial BAX, cytochrome c (Cyt c), ZIKV E and cleaved caspase 3 (CASP 3), 48 hpi. The white scale bar represents 10 μm. Right panel series show magnified details of selected cells from the ×200 microscopic field (white square). Arrows indicate (a): an infected cell (stained for ZIKV E) and (b): an infected and apoptotic cell (stained for ZIKV E and with mitochondrial localization of BAX or Cytosolic Cyt c or cleaved CASP3. (**B**) Percentage of A549 infected cells co-immunolabeled for ZIKV E and for active mitochondrial BAX, among the ZIKV E positive cells were determined at 24, 48 and 72 hpi. (**C**) Percentage of A549 infected cells co-immunolabeled for ZIKV E and for cytosolic Cyt c, among the ZIKV E positive cells were determined at 24, 48 and 72 hpi. (**D**) Percentage of A549 infected cells immunostained with anti-cleaved CASP 3 antibody among the ZIKV E positive cells were followed at 24, 48 and 72 hpi. (**E**) The infectious viral particles were collected from infected cell culture supernatants at 24, 48, 72 and 96 hpi and titrated. Values represent the mean and standard deviation of three independent experiments.

3.2. ZIKV Infected Cells Display Resistance to Apoptosis Inducers

Apoptosis can be initiated by extrinsic or intrinsic pathways, the former being mediated by death receptors located on the cell surface, the latter being driven by various cellular stresses such as DNA damage. Activation of apoptosis-initiating caspases 8 or 9 results in mitochondrial outer membrane permeabilization (MOMP) via oligomerization and insertion of the proapoptotic factors BAX/BAK in the mitochondria of type II cells such as epithelial cells [26].

The late onset of apoptosis in infected cells led us to postulate that ZIKV may modulate the apoptotic response of the cell, delaying it through transient inhibition. To test this hypothesis, we investigated whether ZIKV could counteract the effect of death inducers added during the infection time course. We induced apoptosis through extrinsic and intrinsic pathways and tested the addition of the inducer at 2 h prior-to and 2 h post infection (Figures 3 and 4).

3.2.1. ZIKV Provides Protection Against Death Receptor Mediated Cell Death

To drive an extrinsic apoptosis, we induced the TNF-receptor using its ligand, TNF-alpha (TNFα), inhibiting the cytoprotective NFkb response by the addition of the translation inhibitor cycloheximide (CHX). Between 6 and 8 h of treatment leads to an estimated 20% cell mortality in A549 cells, when counting BAX positive cells (Figure 3A). No variation in the percentage of cells engaged in early apoptosis (BAX+) 8 h after the onset of treatment was observed when TNFα/CHX was added 2 h prior to ZIKV infection.

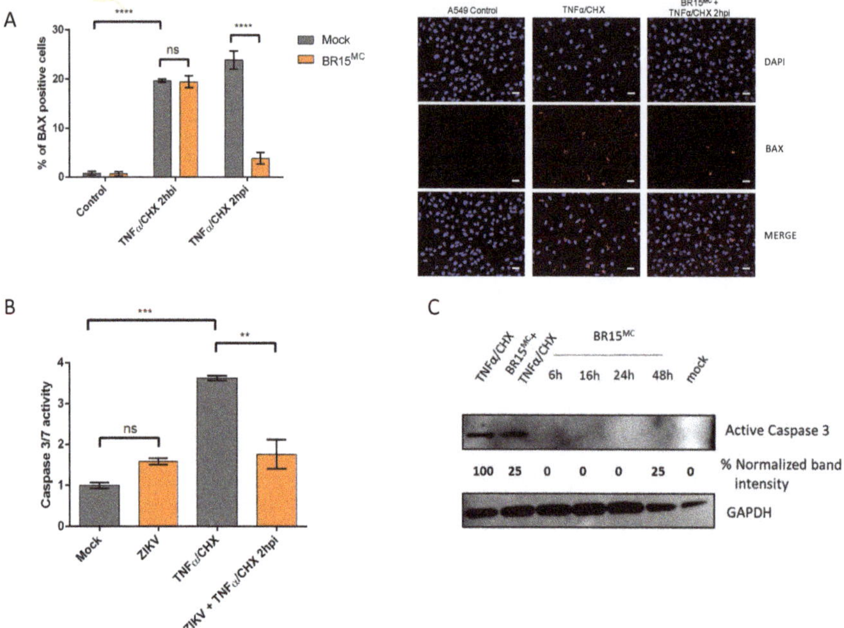

Figure 3. BR15^MC provides a protection against extrinsically induced cell death. A549 cells were infected with BR15^MC at MOI of 1 for 8 h and treated with TNFα and cycloheximide (TNFα/CHX) 2 h post infection (2 hpi) or 2 h before infection (2 hbi). (**A**) The percentage of A549 cells immunolabeled with anti-BAX antibody was enumerated from images representative of immunofluorescence observations (panels on the right, white scale bar: 10μm). (**B**) Caspase 3/7 activity was measured after TNFα/CHX treatment 2 hpi and normalized to mock treated and infected cells. (**C**) Immunoblot of active caspase 3 during TNFα treatment 2 hpi and infection time course with BR15^MC. Active caspase 3 band intensity was normalized with GAPDH. Western blotting (WB) is representative of three independent experiments. Values represent the mean and standard deviation of three independent experiments. Data were analyzed by a one-way ANOVA test with post-hoc Tukey's test (** $p < 0.01$; *** $p < 0.001$, **** $p < 0.0001$, ns = not significant).

Conversely, if TNFα/CHX was added 2 h post infection, a drastic and significant drop in the percentage of cells engaged in apoptosis was observed. This drop was corroborated by the measure of caspase 3 activity (Figure 3B). It is worth recalling that in this lapse of time, virally induced apoptosis is undetectable and therefore is unlikely to interfere with the quantification of cell death induced by the action of TNFα/CHX. Moreover, in Western blots, detectable levels of the active form of caspase 3 in BR15^{MC}-infected cells were seen to be reduced by around 75% after treatment with TNFα/CHX (Figure 3C).

3.2.2. ZIKV Provides Protection Against Intrinsically Induced Cell Death

Intrinsically induced apoptosis was stimulated by chemical DNA damage. To do this, we used the genotoxic agent etoposide, a topoisomerase-II inhibitor, which causes chromosome breaks during DNA replication. Overnight treatment with etoposide (16–18 h) resulted in a percentage of cells with a mitochondrial BAX among the remaining adherent cells that was between 7% and 12%, depending on the experiment (Figure 4). Similar to the induction of cell death by TNFα/CHX, although the effects are slightly more modest, apoptosis produced in A549 cells after 16 h of etoposide treatment was significantly reduced in the case where ZIKV was added 2 h post-infection (Figure 4A). In addition, both BR15, the molecular clone of ZIKV (Figure 4A) and a clinical isolate of the epidemic strain PF13 (Figure 4B) were capable to interfere with apoptosis induction by etoposide.

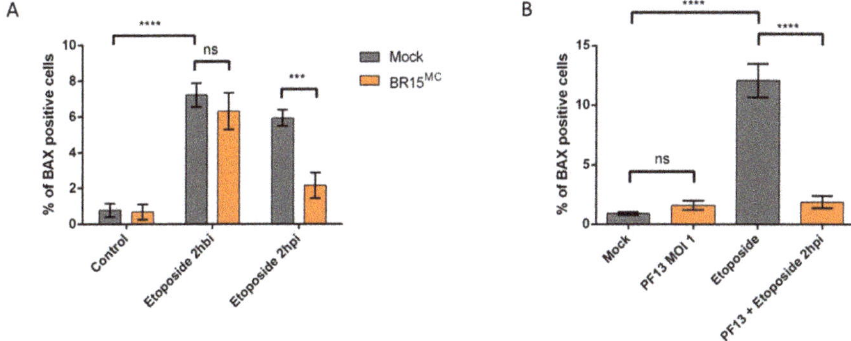

Figure 4. ZIKV provides a protection against intrinsically induced cell death by etoposide. (**A**) A549 cells were infected with BR15^{MC} at MOI of 1 for 8 h and treated with etoposide 2 h before infection (2 hbi) or 2 h post infection (2 hpi). The percentage of A549 cells immunostained with anti-BAX antibody was followed. (**B**) A549 cells were infected with PF13 at MOI of 1 for 8 h and treated with etoposide for 16 h, 2 hpi. The percentage of A549 cells immunostained with anti-BAX antibody was followed. Values represent the mean and standard deviation of three independent experiments Data were analyzed by a one-way ANOVA test with post-hoc Tukey's test (*** $p < 0.001$, **** $p < 0.0001$, ns = not significant).

We also tested the effect of the cell line used for infection by repeating the procedure in Vero cells, in this instance using blasticidin to induce intrinsic apoptosis in response to the inhibition of translation instead of DNA damage with etoposide. Following cell death by MTT assay and caspase 3/7 activity again suggested that ZIKV infection represses apoptosis, regardless of cell type and death inducers (Figure 5A,B).

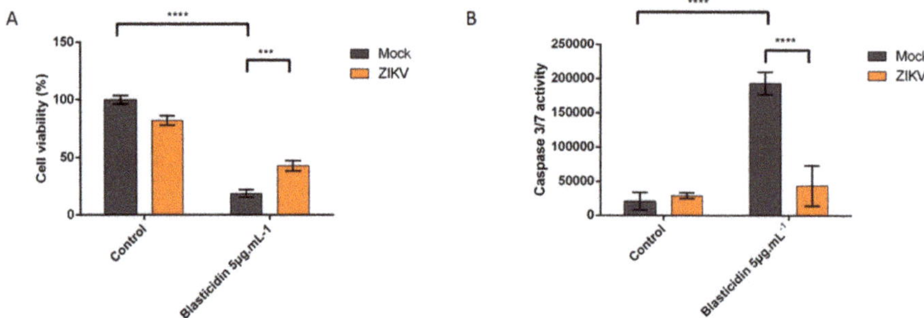

Figure 5. ZIKV provides a protection against intrinsically induced cell death by blasticidin. Vero cells were infected with ZIKV-PF13 at MOI of 1 for 8 h and treated with blasticidin for 16 h, followed by viability assay (MTT) (**A**) or caspase 3/7 activity (**B**). Values represent the mean and standard deviation of three independent experiments. Values represent the mean and standard deviation of three independent experiments Data were analyzed by a one-way ANOVA test with post-hoc Tukey's test (** $p < 0.01$; *** $p < 0.001$, **** $p < 0.0001$, ns = not significant).

In order to ensure that the ZIKV mediated repression of apoptosis was conserved between ZIKV strains, we looked at the effect of apoptosis induction with etoposide after infection with the epidemic clinical isolate PF13 (ZIKV-PF13) but also with a molecular clone of historical strain from African lineage, MR766-NIID isolated in Uganda in 1947 (MR766[MC]; Supplemental Figure S4). ZIKV infection resulted in reduced apoptosis for all tested strains. Convergent measurements of several parameters that establish the death rate and degree of protection confirm that ZIKV interferes with the achievement of apoptosis in response to a death signal.

3.3. Apoptosis is Repressed Through ZIKV Replication

Our data suggest that protection against an exogenous induced apoptosis is acquired once the virus has entered the cells and started its multiplication cycle. In order to determine the contribution of the replicative process in the protective mechanism, we exploited the "replicon" systems.

3.3.1. Cells Stably Expressing a ZIKV Replicon are Protected from Intrinsically and Extrinsically Mediated Apoptosis

Since both epidemic Asian (BR15[MC]) and historical African (MR766[MC]) strains of ZIKV were found to be able to control apoptosis, we investigated the role of viral replication in the mechanism of apoptosis repression using a previously established MR766 replicon system in HEK-293 cells [25]. The Rep ZIKV-GFP cells have a self-replicating RNA encoding the viral NS proteins from MR766-NIID, a puromycin resistance gene to facilitate selection and GFP as a reporter. HEK-293 cells stably transfected with a plasmid encoding a GFP reporter gene and puromycin resistance without any viral material was used as a control (HEK CT; Figure 6). Apoptosis was induced either with TNFα/CHX, with etoposide or with blasticidin. We monitored apoptosis and in particular measured caspase 3/7 activity 6h post addition of TNFα/CHX or 16 h post addition of etoposide or blasticidin. We compared the values related to caspase 3/7 activity in Rep ZIKV-GFP with those obtained with the control cells (HEK CT; Figure 6A).

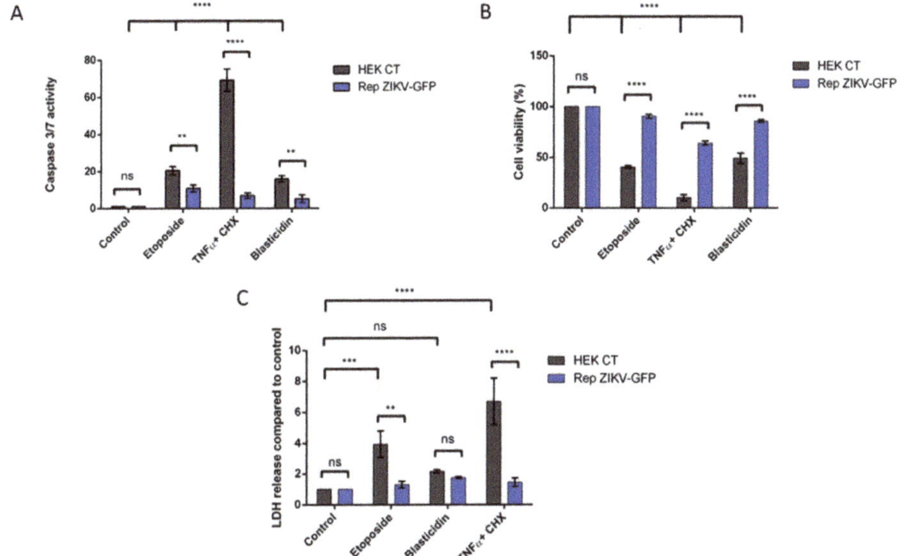

Figure 6. ZIKV replicon-expressing cells are protected from apoptosis. HEK-293 cells stably expressing a ZIKV replicon (Rep ZIKV-GFP) are protected from intrinsically and extrinsically mediated apoptosis. HEK-293 cells with Rep ZIKV-GFP were treated with TNFα and CHX for 6 h, or treated with etoposide 10 μM or blasticidin 25 μg/mL^{-1} for 16 h and analyzed for: caspase 3/7 activity (**A**), cell viability (MTT assay) (**B**) and released LDH activity (**C**). Values represent the mean and standard deviation obtained with three different clones of Rep ZIKV-GFP. Values represent the mean and standard deviation of three independent experiments. Data were analyzed by a one-way ANOVA test with post-hoc Tukey's test (** $p < 0.01$; *** $p < 0.001$, **** $p < 0.0001$, ns = not significant).

Monitoring apoptosis by caspase 3/7 activity (Figure 6A), together with the measure of cell viability (Figure 6B) and released LDH activity (Figure 6C) demonstrated that ZIKV replicon resulted in a significant reduction in the indicators of cell death. An inhibition of apoptosis was not observed in the case of the cell control, selected for GFP and puromycin resistance (HEK CT) as well as in untransfected HEK-293 cells (data not shown). It can be concluded that resistance to puromycin or the presence of a GFP encoding gene were not responsible for a protective effect.

It can therefore be proposed that the presence of an autonomous replication of a viral RNA associated with the expression of ZIKV NS proteins makes cells resistant to several extrinsic and intrinsic apoptotic inducers.

3.3.2. A549 Cells Transiently Expressing a ZIKV Replicon are Protected from Different Intrinsically Induced Apoptosis

To address a protective effect of the viral replication in a system consistent with the one in which the effect was revealed through infection, we adapted the ISA method to obtain A549 cells transiently expressing a ZIKV replicon with the non-structural sequences from BeH819015 (REP BR15). We used A549 cells transfected with an incomplete set of amplicons as a negative control (REP-NEG; Figure 7A). Apoptosis was induced 48 h after amplicon transfection either by etoposide treatment (Figure 7B) or by DNA damage through exposure to UV light (Figure 7C).

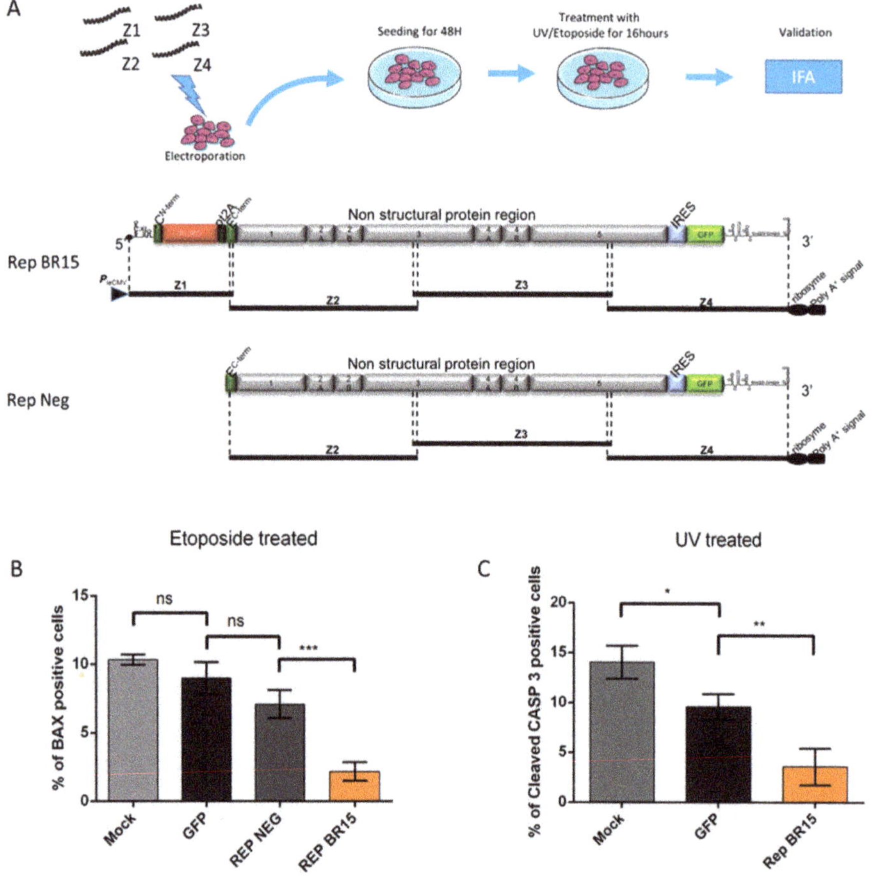

Figure 7. A549 cells transiently expressing a ZIKV replicon are protected against cell death by apoptosis. A549 cells were transfected with ZIKV amplicons (Z genomic overlapping fragments) for production of REP BR15 (Z1, 2, 3 and 4) and REP NEG (Z2, 3 and 4) or with pEGFP-N1 (**A**). At 48 h after transfection A549 cells were treated with etoposide for 16 h (**B**) or UV at 400 mJ (**C**). The percentage of A549 cells immunostained with anti-BAX antibody or anti-cleaved CASP 3 was monitored. Values represent the mean and standard deviation of three independent experiments. Data were analyzed by a one-way ANOVA test with post-hoc Tukey's test (* $p < 0.05$; ** $p < 0.01$; *** $p < 0.001$, ns = not significant).

In cells treated with etoposide, REP BR15 expressing cells showed approximately half the number of apoptotic cells of the replicon control (REP NEG; Figure 7B). The percentage of dying cells was even lower in cells expressing REP BR15 after UV treatment (Figure 7C). Thus, REP BR15 was able to confer resistance to apoptosis. This resistance was also observed in an experiment conducted with A549 cells transiently expressing a ZIKV replicon from the MR766 strain of ZIKV (Figure S5). All together these results suggest that replication of the ZIKV was capable of inhibiting apoptosis.

3.4. ZIKV Promotes an Anti-Apoptotic Prevailing Status in Infected Cells Through the Bcl-2 Family Protein

The long delay in the onset of viral apoptosis induced by ZIKV infection, particularly in the case of Asian viral strains responsible for current epidemics, and the demonstration that, when the viral RNA is present and replicating there is an inhibition of apoptotic induction, suggest that cells have acquired with the virus a status in which anti-apoptotic activity prevails. The protective effect acquired with ZIKV could be at the level of convergence of the intrinsic and extrinsic pathways. As we followed BAX relocalization we could argue in favor of protection around the mitochondria events and the control of the outer membrane permeabilization (OMP). A prominent anti-apoptotic factor involved in the regulation of early apoptosis, by operating mainly at the mitochondrial level for the control of its permeabilization is Bcl-2 and the related Bcl-XL protein [27]. To identify to which extent these anti-apoptotic factors play a role in the protection provided by ZIKV, we examined the effect of ABT-737 on cell death outcomes, with or without BR15^{MC} (Figure 8A,B). ABT-737 is a BH3 mimetic molecule that can bind to the hydrophobic groove of the members of the anti-apoptotic Bcl-2 protein family, Bcl-2 and Bcl-xL, and therefore inhibits their activity by shifting oligomerization mechanisms in favor of BAX/BAK dimerization [28].

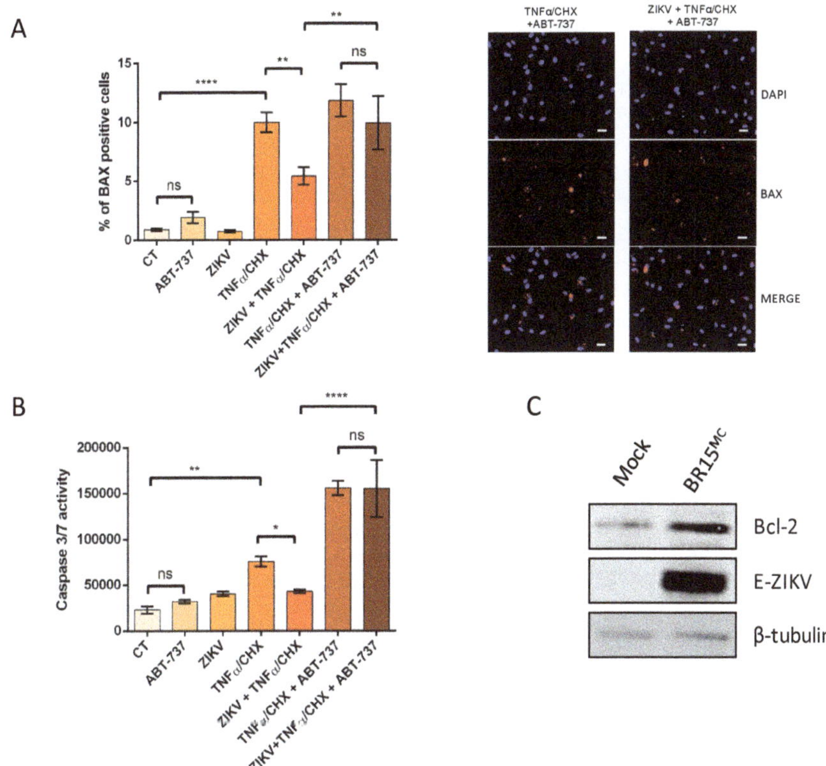

Figure 8. Inhibition of anti-apoptotic Bcl-2 family proteins abrogates the protection mediated by ZIKV. A549 cells were infected with ZIKV at MOI of 1. TNFα and CHX were added 2 h post infection for 6 h with or without ABT-737. A549 cells were immunostained with an anti-BAX antibody (representative images in panels on the right, white scale bar: 10μm) (**A**) and caspase 3/7 activity was followed after treatment (**B**). A549 cells were infected with ZIKV at MOI of 1 and level of Bcl-2 was followed by western blot 24 h post infection (**C**). Values represent the mean and standard deviation of three independent experiments. Data were analyzed by a one-way ANOVA test with post-hoc Tukey's test (* $p < 0.05$; ** $p < 0.01$; *** $p < 0.001$, **** $p < 0.0001$, ns = not significant).

When inducing apoptosis with TNFα/CHX, addition of ABT-737 restored the percentage of ZIKV-infected cells with mitochondrial BAX (Figure 8A) and caspase 3/7 activity (Figure 8B) to levels that were similar to cells that were not infected with ZIKV.

These observations suggest that ABT-737 has counteracted the protective effect acquired with the virus. It can legitimately be deduced that the viro-induced protective effect probably depends on the anti-apoptotic activity of Bcl-2 family proteins. When Bcl-2/Bcl-XL is inhibited, ZIKV no longer allows a quantitative reduction of apoptosis.

As we used cycloheximide or blasticidin in our assays, we could therefore assume that the mechanism implemented by ZIKV did not involve a "de novo" synthesis of proteins. Based on these remarks, we formulated the hypothesis that ZIKV might allow stabilization of the Bcl-2 protein over time. We performed a western blot assay to measure Bcl-2 protein level upon ZIKV infection. Upon ZIKV infection, we observed an increase of Bcl-2 at the protein level (Figure 8C). This stabilization of Bcl-2 could support the anti-apoptotic capacity of ZIKV.

4. Discussion

ZIKV has attracted tremendous attention in the field of medically important flaviviruses, which are responsible of world epidemics that are difficult to control [29]. To date, there is no effective treatment against this emerging pathogen despite significant efforts made to succeed in providing an effective vaccine. The characterization of the interaction modalities between the pathogen and host cells are important in order to better understand the strategies adopted by the virus to be effective in its replication and dispersal of its progeny. Among the responses of infected cells, a high priority must be given to virally induced apoptosis since its completion can significantly interfere with the virus's multiplication cycle and hinder its spread [30]. Moreover, crosstalk between innate immune signaling and cell death pathways and how the viruses are able to manipulate each other are essential for viral clearance or persistence and for the global outcomes of the infection. Many studies support that ZIKV infection induces apoptosis in vivo as well as in vitro [31–33] However, several of them also assert that ZIKV induced cell death could be delayed [17,18]. We therefore wanted to define whether this capacity was specific to the strain of ZIKV responsible for the current epidemic and whether ZIKV was able to interfere with the induction of apoptosis.

Using a molecular clone of the BeH819015 from the Brazilian 2015 outbreak (BR15^{MC}) and its comparison to the clinical isolate from the French Polynesian 2013 outbreak (PF13), our work has highlighted that ZIKV strains from the actual epidemic Asian lineage are particularly inclined to delay the onset of apoptosis in infected human epithelial cells (A549) as well as in human brain glioblastoma–astrocytoma cells (U251 MG; Figure 1, Figure 2 and Figure S3). These viruses are also characterized by a rather slow viral growth compared to the molecular clone of the historical strain of ZIKV MR766 from the African lineage [21]. Infection with MR766^{MC} was marked by a higher cytotoxicity over the duration of infection, with 10% of infected cells showing signs of entry into apoptosis as early as 24 h, whereas epidemic strains had no signs of mortality before 48 h. However, despite this behavior of MR766^{MC}, which seems more aggressive, it must be admitted that the mortality rate among infected cells remained limited until 48 h post infection (Figure 2 and Figure S4). We proposed from these results that both viruses had the ability to interfere with the onset of apoptosis even though faster and more effective viral growth in the early hours of infection for MR766^{MC} than for BR15^{MC} had resulted in faster and more cytopathic effects. In both cases, the maximum mortality was only recorded when the virion production had reached its highest titer. This observation suggests a manipulation of apoptosis orchestrated by ZIKV in order to give it enough time to complete its entire production cycle. This phenomenon has already been described for other flaviviruses such as DENV, JEV and WNV as they can delay apoptosis through activation of the phosphatidylinositol 3-kinase (PI3K) and Akt pathway [16,34].

In a rather unexpected way, our work mainly shows that in the early stages of infection, ZIKV infection provides a solid protection against an exogenous induced cell death. We provided supporting evidence that protection is acquired both with the Asian epidemic strains and with the African strain (Figure 3, Figure 4, Figures S4 and S5). This protection is effective against apoptosis mediated by an extrinsic death inducer (TNFα) as well as by an intrinsic signal (provided by the action of etoposide or blasticidin). Resistance to these induction modes has also been found to characterize cells expressing replicons, either HEK 293 cells stably expressing a MR766 replicon or A549 cells transiently expressing BR15 replicons or MR766 replicons (Figure 6, Figure 7 and Figure S5). The data obtained with the use of these replicons are in support of a greater protection granted by BR15. This is consistent with the data obtained with the whole virus that is responsible for the longest delay in apoptosis entry. Cell death inhibition ability acquired with the ZIKV replicons would imply that the single presence of a viral RNA leading to the production of the NS proteins and allowing its self-replication is the driving force behind the protection acquired against apoptosis.

This discovery is rather unusual when one considers the literature mainly in favor of pro apoptotic functions for the NS proteins [35]. However, studies have also shown that some NS may help ZIKV to evade antiviral immunity and cell death. NS2B in particular was proposed to be responsible for blocking (RIG-I)-like receptors triggered apoptotic cell death [36]. A thorough identification of the viral protein responsible for an inhibitory effect needs to be confirmed and further investigated. While our work suggests a possible role for NS proteins in the control of apoptosis, we cannot rule out the possibility that structural proteins may also act. Anti-apoptotic activities have been previously reported in relation to the capsid [16,18]. It should be noted that in the construct we used to generate the replicons, the polyprotein produced retains the first 33 amino acids of the capsid. It would be interesting to see if this part of the structural protein, released after the cleavage of the GFP has a role in protection. We could not also exclude a role of the viral RNA by itself. Recent work has also investigated the effects of viral RNAs, as flaviviruses are known to produce multiple small RNAs that may have interference activities in the cell physiology. It was recently described that recent epidemic Asian lineage display more negative-strand replicative intermediates than the historical African strain [37]. This important production could be a key element in the search for which viral determinants are crucial in the control of apoptosis. We also know how important are the viral genomic 3′ UTR regions and the Subgenomic Flavivirus RNA (sfRNA) produced, in the implementation of cellular responses to infection [38]. It remains to be discovered which factor associated with the replication process of ZIKV viral RNA is involved in the protection mechanism. What we already know from our study is that this mechanism required the Bcl-2 family protein as ABT-737 abrogates the protection acquired against apoptosis by ZIKV infection (Figure 8). The anti-apoptotic capacity of the pro-survival Bcl-2 proteins is known to depend mainly on their ability to sequester pro-apoptotic proteins by binding their BH3 domains. A decrease in Bcl-2 leads to the disruption of associated pro-survival and pro-apoptotic Bcl-2 proteins and will promote apoptosis whereas overexpression of Bcl-2 will inhibit mitochondrial OMP [39,40]. A control of the stability and degradation of Bcl-2 is therefore a key in the subtle balance that takes place between pro and anti-apoptotic suits to determine the cell's fate [40]. A previous study showed the importance of Bcl-xL for cells survival, in deficient Bcl-2 cells during ZIKV infection [41], but here we could not exclude a role for Bcl-2 and/or Bcl-xL protein as our model expressed these two anti-apoptotic proteins. Based on the fact that Bcl-2 family protein was involved in ZIKV infection with a Bcl-2 protein level quantitatively maintained (Figure 8), we formulated the hypothesis that ZIKV might allow a stabilization of the Bcl-2 protein over time and inhibit MOMP formation (Figure 9). The mechanism by which the virus allows the Bcl-2 stabilization and blocks apoptosis needs further investigation.

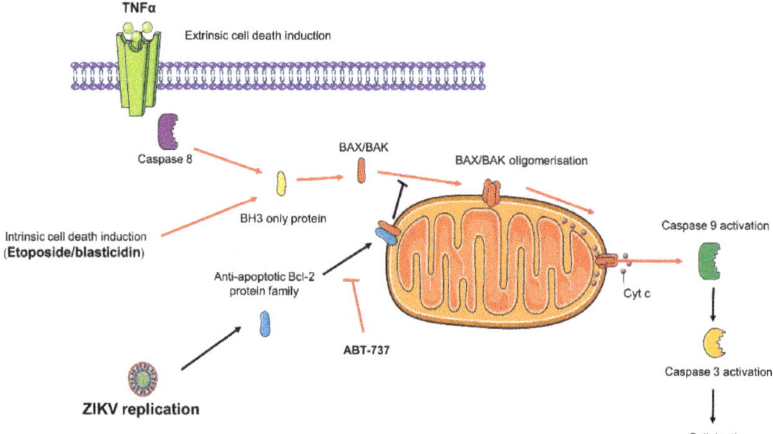

Figure 9. Model depicting the protective action of ZIKV infection against apoptosis. Intrinsic or extrinsic activation of cell death occurs through the formation of mitochondrial outer membrane pore (MOMP) via the BAX/BAK complex. The anti-apoptotic family Bcl-2 members Bcl-2/Bcl-XL interfere with the complex formation by sequestering BAX via their BH3 domain. ZIKV replication interferes with BAX relocalization at the mitochondria. ABT-737, a BH3 mimetic, which inhibits Bcl-2/Bcl-XL, abrogates the inhibition induced by ZIKV. Indeed, the virus delays apoptosis during infection by modulating the homeostasis of Bcl-2/Bcl-XL.

Supplementary Materials: The following are available online at http://www.mdpi.com/2073-4409/8/11/1338/s1, Figure S1. IF imaging to validate markers of apoptosis in A549 cells. Figure S2. Alphavirus RRV induces early and massive apoptosis compared to Zika virus. Figure S3. ZIKV-PF13 does not cause significant activation of apoptosis until late in infection in U251MG cells. Figure S4. ZIKV-MR766 does not cause significant activation of apoptosis until late in infection and ZIKV-MR766 is able to control cell death. Figure S5. A549 cells transiently expressing a ZIKV-MR766 replicon are protected against cell death by apoptosis.

Author Contributions: J.T., E.F., W.V., P.K.-T. designed the research J.T., E.F., W.V., P.K.-T. performed the research; J.T., E.F., P.D., W.V., P.K.-T. contributed new reagents/analytic tools; all authors analyzed the data; J.T., E.F., W.V., P.K.-T. wrote, revised and edited the manuscript.

Funding: This work was supported by the Federation BioST from Reunion Island University (Zikapone project), the ZIKAlert project (European Union-Région Réunion program under grant agreement n° SYNERGY: RE0001902). E.F. holds a fellowship from the Regional Council of Reunion Island (European Union-Région Réunion program under grant agreement n° SYNERGY: RE0012406). J.T. has a PhD degree scholarship from La Réunion Island University (Ecole Doctorale STS), funded by the French ministry MEESR.

Acknowledgments: We thank the members of PIMIT and DéTROI laboratories for helpful discussions. We are grateful to David A. Wilkinson for the critical review of the manuscript and Valérie Ferment-Van de Moortele for the English spelling. We also thank Valérie Choumet for providing us the rabbit anti-ZIKV EDIII antibodies. Servier Medical art provided the icons used in Figures 7 and 9.

Conflicts of Interest: The authors declare no conflict of interest.

References

1. Paixão, E.S.; Barreto, F. History, Epidemiology, and Clinical Manifestations of Zika: A Systematic Review. *Am. J. Public Health* **2016**, *106*, 606–612.

2. Cao-Lormeau, V.-M.; Blake, A.; Mons, S.; Lastère, S.; Roche, C.; Vanhomwegen, J.; Dub, T.; Baudouin, L.; Teissier, A.; Larre, P.; et al. Guillain-Barré Syndrome outbreak associated with Zika virus infection in French Polynesia: A case-control study. *Lancet* **2016**, *387*, 1531–1539. [CrossRef]

3. Lover, A.A. Zika virus and microcephaly. *Lancet Infect. Dis.* **2016**, *16*, 1331–1332. [CrossRef]

4. McKenzie, B.A.; Wilson, A.E.; Zohdy, S. Aedes albopictus is a competent vector of Zika virus: A meta-analysis. *PLoS ONE* **2019**. [CrossRef]

5. Kraemer, M.U.; Sinka, M.E.; Duda, K.A.; Mylne, A.Q.; Shearer, F.M.; Barker, C.M.; Moore, C.G.; Carvalho, R.G.; Coelho, G.E.; Van Bortel, W.; et al. The global distribution of the arbovirus vectors Aedes aegypti and Ae. albopictus. *eLife* **2015**. [CrossRef] [PubMed]

6. Haddow, A.D.; Schuh, A.J.; Yasuda, C.Y.; Kasper, M.R.; Heang, V.; Huy, R.; Guzman, H.; Tesh, R.B.; Weaver, S.C. Genetic Characterization of Zika Virus Strains: Geographic Expansion of the Asian Lineage. *PLoS Negl. Trop. Dis.* **2012**. [CrossRef] [PubMed]

7. Giovanetti, M.; Milano, T.; Alcantara, L.C.J.; Carcangiu, L.; Cella, E.; Lai, A.; Lo Presti, A.; Pascarella, S.; Zehender, G.; Angeletti, S.; et al. Zika Virus spreading in South America: Evolutionary analysis of emerging neutralizing resistant Phe279Ser strains. *Asian Pac. J. Trop. Med.* **2016**, *9*, 445–452. [CrossRef] [PubMed]

8. Apte-Sengupta, S.; Sirohi, D.; Kuhn, R.J. Coupling of replication and assembly in flaviviruses. *Curr. Opin. Virol.* **2014**, *9*, 134–142. [CrossRef] [PubMed]

9. Hamel, R.; Dejarnac, O.; Wichit, S.; Ekchariyawat, P.; Neyret, A.; Luplertlop, N.; Perera-Lecoin, M.; Surasombatpattana, P.; Talignani, L.; Thomas, F.; et al. Biology of Zika Virus Infection in Human Skin Cells. *J. Virol.* **2015**, *89*, 8880–8896. [CrossRef]

10. Mehrbod, P.; Ande, S.R.; Alizadeh, J.; Rahimizadeh, S.; Shariati, A.; Malek, H.; Hashemi, M.; Glover, K.K.M.; Sher, A.A.; Coombs, K.M.; et al. The roles of apoptosis, autophagy and unfolded protein response in arbovirus, influenza virus, and HIV infections. *Virulence* **2019**, *10*, 376–413. [CrossRef]

11. Roulston, A.; Marcellus, R.C.; Branton, P.E. Viruses and Apoptosis. *Annu. Rev. Microbiol.* **1999**, *53*, 577–628. [CrossRef] [PubMed]

12. Krejbich-Trotot, P.; Denizot, M.; Hoarau, J.-J.; Jaffar-Bandjee, M.-C.; Das, T.; Gasque, P. Chikungunya virus mobilizes the apoptotic machinery to invade host cell defenses. *FASEB J. Off. Publ. Fed. Am. Soc. Exp. Biol.* **2011**, *25*, 314–325. [CrossRef] [PubMed]

13. Li, M.-L.; Stollar, V. Alphaviruses and apoptosis. *Int. Rev. Immunol.* **2004**, *23*, 7–24. [CrossRef] [PubMed]

14. Okamoto, T.; Suzuki, T.; Kusakabe, S.; Tokunaga, M.; Hirano, J.; Miyata, Y.; Matsuura, Y. Regulation of Apoptosis during Flavivirus Infection. *Viruses* **2017**. [CrossRef] [PubMed]

15. Bhuvanakantham, R.; Cheong, Y.K.; Ng, M.-L. West Nile virus capsid protein interaction with importin and HDM2 protein is regulated by protein kinase C-mediated phosphorylation. *Microbes Infect.* **2010**, *12*, 615–625. [CrossRef] [PubMed]

16. Urbanowski, M.D.; Hobman, T.C. The West Nile Virus Capsid Protein Blocks Apoptosis through a Phosphatidylinositol 3-Kinase-Dependent Mechanism. *J. Virol.* **2013**, *87*, 872–881. [CrossRef] [PubMed]

17. Frumence, E.; Roche, M.; Krejbich-Trotot, P.; El-Kalamouni, C.; Nativel, B.; Rondeau, P.; Missé, D.; Gadea, G.; Viranaicken, W.; Desprès, P. The South Pacific epidemic strain of Zika virus replicates efficiently in human epithelial A549 cells leading to IFN-β production and apoptosis induction. *Virology* **2016**, *493*, 217–226. [CrossRef]

18. Limonta, D.; Jovel, J.; Kumar, A.; Airo, A.M.; Hou, S.; Saito, L.; Branton, W.; Ka-Shu Wong, G.; Mason, A.; Power, C.; et al. Human Fetal Astrocytes Infected with Zika Virus Exhibit Delayed Apoptosis and Resistance to Interferon: Implications for Persistence. *Viruses* **2018**. [CrossRef]

19. Van der Linden, V.; Pessoa, A.; Dobyns, W.; Barkovich, A.J.; van der Linden Júnior, H.; Filho, E.L.R.; Ribeiro, E.M.; de Carvalho Leal, M.; de Araújo Coimbra, P.P.; de Fátima Viana Vasco Aragão, M.; et al. Description of 13 Infants Born During October 2015-January 2016 With Congenital Zika Virus Infection Without Microcephaly at Birth-Brazil. *MMWR Morb. Mortal. Wkly. Rep.* **2016**, *65*, 1343–1348. [CrossRef]

20. Gadea, G.; Bos, S.; Krejbich-Trotot, P.; Clain, E.; Viranaicken, W.; El-Kalamouni, C.; Mavingui, P.; Desprès, P. A robust method for the rapid generation of recombinant Zika virus expressing the GFP reporter gene. *Virology* **2016**, *497*, 157–162. [CrossRef]

21. Bos, S.; Viranaicken, W.; Turpin, J.; El-Kalamouni, C.; Roche, M.; Krejbich-Trotot, P.; Desprès, P.; Gadea, G. The structural proteins of epidemic and historical strains of Zika virus differ in their ability to initiate viral infection in human host cells. *Virology* **2018**, *516*, 265–273. [CrossRef] [PubMed]

22. Nativel, B.; Marimoutou, M.; Thon-Hon, V.G.; Gunasekaran, M.K.; Andries, J.; Stanislas, G.; Planesse, C.; Da Silva, C.R.; Césari, M.; Iwema, T.; et al. Soluble HMGB1 is a novel adipokine stimulating IL-6 secretion through RAGE receptor in SW872 preadipocyte cell line: Contribution to chronic inflammation in fat tissue. *PLoS ONE* **2013**. [CrossRef] [PubMed]

23. Viranaicken, W.; Gasmi, L.; Chaumet, A.; Durieux, C.; Georget, V.; Denoulet, P.; Larcher, J.-C. L-Ilf3 and L-NF90 Traffic to the Nucleolus Granular Component: Alternatively-Spliced Exon 3 Encodes a Nucleolar Localization Motif. *PLoS ONE* **2011**. [CrossRef] [PubMed]

24. Antwerp, D.J.V.; Martin, S.J.; Verma, I.M.; Green, D.R. Inhibition of TNF-induced apoptosis by NF-κB. *Trends Cell Biol.* **1998**, *8*, 107–111. [CrossRef]

25. El Kalamouni, C.; Frumence, E.; Bos, S.; Turpin, J.; Nativel, B.; Harrabi, W.; Wilkinson, D.A.; Meilhac, O.; Gadea, G.; Desprès, P.; et al. Subversion of the Heme Oxygenase-1 Antiviral Activity by Zika Virus. *Viruses* **2018**, *11*. [CrossRef]

26. Kantari, C.; Walczak, H. Caspase-8 and bid: Caught in the act between death receptors and mitochondria. *Biochim. Biophys. Acta* **2011**, *1813*, 558–563. [CrossRef]

27. Gross, A.; McDonnell, J.M.; Korsmeyer, S.J. BCL-2 family members and the mitochondria in apoptosis. *Genes Dev.* **1999**, *13*, 1899–1911. [CrossRef]

28. Oltersdorf, T.; Elmore, S.W.; Shoemaker, A.R.; Armstrong, R.C.; Augeri, D.J.; Belli, B.A.; Bruncko, M.; Deckwerth, T.L.; Dinges, J.; Hajduk, P.J.; et al. An inhibitor of Bcl-2 family proteins induces regression of solid tumours. *Nature* **2005**, *435*, 677–681. [CrossRef]

29. Musso, D.; Gubler, D.J. Zika Virus. *Clin. Microbiol. Rev.* **2016**, *29*, 487–524. [CrossRef]

30. Jorgensen, I.; Rayamajhi, M.; Miao, E.A. Programmed cell death as a defence against infection. *Nat. Rev. Immunol.* **2017**, *17*, 151–164. [CrossRef]

31. Souza, B.S.F.; Sampaio, G.L.A.; Pereira, C.S.; Campos, G.S.; Sardi, S.I.; Freitas, L.A.R.; Figueira, C.P.; Paredes, B.D.; Nonaka, C.K.V.; Azevedo, C.M.; et al. Zika virus infection induces mitosis abnormalities and apoptotic cell death of human neural progenitor cells. *Sci. Rep.* **2016**. [CrossRef] [PubMed]

32. Li, C.; Xu, D.; Ye, Q.; Hong, S.; Jiang, Y.; Liu, X.; Zhang, N.; Shi, L.; Qin, C.-F.; Xu, Z. Zika Virus Disrupts Neural Progenitor Development and Leads to Microcephaly in Mice. *Cell Stem Cell* **2016**, *19*, 120–126. [CrossRef] [PubMed]

33. Ghouzzi, V.E.; Bianchi, F.T.; Molineris, I.; Mounce, B.C.; Berto, G.E.; Rak, M.; Lebon, S.; Aubry, L.; Tocco, C.; Gai, M.; et al. ZIKA virus elicits P53 activation and genotoxic stress in human neural progenitors similar to mutations involved in severe forms of genetic microcephaly and p53. *Cell Death Dis.* **2016**, *7*, e2440. [CrossRef] [PubMed]

34. Lee, C.-J.; Liao, C.-L.; Lin, Y.-L. Flavivirus Activates Phosphatidylinositol 3-Kinase Signaling To Block Caspase-Dependent Apoptotic Cell Death at the Early Stage of Virus Infection. *J. Virol.* **2005**, *79*, 8388–8399. [CrossRef]

35. Liu, J.; Li, Q.; Li, X.; Qiu, Z.; Li, A.; Liang, W.; Chen, H.; Cai, X.; Chen, X.; Duan, X.; et al. Zika Virus Envelope Protein induces G2/M Cell Cycle Arrest and Apoptosis via an Intrinsic Cell Death Signaling Pathway in Neuroendocrine PC12 Cells. *Int. J. Biol. Sci.* **2018**, *14*, 1099–1108. [CrossRef]

36. Wu, Y.; Liu, Q.; Zhou, J.; Xie, W.; Chen, C.; Wang, Z.; Yang, H.; Cui, J. Zika virus evades interferon-mediated antiviral response through the co-operation of multiple nonstructural proteins *in vitro*. *Cell Discov.* **2017**. [CrossRef]

37. Barnard, T.R.; Rajah, M.M.; Sagan, S.M. Contemporary Zika Virus Isolates Induce More dsRNA and Produce More Negative-Strand Intermediate in Human Astrocytoma Cells. *Viruses* **2018**. [CrossRef]

38. Göertz, G.P.; Abbo, S.R.; Fros, J.J.; Pijlman, G.P. Functional RNA during Zika virus infection. *Virus Res.* **2018**, *254*, 41–53. [CrossRef]

39. Kim, H.; Rafiuddin-Shah, M.; Tu, H.-C.; Jeffers, J.R.; Zambetti, G.P.; Hsieh, J.J.-D.; Cheng, E.H.-Y. Hierarchical regulation of mitochondrion-dependent apoptosis by BCL-2 subfamilies. *Nat. Cell Biol.* **2006**, *8*, 1348–1358. [CrossRef]

40. Rooswinkel, R.W.; van de Kooij, B.; de Vries, E.; Paauwe, M.; Braster, R.; Verheij, M.; Borst, J. Antiapoptotic potency of Bcl-2 proteins primarily relies on their stability, not binding selectivity. *Blood* **2014**, *123*, 2806–2815. [CrossRef]

41. Suzuki, T.; Okamoto, T.; Katoh, H.; Sugiyama, Y.; Kusakabe, S.; Tokunaga, M.; Hirano, J.; Miyata, Y.; Fukuhara, T.; Ikawa, M.; et al. Infection with flaviviruses requires BCLXL for cell survival. *PLOS Pathog.* **2018**. [CrossRef] [PubMed]

Article

Syncytiotrophoblast of Placentae from Women with Zika Virus Infection Has Altered Tight Junction Protein Expression and Increased Paracellular Permeability

Jael Miranda [1], Dolores Martín-Tapia [1], Yolotzin Valdespino-Vázquez [2], Lourdes Alarcón [1], Aurora Espejel-Nuñez [2], Mario Guzmán-Huerta [2], José Esteban Muñoz-Medina [3], Mineko Shibayama [4], Bibiana Chávez-Munguía [4], Guadalupe Estrada-Gutiérrez [2], Samuel Lievano [5], Juan Ernesto Ludert [4] and Lorenza González-Mariscal [1,*]

[1] Department of Physiology, Biophysics and Neuroscience, Center for Research and Advanced Studies (Cinvestav), Mexico City 07360, Mexico; jamiguz_316@hotmail.com (J.M.); dolores@fisio.cinvestav.mx (D.M.-T.); lalarcon@fisio.cinvestav.mx (L.A.)

[2] Research Division, Instituto Nacional de Perinatología (INPer) Isidro Espinosa de los Reyes, Mexico City 11000, Mexico; yolotzin@gmail.com (Y.V.-V.); aurora_espnu@yahoo.com.mx (A.E.-N.); mguzmanhuerta@yahoo.com.mx (M.G.-H.); gpestrad@gmail.com (G.E.-G.)

[3] Laboratorio Central de Epidemiología, Instituto Mexicano del Seguro Social, Ciudad de México 02990, Mexico; jose.munozm@imss.gob.mx

[4] Department of Infectomics and Molecular Pathogenesis, Center for Research and Advanced Studies (Cinvestav), Mexico City 07360, Mexico; mineko@cinvestav.mx (M.S.); bchavez@cinvestav.mx (B.C.-M.); jludert@cinvestav.mx (J.E.L.)

[5] Quality division, Obstetrics and Gynecology Hospital No. 4, Mexican Institute of Social Security (IMSS), Mexico City 01090, Mexico; samuel.lievano@yahoo.com.mx

[*] Correspondence: lorenza@fisio.cinvestav.mx; Tel.: +52-55-5747-3966

Received: 9 July 2019; Accepted: 26 September 2019; Published: 29 September 2019

Abstract: The cytotrophoblast of human placenta transitions into an outer multinucleated syncytiotrophoblast (STB) layer that covers chorionic villi which are in contact with maternal blood in the intervillous space. During pregnancy, the Zika virus (ZIKV) poses a serious prenatal threat. STB cells are resistant to ZIKV infections, yet placental cells within the mesenchyme of chorionic villi are targets of ZIKV infection. We seek to determine whether ZIKV can open the paracellular pathway of STB cells. This route is regulated by tight junctions (TJs) which are present in the uppermost portion of the lateral membranes of STB cells. We analyzed the paracellular permeability and expression of E-cadherin, occludin, JAMs –B and –C, claudins -1, -3, -4, -5 and -7, and ZO-1, and ZO-2 in the STB of placentae from ZIKV-infected and non-infected women. In ZIKV-infected placentae, the pattern of expression of TJ proteins was preserved, but the amount of claudin-4 diminished. Placentae from ZIKV-infected women were permeable to ruthenium red, and had chorionic villi with a higher mean diameter and Hofbauer hyperplasia. Finally, ZIKV added to the basolateral surface of a trophoblast cell line reduced the transepithelial electrical resistance. These results suggest that ZIKV can open the paracellular pathway of STB cells.

Keywords: zika virus; flavivirus; tight junctions; claudins; ZO-1; blood-placental barrier; placenta

1. Introduction

The placental syncytiotrophoblast (STB) layer that covers the floating chorionic villi in direct contact with maternal blood in the intervillous space constitutes a major barrier to vertical transmission of parasites and microorganisms. The consequences of STB damage have been studied in first trimester placental explants where the subsyncytial cytotrophoblasts (CTB) becomes infected by *Listeria monocytogenes* if the STB layer is damaged with collagenase [1]. With *Toxoplasma gondii*, the situation is more complex, as multiparasite vacuoles in chorionic villi are present not only immediately below or besides syncytial interruptions, but also at sites of no visible syncytial interruptions, suggesting the possibility of hard to detect breaks in the STB layer or of earlier breaks that had subsequently healed [2]. These observations highlight the importance of considering syncytial permeability upon studying placental vulnerability to infection.

Several viruses cross the uterine-placental interface, infecting the fetus and causing birth defects, including rubella, varicella-zoster, parvovirus B19, human cytomegalovirus, hepatitis E type 1 and Zika (ZIKV) (see [3]). ZIKV is a flavivirus which is transmitted by *Aedes* mosquitoes; it was first identified in 1947 in an African forest, and triggered epidemics in the South Pacific in 2007 and in the Americas in 2015–2017 [4]. ZIKV infection during pregnancy is associated to an array of devastating birth defects known as congenital Zika syndrome, which include microcephaly, brain calcifications, neurological impairment, and retinal damage.

We are interested in exploring the mechanisms that allow ZIKV transplacental transmission in humans. During the development of human placenta, CTB epithelial cells may be differentiated in two ways. Firstly, they aggregate into cellular columns that invade the uterine interstitium and colonize the spiral arterioles, allowing the anchorage of the fetus to the mother and the flow of blood to the placenta. Secondly, CTB form a bilayer in which the cells of the external sheet fuse to form the multinucleated STB that covers the chorionic villi. The STB layer is crucial for the interchange of ions, nutrients, gases, and waste between the fetus and the mother. During pregnancy, as a result of syncytial ruptures or focal degeneration of STB, lateral cell membranes subdivide segments of STB from the surrounding STB continuum. This appears to be a dynamic process where the disconnected parts of the STB eventually fuse after the disintegration of the lateral separating membranes [5].

Tight junctions (TJs) regulate transit through the paracellular pathway of epithelial cells. In the STB, these cell-cell adhesion structures located at the uppermost portion of the lateral membranes that subdivide the STB layer constitute a cornerstone of the blood-placental barrier (BPB) that protects the fetus from toxins and pathogens. TJs in the STB of human placental chorionic villi have been observed by freeze fracture, and their function as paracellular seals has been demonstrated by electron microscopy in thin sections, with the blockade of the transit of electron-dense markers through the paracellular pathway [6–9]. The apical surface of the STB of chorionic villi expresses several TJ proteins, including the integral proteins occludin, claudins -1, -3 and -16, and the adaptor protein ZO-1 [9,10], while claudin-4 is strongly expressed during all trimesters of pregnancy [11], but localizes at the basal membrane of the STB [9].

ZIKV infects cells that strongly express TIM-1 [12], a cell surface phosphatidylserine and phosphatidylethanolamine receptor [13]. ZIKV in humans replicates in the glandular epithelium of the decidua and in decidual cells and infects invasive CTB, the CTB of cell columns, as well as fetal fibroblasts and macrophages known as Hofbauer cells which are present in the parenchyma of floating chorionic villi [12,14,15]. These observations led to the proposal of a ZIKV transmission route that goes from the cells of the basal decidua in the mother to the fetal invading CTB, followed by the infection of cell columns of CTB and Hofbauer cells present in the parenchyma of chorionic villi [12]. Moreover, since the envelope (E) proteins between the dengue virus (DENV) and ZIKV are very similar structurally [16], cross-reactive antibodies are generated that may enhance ZIKV infection [17]. In this respect, it has been observed that ZIKV infection in human placental explants are enhanced by the presence of dengue virus antibodies, suggesting that ZIKV immune complexes could use the neonatal Fc receptor for IgG as a transport system to transcytose across the STB layer to infect Hofbauer cells in the chorionic villi [18]. However, since the clinical severity of maternal ZIKV infection has not being associated with the existence of prior dengue antibodies [19], and not all pregnant women infected with ZIKV have been previously infected with DENV, here, we explore another complementary route of vertical ZIKV transmission. Taking into account that STB cells are not infected by ZIKV [12], we explore the possibility of ZIKV reaching the chorionic mesenchyme via transit through the paracellular route of the STB.

Virus passage through the paracellular route of epithelial and endothelial cells has previously been reported. Thus, after human airway epithelia infection with adenovirus, the viruses are first released to the basolateral surface and then escape to the apical surface. This process involves the binding of the fiber proteins of the adenovirus to its receptor CAR located within TJs, which triggers a disruption of junctional integrity that allows the virus to escape through the paracellular pathway between the cells to reach the apical surface [20]. Another case is that of human immunodeficiency virus (HIV), whose disruption of TJs and adherens junctions (AJs) in oral epithelial cells facilitates the paracellular spread of the herpes simplex virus 1 (HSV1), emerging as a mechanism to explain the rapid development of HSV-associated oral lesions in HIV infected individuals [21]. Likewise, sealing the blood-brain barrier (BBB) is compromised by HIV-1 induced inflammatory cytokines; in this respect, TNF-α has been shown to open the paracellular route for HIV-1 invasion across the BBB [22]. With regards to *Flaviviruses*, the West Nile virus disrupts the BBB in mice inducing an increase in BBB permeability and a reduced expression of TJ and AJ proteins [23]; moreover, in vitro infection of mouse brain endothelia with DENV delocalizes TJ proteins from the membrane to the cytoplasm, reduces the transendothelial electrical resistance, and increases the macromolecule permeability and the paracellular passing of free virus particles [24].

Here, we analyzed the permeability of the paracellular pathway and the molecular composition of TJs in the STB cell layer of placentae derived from ZIKV-infected women. In addition, in the trophoblast cell line BeWo, we observed that basolateral incubation with ZIKV reduces the transepithelial electrical resistance (TER).

Our results indicate that ZIKV infection alters the composition of placental TJs and increases paracellular permeability.

2. Materials and Methods

2.1. Ethics Statement

All subjects gave their informed consent for inclusion before they participated in this study. This study was approved by the Ethics Committee of the Instituto Nacional de Perinatologia (INPer) Isidro Espinosa de los Reyes in Mexico City (Register 212250-1000-10107-01-16) and San Judas Tadeo Hospital in Mexico City.

2.2. Patient Selection and Specimens

Placental tissues were obtained from 4 ZIKV-infected women and 4 control women. In the control women, the absence of ZIKV, Chikungunya virus (CHIKV), and DENV was confirmed by RT-PCR, done in placental tissue at the Central Epidemiological Laboratory of the Mexican Institute of Social Security (IMSS). For the 4 ZIKV-infected women, the RT-PCR diagnosis of ZIKV was made during the acute phase of ZIKV infection. In the newborns of the 4 ZIKV-infected women, detection of ZIKV was confirmed by RT-PCR on neonatal serum, amniotic fluid, and umbilical cord by the Mexican Institute of Diagnosis and Epidemiological Reference (InDRE), and the absence of CHIKV, DENV, and West Nile virus in these samples was confirmed by RT-PCR done by the National Institute of Respiratory Diseases (INER).

The pregnancies of ZIKV-infected women enrolled in this study were closely followed at INPer in Mexico City. A multidisciplinary clinical team, including maternal-fetal clinicians, geneticists, and pediatric neurologists, assessed the ultrasonographic diagnoses of described fetal alterations; prenatal findings were confirmed at birth. Immediately after giving birth by cesarean section, placental samples were fixed and processed as described below.

2.3. Cell Culture

BeWo cells were obtained from the American Type culture Collection (ATCC® CCL-98™) and cultured at 37 °C and 95% air 5% CO_2 in Ham's F12K medium (Cat. N3520, SIGMA-ALDRICH, St. Louis, MO, USA), supplemented with 10% fetal calf serum (Cat. 26140-079, Gibco, Grand Island, NY, USA) and 1% penicillin/streptomycin (Cat. 150770, Invitro, S.A., CDMX, Mexico). Cells were plated on Transwell filters (12-well plate, 0.4 μm, Cat. 3460, Corning, Kennebunk, ME, USA); 24 h later, the monolayers were incubated for 1.5 h with ZIKV at MOI = 1. Then, the monolayers were fixed for 10 min with methanol at −20 °C for claudin-4 immunofluorescence, or fixed in 4% para-formaldehyde in PBS at RT for 10 min and permeabilized with 0.2% Triton X-100 in PBS for occludin and ZO-1 immunofluorescence.

2.4. Immunofluorescence

Immunofluorescence on placental cryostate sections was done as previously reported [25], with the exception that the incubation with the first antibody was done overnight (ON) at 4 °C; the procedure of tissue fixation varied according to the protein to be detected as follows: (1) for E-cadherin, claudins -2 and -7, occludin, and ZO-1, the tissue was fixed in acetone at −20 °C for 5 min; (2) for claudins -3 and -4, and JAMs –B and –C, the samples were fixed in ethanol at 4 °C for 10 min followed by fixation in acetone at −20 °C for 3 min; (3) for ZO-2 and claudins -1 and -5, the tissue was fixed in 4% para-formaldehyde in PBS at RT for 10 min. As primary antibodies, a mouse monoclonal against cytokeratin 18 was employed (Cat. MAB1600, Chemicon International, Temecula, CA, USA; dilution 1:5000), together with one of the following rabbit polyclonal antibodies: anti E-cadherin (Cat. 3195, Cell Signaling Technology, Inc., Danvers, MA, USA; 1:300); anti claudin-1 (Cat. 51–9000, Invitrogen, Camarillo, CA, USA; dilution 1:100); anti claudin-3 (Cat. Ab52231, abcam, Cambridge, MA, USA; dilution 1:100); anti claudin-4 (Cat. 36–4800, Invitrogen, Camarillo, CA, USA; dilution 1:300); anti claudin-5 (Cat. ab15106, Abcam, San Diego, CA, USA; dilution 1:300); anti claudin-7 (Cat. 34–9100, Invitrogen, Camarillo, CA; dilution 1:100); anti occludin (Cat. 71–1500, Invitrogen, Carlsbad, CA, USA; dilution 1:100); anti ZO-1 (Cat. 61–7300, Invitrogen, Carlsbad, CA; dilution 1:300); and anti ZO-2 (Cat. 71–1400, Invitrogen, Carlsbad, CA; dilution 1:100). We also employed mouse monoclonal antibodies anti claudin-2 (Cat. sc-293233, Santa Cruz Biotechnology, Santa Cruz, CA, USA; dilution 1:100); and (1) anti JAM-B (Cat. sc-293496, Santa Cruz Biotechnology, Santa Cruz, CA; dilution 1:100); and a rat monoclonal antibody anti JAM-C (Cat. MCA5935, CRAM18F26, SeroTec, Kidlington, UK; dilution 1:100). As secondary antibodies, we employed a donkey antibody against mouse IgG coupled to Alexa594 (Cat. A21203, Invitrogen, Carlsbad, CA; dilution 1:1000), a donkey antibody against rabbit IgG coupled to Alexa-488 (Cat. 21206, Invitrogen, Carlsbad, CA; dilution 1:1000), and donkey antibody against goat IgG coupled to Alexa-488 (Cat. A21208, Invitrogen, Carlsbad, CA; dilution 1:1000).

2.5. Relative Mean Fluorescence Intensity Measurements

Relative mean fluorescence intensity measurements of AJ and TJ proteins at the STB were obtained using ImageJ (ImageJ 1.52n, NIH, Bethesda, MD, USA, 2019) with the Freehand function. An area named A, surrounding the chorionic villi, was selected. Then, another area named B, of the parenchyma of the same villi, immediately below the region of the STB, was selected. The integrated density feature of ImageJ was used to record pixel intensities of each of these two areas. Then, the integrated density of area A minus that of area B was recorded and compared to the fluorescent signal of the STB. Data were derived from three randomly-selected fields per placenta, and the images shown in the figures were one of the quantitated fields.

Data were derived from three images per placenta.

2.6. Western Blot

Western blots of placental lysates were done following standard procedures, as previously described [25]. As primary antibodies, we employed the following: rabbit polyclonal antibodies anti E-cadherin (Cat. 3195, Cell Signaling Technology, Inc., Danvers, MA; 1:3000); anti claudin-1 (Cat. 51–9000, Invitrogen, Camarillo, CA; dilution 1:1000); anti claudin-2 (Cat. 51–6100, Invitrogen, Camarillo, CA; dilution 1:1000); anti claudin-3 (Cat. Ab52231, Abcam, Cambridge, MA; dilution 1:500); anti claudin-4 (Cat. 36–4800, Invitrogen, Camarillo, CA; dilution 1:4000); anti claudin-5 (Cat. ab15106, Abcam, San Diego, CA; dilution 1:2000); anti claudin-7 (Cat. 34–9100, Invitrogen, Camarillo, CA; dilution 1:500); anti occludin (Cat. 71–1500, Invitrogen, Carlsbad, CA; dilution 1:1000); anti ZO-1 (Cat. 61–7300, Invitrogen, Carlsbad, CA; dilution 1:1000); and anti ZO-2 (Cat. 71–1400, Invitrogen, Carlsbad, CA; dilution 1:500). We also employed mouse monoclonal antibodies anti JAM-C (Cat. sc-515893, Santa Cruz Biotechnology, Santa Cruz, CA; dilution 1:500) and anti-JAM-B (Cat. sc-293496, Santa Cruz Biotechnology, Santa Cruz, CA; dilution 1:500). As secondary antibodies, we employed goat antibodies against rabbit IgG coupled to peroxidase (Cat. 6111620, Invitrogen, Camarillo, CA; dilution 1:20,000), or against mouse IgG coupled to peroxidase (Cat. 626520, Invitrogen, Camarillo, CA; dilution 1:10,000).

2.7. Histochemical Staining

Chorionic villi derived from the placentae of control and ZIKV-infected women were fixed in 10% para-formaldehyde, embedded in paraffin, and cut in 8 μm sections. Then sections were stained with hematoxylin and eosin, or with Masson's trichrome stain, following standard protocols [26]. The identification of Hofbauer cells in these slides was done with a rat monoclonal antibody against CD68 (Cat. ab53444, Abcam, San Diego, CA; dilution 1:1,000).

2.8. Transmission Electron Microscopy (TEM)

The placental tissues of control and ZIKV-infected women were fixed in 2.5% glutaraldehyde in 0.1 M sodium cacodylate buffer for 1 h at room temperature (RT). Then, the tissue was incubated for 20 h at 4 °C in 2.5% glutaraldehyde in 0.1 M sodium cacodylate buffer, washed thrice with 0.1 M sodium cacodylate buffer, and incubated for 1 h at RT with 1% osmium tetroxide in 0.1 M sodium cacodylate buffer containing 0.5 mg/mL of ruthenium red, a marker of the paracellular pathway. Samples were then processed for TEM, as previously described [27].

2.9. Measurement of Transepithelial Electrical Resistance (TER)

The transepithelial electrical resistance of BeWo cells plated at confluence on Transwell filters (12-well plate, 0.4 μm, Cat. 3460, Corning, Kennebunk, ME) was continuously measured from each insert using the automated cell monitoring system, cellZscope (nanoAnalytics GmbH, Munster, Germany). TER values were obtained using the cellZscope software version 4.3.1. Twenty-four hours after plating, when the monolayers had achieved a steady state value of TER, cells were infected with ZIKV (strain PRVABC-59-Asian, ATCC® VR-1843™) at a MOI of 1.

2.10. Statistical Analysis

The data derived from control and ZIKV-infected placentae obtained from the relative mean fluorescence intensity images and Western blot densitometry analysis were compared for statistically significant differences using the Mann-Whitney U test. The variances obtained with claudin-4 results were equal among the two groups; therefore, we applied a Student's t-test to include effect size estimations; since both groups had similar standard deviations, we used Cohen's d. For the Western blots of occludin, the data were further analyzed to include effect size estimation with Glass'delta, since each group had a different standard deviation. The Graphs were generated with GraphPad Prism 5.01 software (GraphPad Prism 5.01, La Jolla, CA, USA, 2007).

3. Results

3.1. The Trophoblast of Chorionic Villi in Human Placentae Expressed E-cadherin and a Wide Array of TJ Proteins

In this study, we analyzed the expression of TJ proteins in the placentae from control and ZIKV-infected women. The clinical data and perinatal outcomes of control and ZIKV-infected women during pregnancy and their newborns are summarized in Tables 1 and 2. With regards to the newborn of ZIKV-infected donor D8, that carries a 22q11 deletion, all the reported phenotypic characteristic of the child conform with 22q11 deletion syndrome.

Table 1. Clinical data of women with ZIKV during pregnancy and physical findings of their newborns.

			Mother		
Donor	**Age (years)**	**Birth (GW)**	**Zika Symptoms Onset (GW)**	**Rash**	**Pregnancy Complications**
D1	31	36.6	11.5	Yes	Preeclampsia
D5	35	35.1	7.6	Yes	Preeclampsia
D8	30	36.0	13.0	Yes	None
D10	29	36.0	15.0	Yes	None
			Newborn		
Sex	**Weight (kg)**	**Height (cm)**	**Apgar**	**HC (cm)**	**Outcome**
F	2.84	47	8/9	33.0	Right pulmonary cystic adenomatoid malformation
F	2.92	47	6/9	36.0	Ventriculomegaly, macrocephaly, hydrocephaly, and hip dysplasia
F	3.00	48	8/9	33.8	22q11 deletion syndrome with mielomeningocele, hypotonia and right aortic arch
F	3.03	50	8/9	34.0	Brachycephaly, low-set ears, short neck, and widely-spaced nipples

GW, Gestational week; HC, Head circumference; F, Female.

Table 2. Clinical data of control women during pregnancy and physical findings of their newborns.

		Mother	
Donor	Age (years)	Birth (GW)	Pregnancy complications
D9	32	38.0	Preeclampsia
D17	30	38.8	Preeclampsia
D18	23	40.0	None
D19	34	36.0	None
		Newborn	
Sex	Weight (kg)	Height (cm)	Apgar	HC (cm)	Outcome
M	2.20	50	8/9	35.0	Healthy
F	2.90	49	8/9	34.0	Healthy
M	3.20	51	8/9	36.0	Healthy
F	2.97	50	9/9	35.0	Healthy

GW, Gestational week; F, Female; M, Male; HC, Head circumference.

In the trophoblast of chorionic villi in placentae derived from control and ZIKV-infected women during pregnancy, we explored the expression of the AJ protein E-cadherin and of the following TJ proteins: claudins -1, -2, -3, -4, -5, -7, and -10; JAMs -A, -B, and -C; occludin, ZO-1, and ZO-2 (Table 3). All these proteins, with the exception of claudins -2, -10, and JAM-A (data not shown), were expressed in the STB of chorionic villi.

Table 3. Expression of E-cadherin and TJ proteins in the trophoblast of placental chorionic villi derived from control and ZIKV-infected women.

Protein	STB Distribution	Abundance in Chorionic Vessels	ZIKV-infected vs. Control Placentae
E-cadh	basal	−	=
Cl-1	apical/basal	+++	=
Cl-2	nd	−	
Cl-3	apical/basal	+++	=
Cl-4	basal	+++	↓
Cl-5	apical/basal	+++	=
Cl-7	basal	±	=
Cl-10	nd	−	
JAM-A	nd	−	
JAM-B	apical/basal	−	=
JAM-C	bd	+++	=
Occ	apical/basal	+++	=
ZO-1	apical/basal	+++	=
ZO-2	apical/basal	±	=

BD, barely detected; ND, not detected. −, absent; ±, barely detected; +++ highly abundant; =, no change; ↓, decrease.

3.2. E-cadherin Stained the Basal Membrane of the STB and Its Expression was not Affected in ZIKV-Infected Placentae

In chorionic villi of human placentae, E-cadherin was observed not at the apical surface of the STB layer in contact with the intervillous space, which in situ is occupied by maternal blood, but in the basal membrane of the STB in contact with remnants of the CTB layer and the chorionic mesenchyma (Figure S1a). In this and most of the subsequent images, the trophoblast layer was identified with cytokeratin 18, a canonical marker of epithelial cells [28]. The intensity of the fluorescent signal of E-cadherin in the STB layer was not affected in placentae derived from ZIKV-infected women (Figure S1b). Likewise, by Western blot, no difference in the amount of E-cadherin was observed in placental lysates from control and ZIKV-infected women (Figure S1c,d).

3.3. The Expression of Claudin-1 Slightly Increased in Placentae Derived from ZIKV-Infected Women

By immunofluorescence, we observed that claudin-1 stained fetal endothelia within the chorionic villi parenchyma as well as the STB layer. The relative mean fluorescence intensity of the signal in the STB layer slightly increased in placentae from ZIKV-infected women in comparison to control placentae, albeit not at a significant level (Figure S2a,b). By Western blot, we observed that the amount of claudin-1 was not significantly higher in placentae derived from ZIKV-infected women in comparison to control placentae (Figure S2c,d).

3.4. The Expression of Claudin-3 was Strong in Placental Vessels and Faint at the STB Layer of Both Control and ZIKV-Infected Placentae

Claudin-3 was strongly expressed in the vessels of the parenchyma in chorionic villi, while staining at the STB cell layer was weak in both control and ZIKV-infected placentae (Figure S3a,b). By Western blot, we detected no difference in the amount of claudin-3 between control and ZIKV-infected placentae (Figure S3c,d).

3.5. The Expression of Claudin-4 at the Basolateral Membrane of STB Diminished in ZIKV-Infected Placentae

In the chorionic villi of control human placentae, claudin-4 was observed at the basolateral membrane of the STB in contact with the chorionic parenchyma. The same pattern was observed in the placentae from ZIKV-infected women; however, in this case, the fluorescent signal at the STB layer was less intense (Figure 1a,b). Western blot analysis confirmed the decreased expression of claudin-4 in ZIKV-infected placentae in comparison to those of the control (Figure 1c,d).

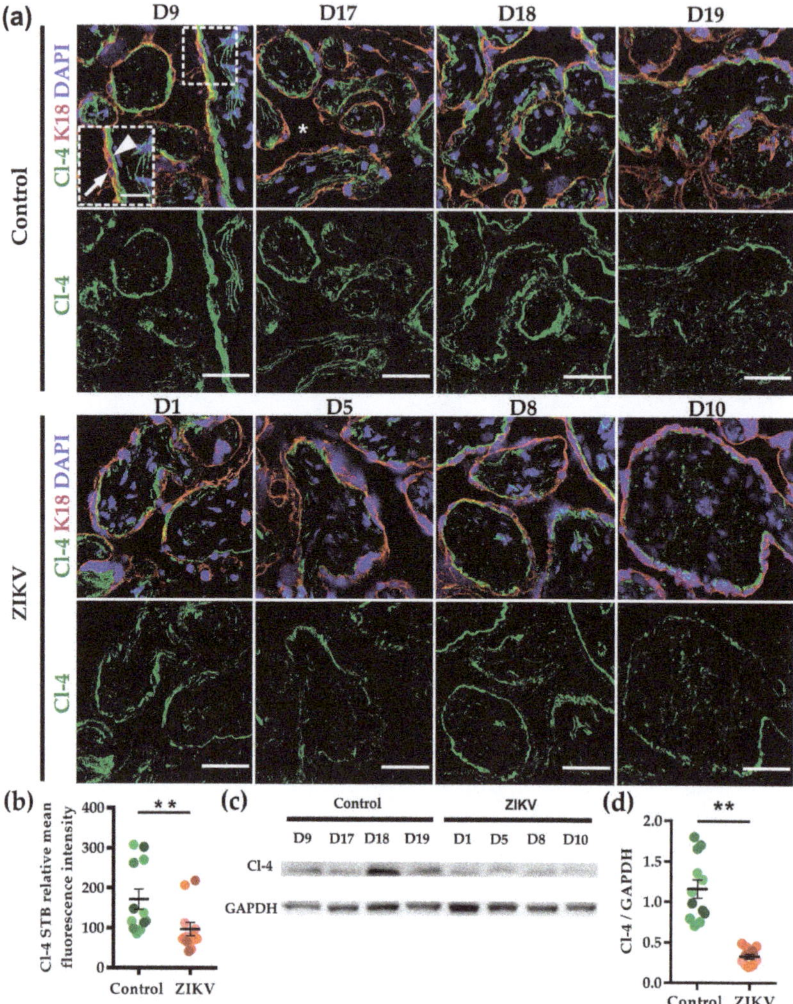

Figure 1. Claudin-4 present at the basal surface of the STB layer diminished in ZIKV-infected placentae. (**a**) Frozen sections of human placentae derived from women infected or not (control) with ZIKV were processed for immunofluorescence with a rabbit antibody against claudin-4 (Cl-4) and a mouse antibody anti cytokeratin 18 (K18). DNA of nuclei was stained with DAPI. Apical surface of STB cell layer (arrow); basolateral surface of STB cell layer (arrowhead); intervillous space (asterisk). Bar, 50 μm; magnification bar, 25 μm. (**b**) Measurements of mean fluorescence intensity of the trophoblast layer were done on three independent images from each condition. Since the variances were equal among the two groups, we applied a Student's *t*-test. To include effect size estimation, since both groups have similar standard deviations, we used Cohen's *d*. The three values obtained per donor are represented by dots with the same color. ** *p* = 0.0054, Cohen's *d* = 3 indicating that 99.9% of the values from the ZIKV group are below the mean value of the control group. (**c**) Representative Western blot of three independent experiments. Glyceraldehyde 3-phosphate dehydrogenase (GAPDH) was employed as loading control. (**d**) Densitometric analysis of Western blots. The three values obtained per donor are represented by dots with the same color. ** *p* = 0.00092 Cohen's *d* = 2.67 indicating that 99.9% of the values from the ZIKV group are below the mean value of the control group.

3.6. Claudin-5 Strong Expression in the Vessels of Chorionic Villi and Faint Stain in STB Cells Did not Change with ZIKV Infection

Claudin-5 was previously identified as an endothelial claudin [29], and accordingly, strongly stained the vessels within the chorionic parenchyma. In addition, claudin-5 faintly stained the STB cell layer in both ZIKV-infected and in control placentae (Figure S4a,b). Western blot analysis also revealed that ZIKV infection induced no change in the amount of claudin-5 expressed in the placentae (Figure S4c,d). The Western blot was done with a placental lysate, and hence, contains claudin-5 from both the vessels and the STB. In the case of claudin-5, this situation is particularly relevant, as this claudin is more abundant in vessels than in the STB cell layer.

3.7. Claudin-7 Expression at the Basal Surface of STB Cells was not Altered by ZIKV Infection

In the chorionic villi of control and ZIKV-infected placentae, claudin-7 was preferentially expressed in the basal membrane of STB, although a faint staining was also detected at the apical surface (Figure S5a,b). Immunofluorescence quantitation of claudin-7 at the STB cell layer and Western blot analysis showed no change in claudin-7 abundance in ZIKV-infected placentae (Figure S5c,d).

3.8. Occludin Expression in the STB was Low and not Affected by ZIKV Infection

In chorionic villi, occludin was strongly expressed in parenchymal vessels and faintly stained the STB cell layer in both control and ZIKV-infected placentae (Figure 2a,b). Western blot analysis showed a decrease in occludin expression in response to ZIKV infection (Figure 2c,d). These Western blots, however, most likely reflect the amount of occludin present in endothelial TJs rather than in the STB cell layer.

3.9. The Expression of ZO-1 was Higher in Chorionic Vessels than in the STB and was not Affected by ZIKV Infection

ZO-1 strongly stained the vessels in the chorionic villi and delineated the STB layer in a moderate manner, although in some placentae from both control (D19) and ZIKV-infected women (D10), labeling at the STB was more intense (Figure S6a,b). Western blot analysis revealed a similar amount of ZO-1 in both group of placentae, and most likely reflects the content of ZO-1 at endothelia, due to the higher expression of this protein in vessels in comparison to the STB (Figure S6c,d).

3.10. In Chorionic Villi, the Expression of ZO-2 in the STB was Much Higher than in the Mesenchymal Vessels

ZO-2 in the chorionic villi of human placentae was clearly expressed in the STB, while it was barely present in the vessels of the mesenchyme. By immunofluorescence, only in one ZIKV-infected placentae (D8), the amount of ZO-2 increased in comparison to the rest of the placentae (Figure S7a,b) and the Western blot analysis showed no change in ZO-2 expression induced by ZIKV infection (Figure S7c,d).

3.11. ZIKV Infection Had no Impact on the Expression of JAMs -B and -C in the STB of Chorionic Villi

JAM-B stained with a similar intensity the STB cell layer of both control and ZIKV-infected placentae. Dots of JAM-B were also present in the chorionic mesenchyme, but not along the TJs of chorionic vessels, which were conspicuously stained with ZO-1 (Figure S8a,b). By Western blot, similar amounts of JAM-B were found between control and ZIKV-infected placentae (Figure S8c,d). In contrast, JAM-C was not observed at the STB cell layer and profusely stained the endothelial vessels in the chorionic mesenchyme. In the control and ZIKV-infected placentae, the same JAM-C staining pattern was observed (Figure S9a,b), and no change in the amount of JAM-C was detected by Western blot after ZIKV infection (Figure S9c,d).

In summary, the exploration of E-cadherin and a wide variety of TJ proteins revealed that E-cadherin and claudins -4 and -7 were present in the basal membrane of the STB, while all the other TJ proteins studied localized in both the apical and basal membranes of the STB. The pattern of expression of these proteins was preserved in ZIKV-infected placentae, although in this pathological condition, the amount of claudin-4 diminished in the STB.

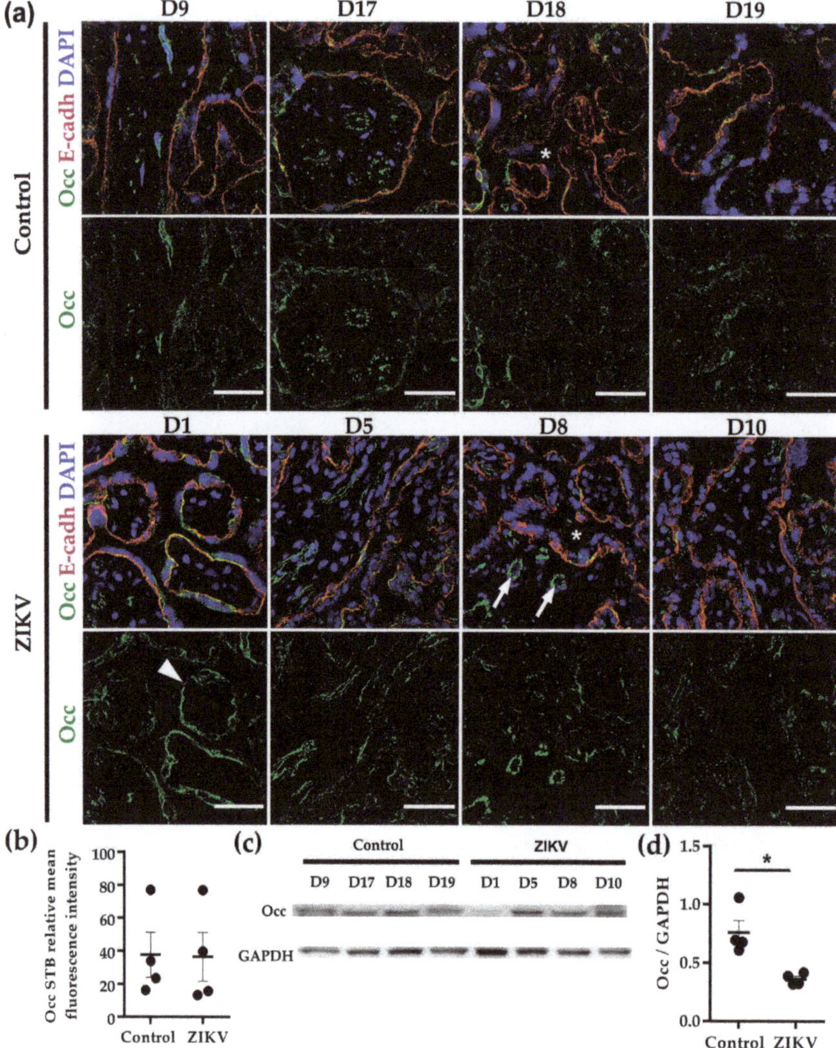

Figure 2. Occludin expression at chorionic vessels is stronger than in the cell layer. (**a**) Frozen sections of human placentae derived from women infected or not (control) with ZIKV were processed for immunofluorescence with a mouse antibody against occludin (Occ) and a rabbit antibody anti E-cadherin (E-cadh). DNA of nuclei were stained with DAPI. Chorionic vessels (arrows); STB (arrowhead); intervillous space (asterisks). Bar, 50 μm. (**b**) Measurements of mean fluorescence intensity of trophoblast layer were done on three independent images from each condition. (**c**) Representative Western blot of three independent experiments. GAPDH was employed as loading control. (**d**) Densitometric analysis of Western blots. * $p = 0.0286$. The data were further analyzed to include effect size estimation with Glass'delta since each group had a different standard deviation. The value of Glass'delta = 1.955 indicates that 99.9% of the values from the ZIKV group are below the mean value of the control group.

3.12. The STB Layer of Placentae of ZIKV-Infected Women is Permeable to Ruthenium Red

Since claudin-4 functions as a cationic barrier [30] or an anion pore [31] that increases TER in cationic and anion selective cell lines and decreases paracellular permeability in the cationic selective cell line MDCK [32], we next determined whether the TJs of ZIKV-infected placentae were leaky. For this purpose, placental tissue was fixed and processed for TEM in the presence of ruthenium red, an electron-dense paracellular marker. In all the ZIKV-infected placentae, we observed ruthenium red staining in the paracellular pathway bellow the TJ region in the STB cell layer, in contrast to control placentae, where, as we had previously shown [9], staining was restricted to the apical membrane of the STB (Figure 3).

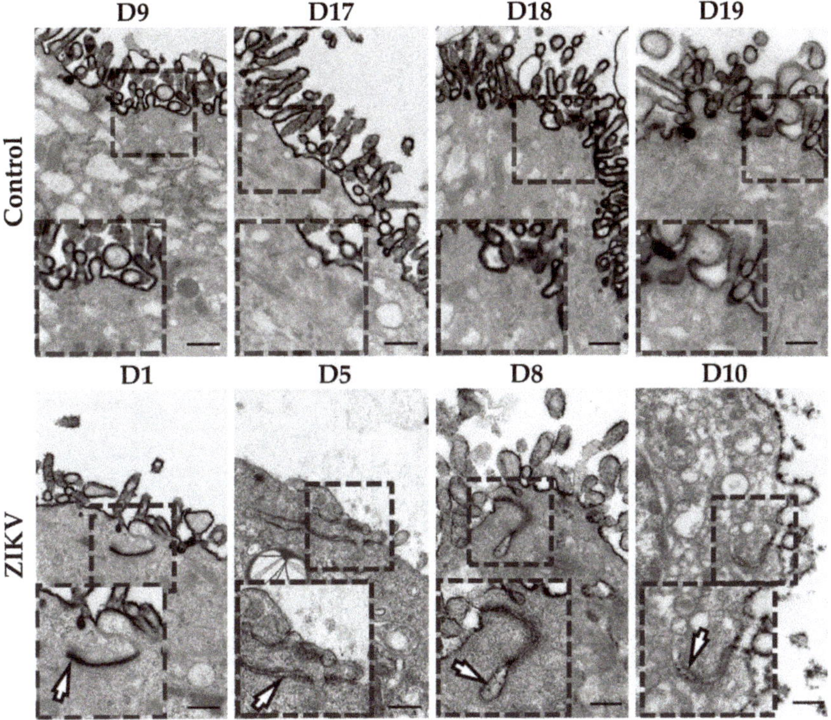

Figure 3. The STB in the placentae of ZIKV-infected women was permeable to ruthenium red. Placental tissue was fixed and processed for TEM in the presence of ruthenium red. Ruthenium red staining in the paracellular pathway (arrows). Bar, 1 μm.

3.13. Hofbauer Cell Hyperplasia, an Increased Diameter of Microvilli and Intravillous Calcifications were Observed in ZIKV-Infected Placentae

The chorionic villi derived from women infected with ZIKV during pregnancy displayed several histological alterations, including: (1) Hofbauer cell hyperplasia, evaluated by counting the number of CD68+ cells in the parenchyma of floating chorionic villi stained with hematoxylin (Figure 4a,b); (2) intravillous calcifications observed in hematoxylin and eosin-stained samples, which show a tendency to increase in ZIKV placentae in comparison to control, but whose difference is not statistically significant (Figure 4c,d); and (3) a higher diameter of chorionic microvilli in hematoxylin and eosin-stained samples (Figure 4e,f).

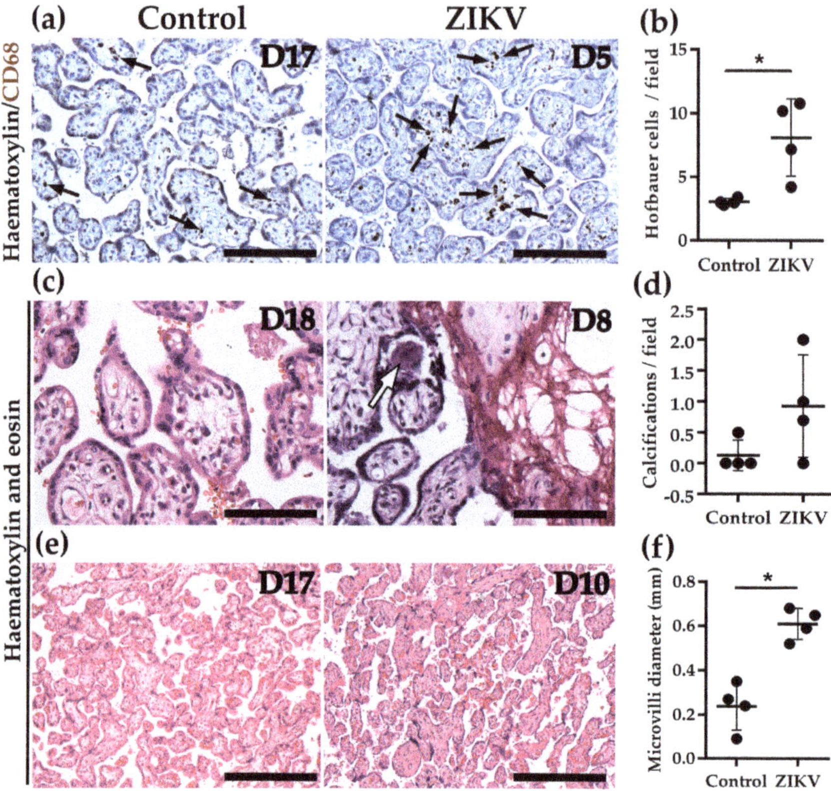

Figure 4. The chorionic villi of ZIKV-infected placentae display Hofbauer cell hyperplasia and a higher mean diameter. (**a**) Hofbauer cells (arrows) in chorionic villi were detected with an antibody against CD68 in slides stained with haematoxylin. Bar, 100 μm. (**b**) The number of Hofbauer cells was evaluated counting CD68+ positive cells in five optical fields per placenta. Each dot corresponds to the mean value of Hofbauer cells/field present in each placenta. * $p = 0.0294$. (**c**) Calcification (arrow) present in a chorionic villus detected in a slide stained with haematoxylin and eosin. Bar, 50 μm. (**d**) Calcifications were counted in five optical fields per placenta. Each dot corresponds to the mean value of calcifications/field present in each placenta. (**e**) The diameter of chorionic microvilli is higher in ZIKV-infected placentae than in the control condition. Bar, 200 μm. (**f**) The diameter of chorionic villi was measured using the image analysis software Zen (version ZEN 2.3 lite, Carl Zeiss Microscopy, Jena, Germany) in five optical fields of samples from placentae stained with haematoxylin and eosin. Each dot corresponds to the mean diameter of chorionic villi per placenta. * $p = 0.0286$.

Some placentae infected with ZIKV also displayed chorionic villi edema, heterogeneous maturation of chorionic villi, characterized by the co-existence of villi with different diameters; increased syncytial knots due to premature aging; and karyorrhexis, the irregular distribution of chromatin in the cytoplasm due to the destructive fragmentation of the nucleus of dying cells (Figures S10–S13). In contrast, tissue sections of chorionic villi from the placentae of control women displayed a homogeneous maturation of chorionic villi, an absence of inflammatory cells, and a parenchyma without abundant Hofbauer cells (Figures S14–S17).

We also performed a Masson's trichrome stain in the chorionic villi of D1 placenta from a ZIKV positive woman in order to detect pathological changes involving the connective tissue. Supplemental

Figure S18 reveals perivascular fibrosis and abundant intravilli collagen, as well as mesenchymal edema and karyorrhexis, which were previously observed with the hematoxylin and eosin stain.

3.14. ZIKV Added to the Basolateral Surface of the Trophoblast-Derived Cell Line BeWo Reduces the Transepithelial Electrical Resistance and Claudin-4 Expression

To further explore the effect of ZIKV on TJs of the trophoblasts, we incubated ZIKV at a MOI of 1 with the trophoblast-derived choriocarcinoma cell line BeWo cultured on Transwell filters. We worked with BeWo cells because they constitute a human cell culture model of placental villous trophoblast, and because the STB layer of human placenta is poorly susceptible to infection by ZIKV [33]. Figure 5a shows that the apical administration of ZIKV immediately increases TER by 34%, reaches 40% above control after 6 h, and diminishes to 27% above control after 24 h. Instead, the administration of ZIKV to the basolateral surface of BeWo cells induced a fast drop of TER of 20% in comparison to control monolayers at 1.5 h. However, after 10 h, the values of TER had recovered and were undistinguishable from those of control monolayers. Hence, these results reveal that ZIKV in contact with the basolateral surface of trophoblast cells is able to reduce the degree of sealing of TJs.

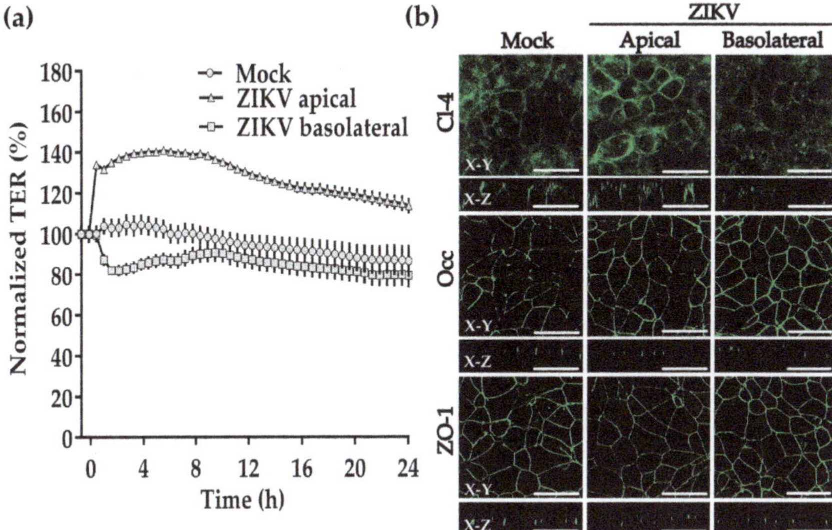

Figure 5. ZIKV added to the basolateral surface of BeWo cells transiently diminishes the transepithelial electrical resistance and claudin-4 expression. Confluent monolayers of BeWo cells that had achieved a stable value of TER were incubated with ZIKV (MOI = 1) added to the apical or basolateral surface. (**a**) TER was continuously measured in the cellZscope system in three inserts per condition. Results are shown with the corresponding standard deviation. (**b**) Immunofluorescence for claudin-4, occludin and ZO-1 in BeWo monolayers done after 1.5 h of incubation with ZIKV. Bar, 50 μm. X-Y, en face view; X-Z, lateral view.

By immunofluorescence, we observed no change in the expression of occludin or ZO-1 in BeWo cells, 1.5 h after the addition of ZIKV to the apical or basolateral surfaces (Figure 5b). Instead, the expression of claudin-4 diminishes 1.5 h after ZIKV is added to the basolateral surface. In addition, while occludin and ZO-1 concentrate as spots at the TJ region in immunofluorescence z-sections, claudin-4 distributes along the basolateral membrane, as had been previously observed in intestinal Caco-2 cells [34] These results thus confirm the capacity of ZIKV to diminish the expression of claudin-4 in trophoblasts.

4. Discussion

TJs in the STB are a cornerstone of the BPB that protects the fetus from toxins and pathogen infections. Given the devastating consequences of ZIKV fetal infection, in this work, we explored the expression of E-cadherin and TJ proteins in the STB of human placentae of women infected with ZIKV. By immunofluorescence, we observed that E-cadherin, claudins -1, -3, -4, -5, and -7; JAM-B; occludin, ZO-1, and ZO-2 were present in the STB of chorionic villi from both control and ZIKV-infected women. Only JAM-A and claudins -2 and -10 (data not shown) were not detected in these tissues. In contrast, in mice, a strong induction of claudin-10 was observed in the decidua at pregnancy day 4.5, although in humans, claudin-10 was not detected in first trimester decidual cells [35].

Claudins -1, -3, and -5, JAM-B, occludin, ZO-1, and ZO-2 are present in both the apical and basal surfaces of the STB layer, whereas E-cadherin and claudins -4 and -7 concentrate at the basal surface of the STB in contact with the underlying CTB or chorionic parenchyma. This pattern of expression was not altered by ZIKV infection.

The TJ proteins that were more conspicuously expressed in the STB layer were claudins -1, -3, -4, -7, JAM-B, occluding, and ZO-2. These results agree with previous observations showing strong expression of claudin-4 in trophoblastic cells during all trimesters of human pregnancy [11] and with our previous results with occludin and claudins -1, -3, and -4 in the placentae of control women and with preeclampsia [9]. In addition, in mice placentae, a real-time PCR study revealed a higher level of expression of mRNA for claudins -1, -2, -4, and -5 in comparison to all other claudins in the family, while a Western blot revealed an increased expression of claudin-4 and -5 and a decrease in the content of claudin-2 as pregnancy advances from day 12 to 20 [36].

Here, we observed that both occludin and ZO-1 delineated the STB layer in a moderate manner in comparison to the strong staining observed in the chorionic vessels. Therefore, the decreased content of occludin detected by Western blot in the placentae from ZIKV-infected women is most likely due to a reduction in the amount of occludin present in endothelia, particularly since the immunofluorescence signal of occludin at the STB layer did not decrease in the placentae from ZIKV-infected women. Therefore, we think that the diminished expression of occludin in the endothelia of the chorionic parenchyma does not contribute to the leakiness observed in the STB layer of chorionic villi in placentae derived from ZIKV-infected women.

These observations, however, do not minimize the importance of ZO-1 and occludin for a healthy STB, particularly since in hydatidiform moles, characterized by hyperplasia of the trophoblastic tissue and distention of the chorioninc villi by fluid, the expression of ZO-1 and occludin is downregulated and their distribution in the STB changes from the cell borders to the cytoplasm [10]. In addition, ZO-1 appears to play a crucial role in the fusion of trophoblastic cells into a syncytium, as ZO-1 expression at intercellular boundaries decreases during fusion; the treatment of primary human trophoblastic cells in culture with ZO-1 siRNA blocks this process [37].

The expression of occludin in fetal vessels of human placenta is observed mainly at term [10], and occurs in the secondary chorionic villi in large and intermediate vessels but not in terminal exchange vessels, as we and others have shown [9,38]. Instead, the expression of *ZO-1* in fetal endothelia has been observed throughout gestation [10] and among the whole placental vascular tree [38]. In this respect, it is interesting to note that one of the reasons that ZO-1 knock out mice are embryonic lethal is due to abnormal angiogenesis in the yolk sac [39].

With regards to ZO-2, it is interesting to note that in contrast to *ZO-1*, this protein was abundantly expressed in the STB layer of human placentae but barely detectable in chorionic vessels. Unlike ZO-1 knock out mice, ZO-2 knock out mice do not die due to alterations in angiogenesis [39]. Instead, ZO-2 knock out mice are embryonic lethal due to defects in the development of the extraembryonic tissue. This was demonstrated when the injection of ZO-2$^{-/-}$ embryonic stem cells into wild type blastocysts generated viable ZO-2 chimera mice [40].

Our results indicate a reduced expression of claudin-4 in BeWo cells when ZIKV was added to the basolateral surface and in the STB layer of chorionic villi from ZIKV-infected women. The change observed in the chorionic villi cannot be attributed to the preeclampsia present in two of the ZIKV-infected donors, because the placentae of ZIKV-infected women without preeclampsia also exhibited a reduced claudin-4 expression. In addition, a histochemistry study reported a slight increase in *claudin-4* expression in the trophoblasts of preeclamptic placentae [11]. Finally, by Western blot, we previously observed that the amount of claudin-4 in the chorionic villi does not vary with preeclampsia [9].

The differences in the gestational age between the ZIKV-infected placentae and the controls arose because women infected with ZIKV were subjected to earlier cesarean sections due to their high-risk pregnancies in order to better protect fetal health, while healthy women in the control group had cesarean sections at the expected time for full-term pregnancies. Nevertheless, the differences in claudin expression here observed cannot be ascribed to the variation in gestational age between the groups, because in humans, there are no differences in the level of expression of claudin-4 in trophoblastic cells of chorionic villi, between the three trimesters of pregnancy [11].

The combination and mixing ratios of claudin species determines the barrier properties of TJ strands [41], and the alteration of a single type of claudin can significantly alter the permeability and transepithelial electrical resistance of a tissue [30,32]. Therefore, the decrease in claudin-4 expression in placentae from ZIKV-infected woman may have a big impact in the paracellular transit through the STB, particularly since this claudin functions as a cation barrier [30,32] or an anion pore [31]. Accordingly, a significant decrease in permeability and an increase in TER have been observed in MDCK cells after claudin-4 transfection [30,32].

Claudin-4 in the renal collecting duct interacts with claudin-8, and their association is required to form a paracellular chloride channel [31]. Therefore, it may be important in the future to determine whether claudin-8 is expressed together with claudin-4 in the STB layer in human placentae, and whether its expression is altered in ZIKV-infected placentae.

Claudin-4 is a critical claudin for the establishment of a permeability barrier to protect the developing embryo. When the blastocyst enters the uterus, the process of implantation and placentation starts. The first contact is established between the blastocyst trophoectoderm and the uterine epithelium. Once the blastocyst attaches, the process of decidualization is triggered, involving the stromal epithelial transition in which uterine stromal cells differentiate into decidual cells surrounding the implanting blastocyst. In this event, the trophoectoderm acts as a stimulus for the creation of a TJ permeability barrier in stromal cells that protects the embryo from the passage of injurious maternal immunoglobulins [42,43]. In rat stromal cells of the uterus, we have observed that claudin-1 is present in all gestational days, and that ZO-1 appears until day 6, albeit at both implantation and non-implantation sites, while claudins -3 and -4 appear until gestational day 7 and only at implantation sites [44], reinforcing the view of claudin-4 as an important TJ proteins for embryonic development. Moreover, in rat uterus, by the time of implantation of the blastocyst at gestational day 6, when the network of TJ strands increases 3-fold in depth along the lateral plasma membrane and displays more branches and interconnections with neighboring strands [45,46], claudin-4 is detected for the first time at the basolateral membrane of uterine epithelial cells [44]. Similarly, in human endometrium, an increase in claudin-4 mRNA is found during the implantation window [47–49], thus suggesting a critical role of claudin-4 during implantation.

The role of claudin-4 in human placenta is highlighted by the observation that its expression increases in hydatidiform moles and in maternal diabetes [11]. In human placentae derived from assisted reproductive technology, claudin-4 mRNA diminishes, and this change is accompanied by an increase in claudin-8 mRNA [50]; while in mice placentae, claudin-4 augments as gestation advances from days 12 to 20 and after the administration of the estrogen receptor antagonist ICI 182,780 and the progesterone receptor antagonist RU-486 [36].

How ZIKV alters the Cl-4 expression of STB without actively replicating in these cells remains an open question. It has been observed that a wide array of viruses use integral proteins located at the apical junctional complex (AJC) of epithelial cells (for review see [51]) as cellular receptors, including for example hepatitis C virus, that associates to claudins -1, -6, and -9 [52,53]. The use of such proteins is important for the entry of viruses into epithelial cells, but also implies the disruption of the AJC that compromises the integrity of the epithelial barrier in consequence. Thus, West Nile virus specifically induces the endocytosis of claudin-1 and JAM-A [54], and in rotavirus, the VP8 protein, generated from the proteolytic cleavage of spike protein VP4, opens the TJs of epithelial cells in a reversible and dose-dependent manner [55]. VP8 has several segments with high similarity to domains present in the extracellular loops of claudins and occludin; hence, it was proposed that VP8 opened the TJs by competing with the homotypic interactions established among the extracellular domains of certain claudins and occludin. Another interesting protein in this respect, derived from a microorganism, is *Clostridium perfringens* enterotoxin (CPE). This toxin opens the paracellular barrier due to the selective removal of claudins -4 and -3 from the TJ [56,57]. In the case of ZIKV, the effect of structural E and M proteins on TJ proteins has not yet been explored.

The observation that only the basolateral exposure of ZIKV reduced TER and claudin-4 expression of BeWo monolayers is not surprising, as a similar situation had been observed with several viruses. Thus, the adenovirus fiber protein can only access its CAR receptor at the apical junctional complex and open the TJs when added to the basolateral surface [20]; HSV-1 only infected epithelial cells if added to the basolateral surface or if depletion of extracellular calcium had weakened the strength of the AJC to allow the virus to access its nectin receptor [58,59]; and hepatitis C virus first localizes with the epidermal growth factor receptor at the basolateral membrane and then accumulates at the TJ and associates to claudin-1 and occludin [60,61]. These results suggest that ZIKV passed from the maternal basal decidua to the fetal invading CTB and the cell columns of CTB, or through the transport system facilitated by the neonatal Fc receptor to transcytose across the STB layer, could open the paracellular pathway of the STB layer due to its presence in the parenchyma of chorionic villi that faces the basolateral surface of STB cells. Hence, the opening of the TJ in the STB could occur not as an initial step in the vertical transmission of ZIKV, but as a consequence of chorionic villi infection.

The heterogeneous maturation of chorionic villi, Hofbauer cell hyperplasia, and intravillous calcifications that we observed in ZIKV-infected placentae have also been reported in other studies [15,62–64]. In this respect, alterations in Hofbauer cells homeostasis are known to be associated with placental pathologies involving infection, inflammation, and inadequate placental development [65]. With regards to the diameter of chorionic villi, as the third trimester of pregnancy advances, stem villi branch into distal villi; in consequence, the diameter of chorionic villi decreases [66]. Thus, the higher diameter of chorionic villi observed in placentae from ZIKV-infected women suggests villous maldevelopment, although an effect due to the different ages of ZIKV and control placenta cannot be disregarded. Nevertheless, it should also be mentioned that this effect could be related to the altered expression of claudin-4, since in both mice and humans, during placental development, frizzled 5 induces the disassociation of cell junctions for chorion branching initiation through the downregulation of ZO-1, claudin-4 and claudin-7 in trophoblast cells [67]. Therefore, another important aspect to study in the future in ZIKV-infected placentae could be the expression of frizzled 5.

In summary, our results indicate that the chorionic villi of placentae from women infected with ZIKV display Hofbauer cell hyperplasia, an increased diameter of microvilli and intravillous calcifications, while the study of the STB layer of these placentae shows a decreased expression of claudin-4 and ruthenium red permeability, suggesting that these placentae are leakier than the normal, control ones (Figure 6). These observations allowed us to propose the paracellular pathway of the STB layer as a route of vertical transmission of ZIKV. However, the observation that ZIKV only reduced the TER of a trophoblast cell line when added to the basolateral surface raises the possibility of seeing the opening of TJs in the STB as a consequence of ZIKV infection of the chorionic villi.

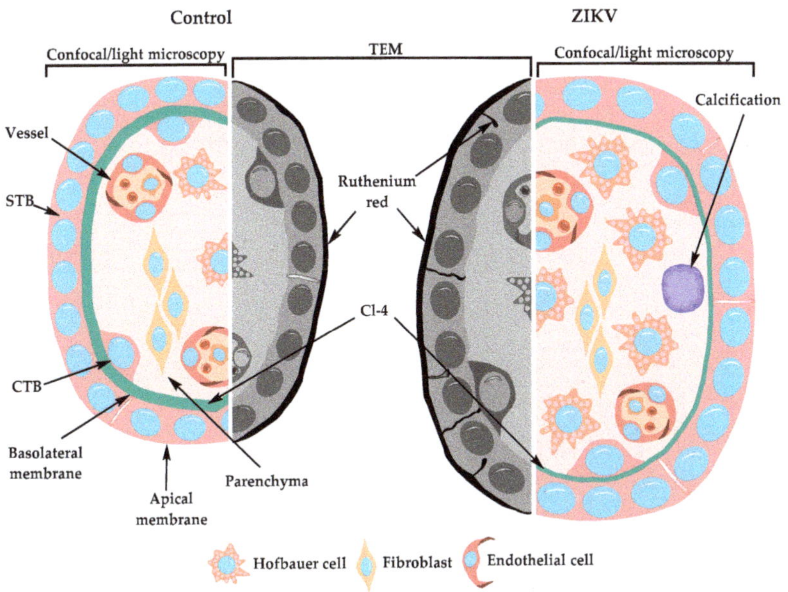

Figure 6. Schematic representation of the changes observed in chorionic villi of ZIKV-infected women. Chorionic villi derived from women infected with ZIKV during pregnancy, displayed several alterations including Hofbauer cell hyperplasia, increased diameter of microvilli, intravillous calcifications, and a STB layer with a diminished expression of claudin-4 and permeable to ruthenium red passage though the paracellular pathway.

Supplementary Materials: The following are available online at http://www.mdpi.com/2073-4409/8/10/1174/s1, Figure S1. E-cadherin presence in the basolateral membrane of the STB was not altered in placental tissue from ZIKV-infected women, Figure S2. The expression of claudin-1 in the STB of chorionic villi in human placenta increased with ZIKV infection, Figure S3: Claudin-3 was strongly expressed in vessels in the chorionic parenchyma, Figure S4. Claudin-5 was strongly expressed in chorionic villi vessels and faintly stained the STB cell layer, Figure S5: Claudin-7 is present in the basolateral surface of STB cells and infection with ZIKV did not alter its expression, Figure S6. ZO-1 strongly stains the chorionic vessels, Figure S7. ZO-2 stained the STB layer of placentae, Figure S8. JAM-B was present in the STB cell layer and its expression was not affected by ZIKV infection, Figure S9: JAM-C expression was abundant in chorionic vessels but scarce in the STB layer, Figure S10: Chorionic villi stained with haematoxylin and eosin in D1 placenta from a ZIKV-infected woman, Figure S11: Chorionic villi stained with haematoxylin and eosin in D5 placenta from a ZIKV-infected woman, Figure S12: Chorionic villi stained with haematoxylin and eosin in D8 placenta from a ZIKV-infected woman, Figure S13: Chorionic villi stained with haematoxylin and eosin in D10 placenta from a ZIKV-infected woman, Figure S14: Chorionic villi stained with haematoxylin and eosin in D9 placenta from the control group, Figure S15: Chorionic villi stained with haematoxylin and eosin in placenta D17 from the control group, Figure S16: Chorionic villi stained with haematoxylin and eosin in placenta D18 from the control group, Figure S17: Chorionic villi stained with haematoxylin and eosin in placenta D19 from the control group, Figure S18: Chorionic villi stained with Masson's trichrome stain in D1 placenta from a ZIKV-infected woman.

Author Contributions: Conceptualization, L.G.-M.; methodology, L.G.-M. and J.M.; formal analysis, J.M. and L.G.-M.; investigation, J.M., Y.V.-V., D.M.-T., L.A., A.E.-N., M.G.-H., J.E.M.-M.; M.S., B.C.-M., J.E.M.-M.; resources, L.G.-M., G.E.-G. and S.L.; data curation, J.M.; writing—original draft preparation, L.G.-M.; writing—review and editing, L.G.-M., J.E.L., G.E.-G., M.S.; validation, L.G.-M.; visualization, L.G.-M. and J.M.; project administration, L.G.-M.; funding acquisition, L.G.-M., J.E.L. and G.E.-G.

Funding: This work was supported by grant C-623/2016 of the Mexican National Council of Science and Technology (Conacyt) to L.G-M., by grant CB-254461 of Conacyt to J.E.L.; by a grant from the Miguel Alemán Valdés Foundation to L.G-M.; by a SEP-Cinvestav grant to L.G-M. and by INPer project 212250-1000-10107-01-16. Jael Miranda was a recipient of a doctoral fellowship from Conacyt (262817) and the Mexiquense Council of Science and Technology (Comecyt, 2018AD0035-1).

Acknowledgments: We would like to thank the technical support of Angélica Silva-Olivares.

Conflicts of Interest: The authors declare no conflict of interest.

References

1. Robbins, J.R.; Skrzypczynska, K.M.; Zeldovich, V.B.; Kapidzic, M.; Bakardjiev, A.I. Placental syncytiotrophoblast constitutes a major barrier to vertical transmission of Listeria monocytogenes. *PLoS Pathog.* **2010**, *6*. [CrossRef] [PubMed]
2. Robbins, J.R.; Zeldovich, V.B.; Poukchanski, A.; Boothroyd, J.C.; Bakardjiev, A.I. Tissue barriers of the human placenta to infection with Toxoplasma gondii. *Infect. Immun.* **2012**, *80*, 418–428. [CrossRef] [PubMed]
3. Pereira, L. Congenital Viral Infection: Traversing the Uterine-Placental Interface. *Annu. Rev. Virol.* **2018**, *5*, 273–299. [CrossRef] [PubMed]
4. WHO. Zika Virus. Available online: https://www.who.int/news-room/fact-sheets/detail/zika-virus (accessed on 6 December 2018).
5. Castellucci, M.; Kaufmann, P. Basic Structure of the Villous Trees. In *Pathology of the Human Placenta*; Benirschke, K., Kaufmann, P., Baergen, R.N., Eds.; Springer: New York, NY, USA, 2006; pp. 50–99.
6. De Virgiliis, G.; Sideri, M.; Fumagalli, G.; Remotti, G. The junctional pattern of the human villous trophoblast. A freeze-fracture study. *Gynecol. Obstet. Investig.* **1982**, *14*, 263–272. [CrossRef] [PubMed]
7. Wang, T.; Schneider, J. Cellular junctions on the free surface of human placental syncytium. *Arch. Gynecol. Obstet.* **1987**, *240*, 211–216. [CrossRef]
8. Metz, J.; Weihe, E.; Heinrich, D. Intercellular junctions in the full term human placenta. *Brain Struct. Funct.* **1979**, *158*, 41–50. [CrossRef] [PubMed]
9. Liévano, S.; Alarcón, L.; Chávez–Munguía, B.; González–Mariscal, L. Endothelia of term human placentae display diminished expression of tight junction proteins during preeclampsia. *Cell Tissue Res.* **2006**, *324*, 433–448. [CrossRef] [PubMed]
10. Marzioni, D. Expression of ZO-1 and occludin in normal human placenta and in hydatidiform moles. *Mol. Hum. Reprod.* **2001**, *7*, 279–285. [CrossRef]
11. Pirinen, E.; Soini, Y. A survey of zeb1, twist and claudin 1 and 4 expression during placental development and disease. *APMIS* **2014**, *122*, 530–538. [CrossRef] [PubMed]
12. Tabata, T.; Petitt, M.; Puerta-Guardo, H.; Michlmayr, D.; Wang, C.; Fang-Hoover, J.; Harris, E.; Pereira, L. Zika Virus Targets Different Primary Human Placental Cells, Suggesting Two Routes for Vertical Transmission. *Cell Host Microbe* **2016**, *20*, 155–166. [CrossRef] [PubMed]
13. Freeman, G.J.; Casasnovas, J.M.; Umetsu, D.T.; DeKruyff, R.H. TIM genes: A family of cell surface phosphatidylserine receptors that regulate innate and adaptive immunity. *Immunol. Rev.* **2010**, *235*, 172–189. [CrossRef] [PubMed]
14. Tabata, T.; Petitt, M.; Puerta-Guardo, H.; Michlmayr, D.; Harris, E.; Pereira, L. Zika Virus Replicates in Proliferating Cells in Explants From First-Trimester Human Placentas, Potential Sites for Dissemination of Infection. *J. Infect. Dis.* **2018**, *217*, 1202–1213. [CrossRef] [PubMed]
15. De Noronha, L.; Zanluca, C.; Azevedo, M.L.V.; Luz, K.G.; Dos Santos, C.N.D. Zika virus damages the human placental barrier and presents marked fetal neurotropism. *Memórias do Inst. Oswaldo Cruz* **2016**, *111*, 287–293. [CrossRef] [PubMed]
16. Sirohi, D.; Kuhn, R.J. Zika Virus Structure, Maturation, and Receptors. *J. Infect. Dis.* **2017**, *216*, S935–S944. [CrossRef] [PubMed]
17. Priyamvada, L.; Quicke, K.M.; Hudson, W.H.; Onlamoon, N.; Sewatanon, J.; Edupuganti, S.; Pattanapanyasat, K.; Chokephaibulkit, K.; Mulligan, M.J.; Wilson, P.C.; et al. Human antibody responses after dengue virus infection are highly cross-reactive to Zika virus. *Proc. Natl. Acad. Sci. USA* **2016**, *113*, 7852–7857. [CrossRef]
18. Zimmerman, M.G.; Quicke, K.M.; O'Neal, J.T.; Arora, N.; Machiah, D.; Priyamvada, L.; Kauffman, R.C.; Register, E.; Adekunle, O.; Swieboda, D.; et al. Cross-Reactive Dengue Virus Antibodies Augment Zika Virus Infection of Human Placental Macrophages. *Cell Host Microbe* **2018**, *24*, 731–742. [CrossRef]
19. Halai, U.-A.; Nielsen-Saines, K.; Moreira, M.L.; De Sequeira, P.C.; Junior, J.P.P.; Zin, A.D.A.; Cherry, J.; Gabaglia, C.R.; Gaw, S.L.; Adachi, K.; et al. Maternal Zika Virus Disease Severity, Virus Load, Prior Dengue Antibodies, and Their Relationship to Birth Outcomes. *Clin. Infect. Dis.* **2017**, *65*, 877–883. [CrossRef]

20. Walters, R.W.; Freimuth, P.; Moninger, T.O.; Ganske, I.; Zabner, J.; Welsh, M.J. Adenovirus Fiber Disrupts CAR-Mediated Intercellular Adhesion Allowing Virus Escape. *Cell* **2002**, *110*, 789–799. [CrossRef]

21. Sufiawati, I.; Tugizov, S.M. HIV-Associated Disruption of Tight and Adherens Junctions of Oral Epithelial Cells Facilitates HSV-1 Infection and Spread. *PLoS ONE* **2014**, *9*. [CrossRef]

22. Fiala, M.; Looney, D.J.; Stins, M.; Way, D.D.; Zhang, L.; Gan, X.; Chiappelli, F.; Schweitzer, E.S.; Shapshak, P.; Weinand, M.; et al. TNF-alpha opens a paracellular route for HIV-1 invasion across the blood-brain barrier. *Mol. Med.* **1997**, *3*, 553–564. [CrossRef]

23. Roe, K.; Kumar, M.; Lum, S.; Orillo, B.; Nerurkar, V.R.; Verma, S. West Nile virus-induced disruption of the blood-brain barrier in mice is characterized by the degradation of the junctional complex proteins and increase in multiple matrix metalloproteinases. *J. Gen. Virol.* **2012**, *93*, 1193–1203. [CrossRef] [PubMed]

24. Velandia-Romero, M.L.; Calderón-Peláez, M.-A.; Castellanos, J.E. In Vitro Infection with Dengue Virus Induces Changes in the Structure and Function of the Mouse Brain Endothelium. *PLoS ONE* **2016**, *11*. [CrossRef] [PubMed]

25. González-Mariscal, L.; Garay, E.; Quirós, M. Identification of Claudins by Western Blot and Immunofluorescence in Different Cell Lines and Tissues. *Adv. Struct. Saf. Stud.* **2011**, *762*, 213–231.

26. Abdelhalim, N.Y.; Shehata, M.H.; Gadallah, H.N.; Sayed, W.M.; Othman, A.A. Morphological and ultrastructural changes in the placenta of the diabetic pregnant Egyptian women. *Acta Histochem.* **2018**, *120*, 490–503. [CrossRef] [PubMed]

27. Ortega-Olvera, J.M.; Winkler, R.; Quintanilla-Vega, B.; Shibayama, M.; Chávez-Munguía, B.; Martín-Tapia, D.; Alarcón, L.; González-Mariscal, L.; Quitanilla-Vega, B. The organophosphate pesticide methamidophos opens the blood-testis barrier and covalently binds to ZO-2 in mice. *Toxicol. Appl. Pharmacol.* **2018**, *360*, 257–272. [CrossRef] [PubMed]

28. Bancher-Todesca, D.; Hefler, L.; Zeisler, H.; Schatten, C.; Husslein, P.; Heinze, G.; Tempfer, C. Placental expression of cytokeratin 18 and serum levels of tissue polypeptide antigen in women with pregnancy-induced hypertension. *Hypertens. Pregnancy* **2001**, *20*, 89–98. [CrossRef] [PubMed]

29. Nitta, T.; Hata, M.; Gotoh, S.; Seo, Y.; Sasaki, H.; Hashimoto, N.; Furuse, M.; Tsukita, S. Size-selective loosening of the blood-brain barrier in claudin-5–deficient mice. *J. Cell Boil.* **2003**, *161*, 653–660. [CrossRef] [PubMed]

30. Van Itallie, C.; Rahner, C.; Anderson, J.M. Regulated expression of claudin-4 decreases paracellular conductance through a selective decrease in sodium permeability. *J. Clin. Investig.* **2001**, *107*, 1319–1327. [CrossRef] [PubMed]

31. Hou, J.; Renigunta, A.; Yang, J.; Waldegger, S. Claudin-4 forms paracellular chloride channel in the kidney and requires claudin-8 for tight junction localization. *Proc. Natl. Acad. Sci. USA* **2010**, *107*, 18010–18015. [CrossRef] [PubMed]

32. Van Itallie, C.M.; Fanning, A.S.; Anderson, J.M. Reversal of charge selectivity in cation or anion-selective epithelial lines by expression of different claudins. *Am. J. Physiol. Physiol.* **2003**, *285*, F1078–F1084. [CrossRef]

33. Bayer, A.; Lennemann, N.J.; Ouyang, Y.; Bramley, J.C.; Morosky, S.; Marques, E.T.D.A.; Cherry, S.; Sadovsky, Y.; Coyne, C.B. Type III Interferons Produced by Human Placental Trophoblasts Confer Protection against Zika Virus Infection. *Cell Host Microbe* **2016**, *19*, 705–712. [CrossRef] [PubMed]

34. Capaldo, C.T.; Farkas, A.E.; Hilgarth, R.S.; Krug, S.M.; Wolf, M.F.; Benedik, J.K.; Fromm, M.; Koval, M.; Parkos, C.; Nusrat, A. Proinflammatory cytokine-induced tight junction remodeling through dynamic self-assembly of claudins. *Mol. Biol. Cell* **2014**, *25*, 2710–2719. [CrossRef] [PubMed]

35. Schumann, S.; Buck, V.U.; Classen-Linke, I.; Wennemuth, G.; Grümmer, R. Claudin-3, claudin-7, and claudin-10 show different distribution patterns during decidualization and trophoblast invasion in mouse and human. *Histochem. Cell Boil.* **2015**, *144*, 571–585. [CrossRef] [PubMed]

36. Ahn, C.; Yang, H.; Lee, D.; An, B.-S.; Jeung, E.-B. Placental claudin expression and its regulation by endogenous sex steroid hormones. *Steroids* **2015**, *100*, 44–51. [CrossRef] [PubMed]

37. Pidoux, G.; Gerbaud, P.; Gnidehou, S.; Grynberg, M.; Geneau, G.; Guibourdenche, J.; Carette, D.; Cronier, L.; Evain-Brion, D.; Malassine, A.; et al. ZO-1 is involved in trophoblastic cell differentiation in human placenta. *Am. J. Physiol. Physiol.* **2010**, *298*, C1517–C1526. [CrossRef]

38. Leach, L.; Lammiman, M.; Babawale, M.; Hobson, S.; Bromilou, B.; Lovat, S.; Simmonds, M. Molecular Organization of Tight and Adherens Junctions in the Human Placental Vascular Tree. *Placenta* **2000**, *21*, 547–557. [CrossRef]

39. Katsuno, T.; Umeda, K.; Matsui, T.; Hata, M.; Tamura, A.; Itoh, M.; Takeuchi, K.; Fujimori, T.; Nabeshima, Y.-I.; Noda, T.; et al. Deficiency of Zonula Occludens-1 Causes Embryonic Lethal Phenotype Associated with Defected Yolk Sac Angiogenesis and Apoptosis of Embryonic Cells. *Mol. Boil. Cell* **2008**, *19*, 2465–2475. [CrossRef] [PubMed]

40. Xu, J.; Anuar, F.; Ali, S.M.; Ng, M.Y.; Phua, D.C.; Hunziker, W. Zona Occludens-2 Is Critical for Blood–Testis Barrier Integrity and Male Fertility. *Mol. Boil. Cell* **2009**, *20*, 4268–4277. [CrossRef] [PubMed]

41. Furuse, M.; Furuse, K.; Sasaki, H.; Tsukita, S. Conversion of Zonulae Occludentes from Tight to Leaky Strand Type by Introducing Claudin-2 into Madin-Darby Canine Kidney I Cells. *J. Cell Boil.* **2001**, *153*, 263–272. [CrossRef]

42. Paria, B.C.; Zhao, X.; Das, S.K.; Dey, S.K.; Yoshinaga, K. Zonula Occludens-1 and E-cadherin Are Coordinately Expressed in the Mouse Uterus with the Initiation of Implantation and Decidualization. *Dev. Boil.* **1999**, *208*, 488–501. [CrossRef]

43. Wang, X.; Matsumoto, H.; Zhao, X.; Das, S.K.; Paria, B.C. Embryonic signals direct the formation of tight junctional permeability barrier in the decidualizing stroma during embryo implantation. *J. Cell Sci.* **2004**, *117*, 53–62. [CrossRef]

44. Martínez-Peña, A.A.; Rivera-Baños, J.; Méndez-Carrillo, L.L.; Ramírez-Solano, M.I.; Galindo-Bustamante, A.; Páez-Franco, J.C.; Morimoto, S.; González-Mariscal, L.; Cruz, M.E.; Mendoza-Rodríguez, C.A.; et al. Perinatal administration of bisphenol A alters the expression of tight junction proteins in the uterus and reduces the implantation rate. *Reprod. Toxicol.* **2017**, *69*, 106–120. [CrossRef] [PubMed]

45. Murphy, C.R.; Swift, J.G.; Mukherjee, T.M.; Rogers, A.W. Effects of ovarian hormones on cell membranes in the rat uterus. *Cell Biophys.* **1981**, *3*, 57–69. [CrossRef] [PubMed]

46. Murphy, C.R.; Swift, J.G.; Mukherjee, T.M.; Rogers, A.W. The structure of tight junctions between uterine luminal epithelial cells at different stages of pregnancy in the rat. *Cell Tissue Res.* **1982**, *223*, 281–286. [CrossRef] [PubMed]

47. Carson, D.D.; Lagow, E.; Thathiah, A.; Al-Shami, R.; Farach-Carson, M.C.; Vernon, M.; Yuan, L.; Fritz, M.A.; Lessey, B. Changes in gene expression during the early to mid-luteal (receptive phase) transition in human endometrium detected by high-density microarray screening. *Mol. Hum. Reprod.* **2002**, *8*, 871–879. [CrossRef] [PubMed]

48. Kao, L.C.; Tulac, S.; Lobo, S.; Imani, B.; Yang, J.P.; Germeyer, A.; Osteen, K.; Taylor, R.N.; Lessey, B.A.; Giudice, L.C. Global gene profiling in human endometrium during the window of implantation. *Endocrinology* **2002**, *143*, 2119–2138. [CrossRef]

49. Riesewijk, A.; Martin, J.; van Os, R.; Horcajadas, J.A.; Polman, J.; Pellicer, A.; Mosselman, S.; Simon, C. Gene expression profiling of human endometrial receptivity on days LH+2 versus LH+7 by microarray technology. *Mol. Human Reprod.* **2003**, *9*, 253–264. [CrossRef]

50. Zhang, Y.; Zhao, N.; Zhao, H.; Cui, Y.; Liu, J. Expression of tight junction factors in human placental tissues derived from assisted reproductive technology and natural pregnancy. *Zhonghua Fu Chan Ke Za Zhi* **2014**, *49*, 125–129. [PubMed]

51. Gonzalez-Mariscal, L. Virus interaction with the apical junctional complex. *Front. Biosci.* **2009**, *14*, 731–768. [CrossRef]

52. Evans, M.J.; Von Hahn, T.; Tscherne, D.M.; Syder, A.J.; Panis, M.; Wölk, B.; Hatziioannou, T.; McKeating, J.A.; Bieniasz, P.D.; Rice, C.M. Claudin-1 is a hepatitis C virus co-receptor required for a late step in entry. *Nature* **2007**, *446*, 801–805. [CrossRef]

53. Zheng, A.; Yuan, F.; Li, Y.; Zhu, F.; Hou, P.; Li, J.; Song, X.; Ding, M.; Deng, H. Claudin-6 and Claudin-9 Function as Additional Coreceptors for Hepatitis C Virus. *J. Virol.* **2007**, *81*, 12465–12471. [CrossRef]

54. Xu, Z.; Waeckerlin, R.; Urbanowski, M.D.; Van Marle, G.; Hobman, T.C. West Nile Virus Infection Causes Endocytosis of a Specific Subset of Tight Junction Membrane Proteins. *PLoS ONE* **2012**, *7*. [CrossRef] [PubMed]

55. Nava, P.; López, S.; Arias, C.F.; Islas, S.; Gonzalez-Mariscal, L. The rotavirus surface protein VP8 modulates the gate and fence function of tight junctions in epithelial cells. *J. Cell Sci.* **2004**, *117*, 5509–5519. [CrossRef] [PubMed]

56. Takahashi, A.; Kondoh, M.; Masuyama, A.; Fujii, M.; Mizuguchi, H.; Horiguchi, Y.; Watanabe, Y. Role of C-terminal regions of the C-terminal fragment of Clostridium perfringens enterotoxin in its interaction with claudin-4. *J. Control. Release* **2005**, *108*, 56–62. [CrossRef] [PubMed]

57. Sonoda, N.; Furuse, M.; Sasaki, H.; Yonemura, S.; Katahira, J.; Horiguchi, Y.; Tsukita, S. Clostridium perfringens enterotoxin fragment removes specific claudins from tight junction strands: Evidence for direct involvement of claudins in tight junction barrier. *J. Cell Biol.* **1999**, *147*, 195–204. [CrossRef] [PubMed]

58. Schelhaas, M.; Jansen, M.; Haase, I.; Knebel-Mörsdorf, D. Herpes simplex virus type 1 exhibits a tropism for basal entry in polarized epithelial cells. *J. Gen. Virol.* **2003**, *84*, 2473–2484. [CrossRef] [PubMed]

59. Marozin, S.; Prank, U.; Sodeik, B. Herpes simplex virus type 1 infection of polarized epithelial cells requires microtubules and access to receptors present at cell-cell contact sites. *J. Gen. Virol.* **2004**, *85*, 775–786. [CrossRef] [PubMed]

60. Mee, C.J.; Grove, J.; Harris, H.J.; Hu, K.; Balfe, P.; McKeating, J.A. Effect of cell polarization on hepatitis C virus entry. *J. Virol.* **2008**, *82*, 461–470. [CrossRef]

61. Baktash, Y.; Madhav, A.; Coller, K.E.; Randall, G. Single Particle Imaging of Polarized Hepatoma Organoids upon Hepatitis C Virus Infection Reveals an Ordered and Sequential Entry Process. *Cell Host Microbe* **2018**, *23*, 382–394. [CrossRef] [PubMed]

62. Rosenberg, A.Z.; Yu, W.; Hill, D.A.; Reyes, C.A.; Schwartz, D.A. Placental Pathology of Zika Virus: Viral Infection of the Placenta Induces Villous Stromal Macrophage (Hofbauer Cell) Proliferation and Hyperplasia. *Arch. Pathol. Lab. Med.* **2017**, *141*, 43–48. [CrossRef] [PubMed]

63. De Noronha, L.; Zanluca, C.; Burger, M.; Suzukawa, A.A.; Azevedo, M.; Rebutini, P.Z.; Novadzki, I.M.; Tanabe, L.S.; Presibella, M.M.; Dos Santos, C.N.D. Zika Virus Infection at Different Pregnancy Stages: Anatomopathological Findings, Target Cells and Viral Persistence in Placental Tissues. *Front. Microbiol.* **2018**, *9*. [CrossRef] [PubMed]

64. Martines, R.B.; Bhatnagar, J.; Keating, M.K.; Silva-Flannery, L.; Muehlenbachs, A.; Gary, J.; Goldsmith, C.; Hale, G.; Ritter, J.; Rollin, D.; et al. Notes from the Field: Evidence of Zika Virus Infection in Brain and Placental Tissues from Two Congenitally Infected Newborns and Two Fetal Losses–Brazil, 2015. *MMWR. Morb. Mortal. Wkly. Rep.* **2016**, *65*, 159–160. [CrossRef] [PubMed]

65. Reyes, L.; Wolfe, B.; Golos, T. Hofbauer Cells: Placental Macrophages of Fetal Origin. In *Macrophages: Origin, Functions and Biointervention*; Kloc, M., Ed.; Springer International Publishing: Houston, TX, USA, 2017; pp. 45–60.

66. Benirschke, K.; Kaufmann, P.; Baergen, R. Classification of Villous Maldevelopment. In *Pathology of the Human Placenta*, 5th ed.; Benirschke, K., Kaufmann, P., Baergen, R., Eds.; Springer: New York, NY, USA, 2006; pp. 491–518.

67. Lu, J.; Zhang, S.; Nakano, H.; Simmons, D.G.; Wang, S.; Kong, S.; Wang, Q.; Shen, L.; Tu, Z.; Wang, W.; et al. A positive feedback loop involving Gcm1 and Fzd5 directs chorionic branching morphogenesis in the placenta. *PLoS Boil.* **2013**, *11*. [CrossRef] [PubMed]

Article

Interplay between Zika Virus and Peroxisomes during Infection

Cheung Pang Wong [1], Zaikun Xu [2], Shangmei Hou [2], Daniel Limonta [2], Anil Kumar [2], Christopher Power [1,3,4,5] and Tom C. Hobman [1,2,4,5,*]

[1] Department of Medical Microbiology & Immunology, University of Alberta, Edmonton, AB T6G 2E1, Canada
[2] Department of Cell Biology, University of Alberta, Edmonton, AB T6G 2H7, Canada
[3] Department of Medicine, University of Alberta, Edmonton, AB T6G 2E1, Canada
[4] Women & Children's Health Research Institute, University of Alberta, Edmonton, AB T6G 1C9, Canada
[5] Li Ka Shing Institute of Virology, University of Alberta, Edmonton, AB T6G 2E1, Canada
* Correspondence: tom.hobman@ualberta.ca; Tel.: +1-780-492-6485

Received: 30 May 2019; Accepted: 12 July 2019; Published: 15 July 2019

Abstract: Zika virus (ZIKV) has emerged as an important human pathogen that can cause congenital defects in the fetus and neurological conditions in adults. The interferon (IFN) system has proven crucial in restricting ZIKV replication and pathogenesis. The canonical IFN response is triggered by the detection of viral RNA through RIG-I like receptors followed by activation of the adaptor protein MAVS on mitochondrial membranes. Recent studies have shown that a second organelle, peroxisomes, also function as a signaling platforms for the IFN response. Here, we investigated how ZIKV infection affects peroxisome biogenesis and antiviral signaling. We show that ZIKV infection depletes peroxisomes in human fetal astrocytes, a brain cell type that can support persistent infection. The peroxisome biogenesis factor PEX11B was shown to inhibit ZIKV replication, likely by increasing peroxisome numbers and enhancing downstream IFN-dependent antiviral signaling. Given that peroxisomes play critical roles in brain development and nerve function, our studies provide important insights into the roles of peroxisomes in regulating ZIKV infection and potentially neuropathogenesis.

Keywords: Zika virus; peroxisomes; innate immune response; interferon; astrocytes; fetal brain

---

1. Introduction

Zika virus (ZIKV) is a mosquito-borne flavivirus whose genome consists of a positive sense single-stranded RNA that encodes three structural proteins (capsid, pre-membrane/membrane and envelope proteins) and seven non-structural proteins (NS1, NS2A, NS2B, NS3, NS4A, NS4B, and NS5) [1,2]. Until recently, ZIKV circulated primarily within Africa and Asia, but the large number of microcephaly cases and other neurological disorders associated with the 2015/2016 pandemic constituted a public health emergency of international concern [3,4]. While intensive research efforts have led to multiple promising vaccine candidates, there are currently no prophylactic or therapeutic antivirals available for this pathogen.

The interferon (IFN) system, a crucial arm of the innate immune system, has been shown to play a role in restricting ZIKV replication and pathogenesis [5–7]. The canonical IFN response is initiated following the detection of viral genomes/transcripts by RNA-sensing helicases such as RIG-I and MDA5 [8,9]. After binding viral RNA, these helicases interact with the adaptor protein MAVS, the bulk of which is localized on mitochondria. The activation of MAVS leads to the phosphorylation of the antiviral transcription factors IFN regulatory factor 3 (IRF3) and NF-κB, followed by transcription of *IFNβ* and *IFNλ* genes [10]. Secreted type I and III IFNs bind to receptors on the cell surface that signal through the JAK/STAT pathway to induce the transcription of IFN-stimulated genes (ISGs), resulting in an antiviral state [11,12]. However, many viruses, including flaviviruses, are known to deploy an array of counter-measures to suppress IFN induction and downstream antiviral signaling [13,14].

In addition to mitochondria, peroxisomes, which are membrane-bound organelles that have well characterized functions in lipid metabolism and regulation of reactive oxygen species [15,16], have recently been shown to play critical roles in antiviral defense. Specifically, activation of MAVS on peroxisomal as well as mitochondrial membranes appears to be important for IFN induction and signaling [17–19]. Evidence indicating that viruses disrupt peroxisome biogenesis began to emerge shortly after, further supporting the importance of peroxisomes in antiviral defense. First, we showed that in cells infected with West Nile (WNV) or Dengue (DENV) viruses, a critical peroxisome biogenesis factor, PEX19, is selectively degraded [20]. This process, which involves the capsid proteins of WNV and DENV, results in reduced levels of peroxisomes and a dampened type III IFN response [20]. Subsequently, it was reported that the NS3-4A protease of hepatitis C virus cleaves MAVS localized on peroxisomes and mitochondria [18,21], whereas the nsp1 protein of porcine diarrhea virus reduces type III IFN induction, in part by reducing peroxisome pools via an unknown mechanism [22]. Finally, human immunodeficiency virus-1 (HIV-1) infection was shown to downregulate peroxisomes by upregulating cellular microRNAs that inhibit the expression of peroxisome biogenesis factors such as PEX2, PEX7, PEX11 and PEX13 [23].

More recently, it was reported that the infection of Vero cells with ZIKV results in a 12% decrease in peroxisome density as well as a 50% loss of the peroxisomal membrane protein PMP70 [24]. It was hypothesized that during ZIKV infection, peroxisomes are consumed and, accordingly, that these organelles are actually required for ZIKV replication. However, this notion contrasts with mounting evidence supporting an antiviral role for peroxisomes [17–22].

Here, we investigated the interplay between ZIKV infection and peroxisomes in primary human fetal astrocytes (HFAs), the most abundant cell type in the brain and potentially a cellular reservoir for ZIKV [25]. Iinfection of HFAs resulted in a dramatic reduction in peroxisomes, regardless of the type of ZIKV strain employed. PEX11B, a biogenesis factor that induces peroxisome proliferation, was found to be a restriction factor for ZIKV. Elevated expression of PEX11B was associated with increased levels of MAVS and enhanced IFN induction and downstream signaling. As peroxisomes are critical for brain development and function [26,27], it is tempting to speculate that the loss of these organelles in HFAs may play a role in the neurological deficits associated with in utero ZIKV infection.

2. Materials and Methods

2.1. Cells and Virus Infection

A549, HEK293T, Vero and U251 cells were purchased from the American Type Culture Collection (Manassas, VA, USA). The cells were cultured in Dulbecco's modified Eagle's medium (DMEM; Gibco; Waltham, MA, USA) supplemented with 100 U/mL penicillin and streptomycin, 1 mM 4-(2-hydroxyethyl)-1-piperazineethanesulfonic acid (HEPES)(Gibco; Waltham, MA, USA), 2 mM glutamine (Gibco; Waltham, MA, USA), 10% heat-inactivated fetal bovine serum (FBS; Gibco; Waltham, MA, USA) at 37 °C in 5% CO_2. Primary human fetal astrocytes (HFAs) were prepared as previously described [28] from 15–19 week aborted fetuses with written consent approved under the protocol 1420 by the University of Alberta Human Research Ethics Board (Biomedical). HFAs were grown in Minimum Essential Media (MEM) (1 g/L Glucose, 15 mM HEPES, Gibco; Waltham, MA, USA) supplemented with 10% FBS, L-glutamine, MEM non-essential amino acids, sodium pyruvate, and 1 g/mL glucose at 37 °C in 5% CO2. PLCal and PRVABC59 strains of ZIKV were kindly provided by Dr. David Safronetz (Public Health Agency of Canada). The Zika virus (strain H/PF/2013, French Polynesia) was kindly provided by Dr. Michael Diamond (Washington University School of Medicine, St. Louis, MO, USA). The Zika virus (strain MR766) was generated from a molecular clone [29] kindly provided by Dr. Matthew J. Evans (Icahn School of Medicine at Mount Sinai, New York, NY, USA). All virus manipulations were performed according to level-2 containment procedures. Virus stocks were generated in C6/36 cells and titrated by plaque assay using Vero cells.

2.2. Plasmids and Transfection

A triple FLAG® (DYKDDDDK)-tagged ZIKV capsid expression construct was generated by polymerase chain reaction (PCR) using a mammalian expression vector encoding the capsid protein [28] as template. The resulting PCR product was then cloned into pcDNA 3.1(-) plasmid using the restriction sites NheI and XhoI. pCMV3-PEX11B was purchased from Sino Biological Inc. Oligonucleotide primers were designed to amplify the desired PEX11B sequence and introduce a myc epitope tag cassette into the 5′ end of the cDNA. The resulting PCR product was then cloned into the lentiviral vector pTRIP-MCS-IRES-AcGFP [30]. All oligonucleotide primers used in this study are shown in Table S1.

For indirect immunofluorescence analysis in U251 cells, transfection of the appropriate expression plasmids was performed using Lipofectamine 2000 (Invitrogen; Carlsbad, CA). Poly(I:C) (Sigma-Aldrich; St. Louis, MO, USA) was transfected into cells using TransIT-LT1 (Mirus Bio; Madison, WI, USA).

2.3. Production of Lentiviral Particles and Transduction of Cells

Pseudotyped lentiviruses were recovered from the media of HEK293T cells transfected with pTRIP-AcGFP plasmids encoding flavivirus capsids or PEX11B, and titered as described [30]. Transduction of U251 cells with recombinant lentiviruses was performed in DMEM containing 3% FBS and 5 µg/mL polybrene for 48 h in 12-well plates.

2.4. Antibodies

Primary antibodies were from the following sources: Mouse monoclonal antibodies against the peroxisomal membrane protein PMP70 and FLAG epitope from Sigma (St. Louis, MO, USA); mouse monoclonal against beta-actin, goat polyclonal antibody against GFP and rabbit polyclonal antibodies to PEX3, PEX7, PEX11B, PEX13, PEX19 and catalase from Abcam (Cambridge, MA, USA); rabbit polyclonal antibody to the tripeptide SKL was produced as described in [23]; goat polyclonal antibody to ZIKV NS5 was produced and purified as described in [31]; mouse monoclonal antibody against MAVS was from Santa Cruz Biotechnology (Santa Cruz, CA, USA); mouse monoclonal antibody against myc from Millipore (Burlington, MA, USA). Donkey anti-mouse IgG conjugated to Alexa Fluor 680, goat anti-rabbit IgG conjugated to Alexa Fluor 680, donkey anti-mouse IgG conjugated to Alexa Fluor 488, donkey anti-goat IgG conjugated to Alexa Fluor 680, chicken anti-goat IgG conjugated to Alexa Fluor 647, and donkey anti-rabbit IgG conjugated to Alexa Fluor 546 were purchased from Invitrogen (Carlsbad, CA, USA).

2.5. Immunoblotting

HFAs or U251 cells collected at designated time points post-infection were washed three times with phosphate-buffer saline (PBS) before lysing with sodium dodecyl sulfate (SDS) Sample Buffer containing β-mercaptoethanol (2%) and 1 unit of Benzonase (Millipore; Burlington, MA, USA) per sample. Proteins in the samples were denatured at 98 °C for 10 min, separated by SDS-PAGE and then transferred to polyvinylidene difluoride membranes for immunoblotting as described [31]. Proteins on the membranes were imaged and analyzed using an Odyssey® CLx Imaging System (LI-COR Biosciences; Lincoln, NE, USA). Relative levels of MAVS, PMP70, PEX3, PEX7, PEX11B, PEX13, PEX19, and catalase (normalized to actin) were determined using Odyssey Infrared Imaging System 1.2 Version software.

2.6. Quantitative Real-Time PCR (qRT-PCR)

Total RNA from HFAs and U251 cells was isolated using the RNA NucleoSpin Kit (Machery Nagel; Bethlehem, PA, USA) and reverse transcribed with random primers (Invitrogen; Carlsbad, CA, USA) and the Improm-II reverse transcriptase system (Promega; Madison, WI, USA) at 42 °C for 2 h. The resulting cDNAs were mixed with the appropriate primers (Integrated DNA Technologies; Coralville, IA) and PerfeCTa SYBR Green SuperMix Low 6-Carboxy-X-Rhodamine (ROX) (Quanta Biosciences; Beverly, MA, USA) and then amplified for 40 cycles (30 s at 94 °C, 40 s at 55 °C and 20 s at 68 °C) in a CFX96 Touch™ Real-Time PCR Detection System. The gene targets and primers used are listed in (Table S2). The ΔCT values were calculated using β-actin mRNA as the internal control. The $\Delta\Delta$CT values were determined using control samples as the reference value. Relative levels of mRNAs were calculated using the formula 2($-\Delta\Delta$CT) [32].

2.7. Co-Immunoprecipitation and Immunoblotting

U251 cells (1×10^6), seeded the day before into 10 cm dishes, were infected with PRVABC59 strain of ZIKV (MOI = 1). At 48 h post-infection, cells were washed three times with PBS before lysing with NP-40 lysis buffer (150 mM NaCl, 2 mM ethylenediaminetetraacetic acid (EDTA), 1% Nonidet P-40, 50 mM Tris-HCl (pH 7.4), 1 mM dithiothreitol) containing CompleteTM protease inhibitors (Roche, Mannheim, Germany) on ice for 30 min. Lysates were clarified at 16,000 g for 20 min in a microcentrifuge at 4 °C. Small aliquots of the clarified lysates were kept for loading controls. The remaining lysates were treated with 20 µg/mL RNase A (Roche; Mannheim, Germany) for 1 h on ice, precleared with protein G-Sepharose beads (Sigma Aldrich; St. Louis, MO, USA) for 1 h at 4 °C before sequential incubation with antibodies overnight and then protein G-Sepharose beads for 2 h at 4 °C. Rabbit IgG was used in parallel as a negative control. Immunoprecipitates were washed three times with NP-40 lysis buffer before the bound proteins were eluted by boiling in SDS Sample buffer. Proteins were separated by SDS-PAGE and transferred to (PVDF) membranes for immunoblotting.

2.8. Cell Viability Assay

Cells were harvested in PBS and processed for cell viability assays using a CellTiter-Glo® Luminescent Cell Viability Assay kit (Promega; Madison, WI, USA).

2.9. Confocal Microscopy

HFAs or U251 cells on coverslips were fixed for 15 min at room temperature with 4% electron microscopy grade paraformaldehyde (Electron Microscope Sciences; Hatfield, PA) in PBS. Samples were washed three times in PBS and incubated in blocking buffer (0.2% Triton X-100 (VWR Internationals; Radnor, PA, USA) and 3% bovine serum albumin (BSA; Sigma Aldrich; St. Louis, MO, USA) in PBS) at room temperature for 1.5 h. Incubations with primary antibodies diluted (1:1000) in blocking buffer (3% BSA and PBS) were carried out at room temperature for 2 h, followed by three washes in PBS containing 0.1% BSA. Samples were then incubated with secondary antibodies in Blocking buffer for 1 h at room temperature, followed by three washes in PBS containing 0.1% BSA. The secondary antibodies were Alexa Fluor 488 donkey anti-mouse, Alexa Fluor 546 donkey anti-rabbit, and Alexa Fluor 647 chicken anti-goat (Invitrogen; Carlsbad, CA, USA). Prior to mounting, samples were incubated with 5 µg/mL DAPI (4′,6-diamidino-2-phenylindole) for 5 min at room temperature before washing in PBS containing 0.1% BSA. Coverslips were mounted onto microscope slides using ProLong Gold antifade reagent with DAPI (Invitrogen; Carlsbad, CA, USA). Samples were examined using an Olympus 1 × 81 spinning disk confocal microscope equipped with a 60×/1.42 oil PlanApo N objective. Confocal images were acquired and processed using Volocity 6.2.1 software.

2.10. Quantification of Peroxisomes

Z-stack images acquired using a confocal microscope were exported from Volocity 6.2.1 as an OEM.tiff file. The exported images were then processed using Imaris 7.2.3 software (Bitplane, Concord, MA, USA). Peroxisomes within polygonal areas that excluded the nucleus were quantified (quality and voxel). Within the selected regions, the absolute intensity of the peroxisomes was determined and then entered into a Microsoft Excel spreadsheet. The data were then analyzed using student's *t*-test. In each cell, peroxisomes were selected based on the absolute pixel intensity in the corresponding channel, and their numbers were then determined. Only those SKL/PMP70-positive structures with volumes between 0.001 and 0.05 μm^3 were included for measurement.

2.11. Statistical Analyses

A paired Student's *t*-test was used for pair-wise statistical comparison. The mean ± standard error of the mean is shown in all bar and line graphs. All statistical analyses were performed using Microsoft Excel software.

3. Results

3.1. ZIKV Infection Decreases the Pool of Peroxisomes in Primary HFAs and U251 Cells

Given the important roles of peroxisomes in antiviral defense and brain development, we investigated how ZIKV infection of the most abundant cell type in the fetal brain affects these organelles. Primary HFAs were infected with four different ZIKV strains including two pandemic strains of Asian lineage; one isolated during the 2015/2016 outbreak in South America (PRVABC59) and one isolated from an outbreak in French Polynesia in 2013 (H/PF/2013). The other two strains included a third contemporary Asian strain (PLCal) isolated from a returning Canadian traveler [33] and the prototype African strain MR766. The effects of ZIKV infection on peroxisomes and peroxisomal proteins in HFAs were assessed by confocal microscopy analysis and immunoblotting, respectively.

Data in Figure 1A show that depending upon the viral strain, the numbers of peroxisomes as identified by staining with anti-PMP70 antibodies were reduced by 60–70% at 48-h post-infection. To rule out the possibility that the apparent loss of peroxisomes in ZIKV-infected HFAs was not due to decreased expression of PMP70 alone, the steady levels of other peroxisome-associated proteins were assessed by immunoblotting at 24- and 48-h post-infection. This included PEX3, PEX7, PEX11B, PEX13 and PEX19, all of which are critical for peroxisome biogenesis. The loss of multiple PEX proteins was evident at 24-h post-infection; however, by 48-h, the reduction in PEX3, PEX7, PEX11B, PEX13 and PEX19 was much more pronounced (Figure 1B). PEX19 was most sensitive to viral infection as its levels were decreased by as much as 90% depending upon the infecting ZIKV strain. Unlike peroxisome biogenesis factors, levels of the peroxisomal matrix protein catalase were not affected by ZIKV infection (Figure 1B). This was not unexpected, however, as a previous study showed that unlike many other cellular proteins that are degraded when not targeted to their proper location, catalase is quite stable in the absence of peroxisomes [23]. Infection with PRVABC59 and PLCal strains was associated with the greatest loss of PEX proteins from HFAs. This did not appear to be due to more robust replication as the infection of HFAs with PRVABC59, which consistently resulted in slightly lower titers than the other three strains (Figure 1C), generally had the greatest effect on peroxisomes (Figure 1A,B).

Given that primary HFAs have a finite lifespan and limited expansion capacity, we assessed whether the human astrocytoma U251 cell line could be used to further examine the interplay between ZIKV infection and peroxisome biogenesis. Similar to HFAs, the infection of U251 cells with PRVABC59 or MR766 strains resulted in a significant loss of peroxisomes and peroxisome biogenesis factors (Figure 2A,B). Moreover, the infection of U251 cells with the pandemic strain PRVABC59 resulted in a greater loss of peroxisomes than MR766.

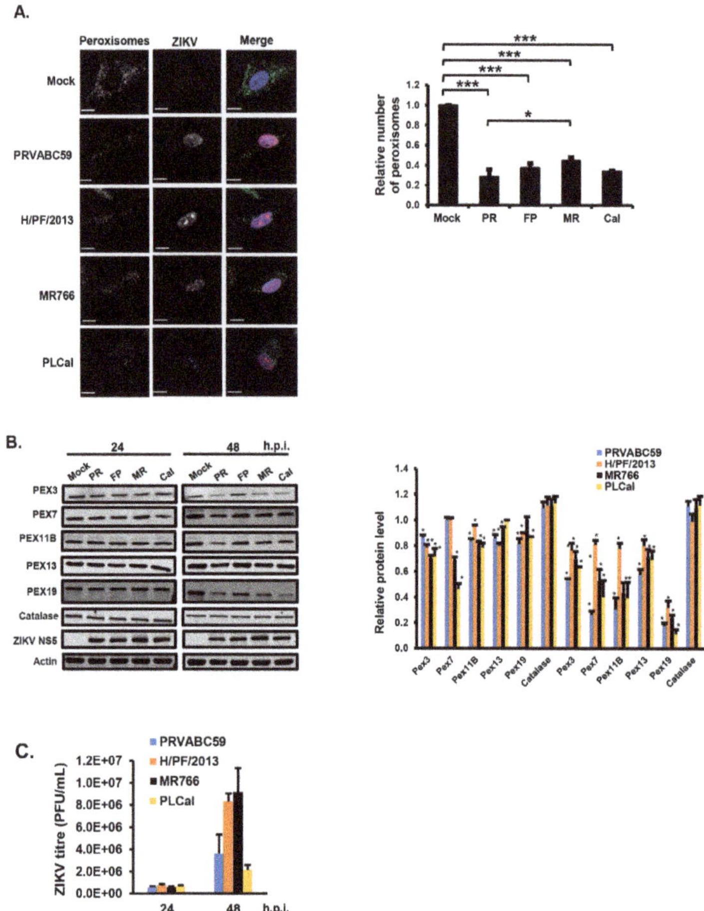

Figure 1. Zika virus (ZIKV) infection decreases the pool of peroxisomes in primary human fetal astrocytes (HFAs). (**A**) HFAs were infected at the multiplicity of infection (MOI) of 3) with ZIKV (PRVABC59 (PR), H/PF/2013 (FP), MR766 (MR) or PLCal (Cal) strains) for 48 h and then processed for confocal microscopy. Peroxisomes were detected using a mouse monoclonal to PMP70 and donkey anti-mouse IgG conjugated to Alexa Fluor 488. Infected cells were detected using a goat polyclonal antibody to ZIKV NS5 and chicken anti-goat IgG conjugated to Alexa Fluor 647. Nuclei were stained with DAPI (4′,6-diamidino-2-phenylindole). Images were obtained using a spinning disc confocal microscope. The number of peroxisomes in mock- and ZIKV-infected cells were determined using Volocity image analysis software. Averages were calculated from three independent experiments, in which a minimum of 20 cells for each sample were analyzed. The average number of peroxisomes in mock-treated cells was normalized to 1.0. Bars represent standard error of the mean. *** $p < 0.001$, * $p < 0.05$. (**B**) HFAs were infected with ZIKV strains (MOI = 5) for the indicated time periods, after which total cell lysates were processed for immunoblot analyses with antibodies to catalase, PEX3, PEX7, PEX11B, PEX13, PEX19, ZIKV NS5 and actin. The relative levels of peroxisomal proteins (compared to actin) from three independent experiments were averaged and plotted. The average levels of peroxisomal proteins in mock-infected cells were normalized to 1.0. Error bars represent standard error of the mean. * $p < 0.05$. (**C**) HFAs were infected with ZIKV (MOI = 1) for the indicated time periods, after which the viral titers in the media were determined by plaque assay. The data are averaged from the results of three independent experiments. Bars represent standard error of the mean.

A.

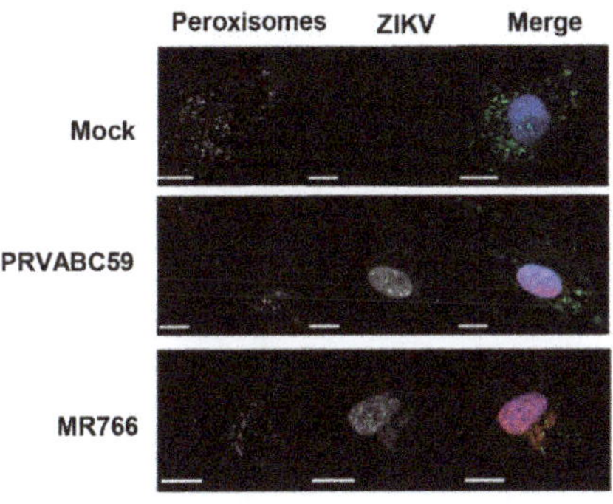

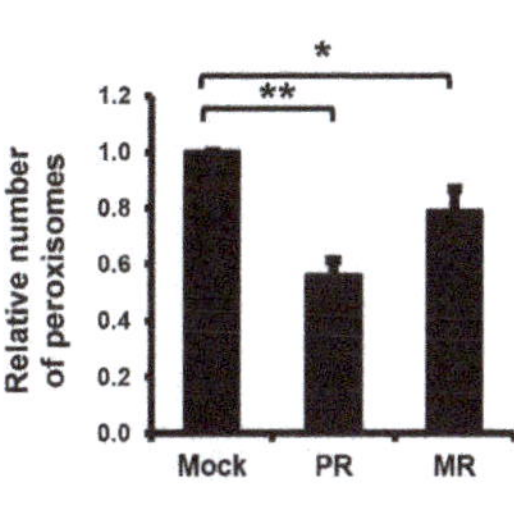

B.

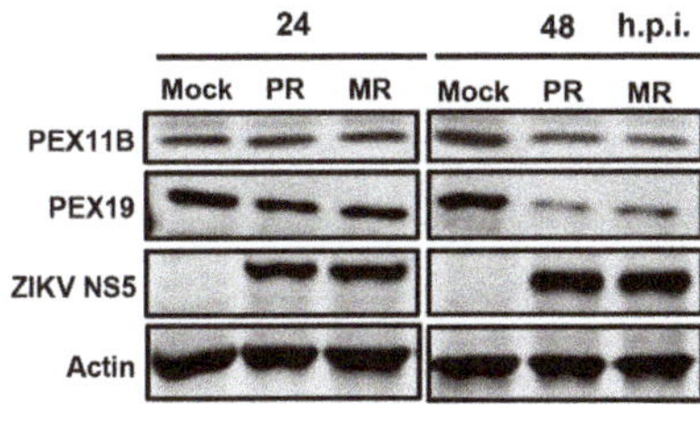

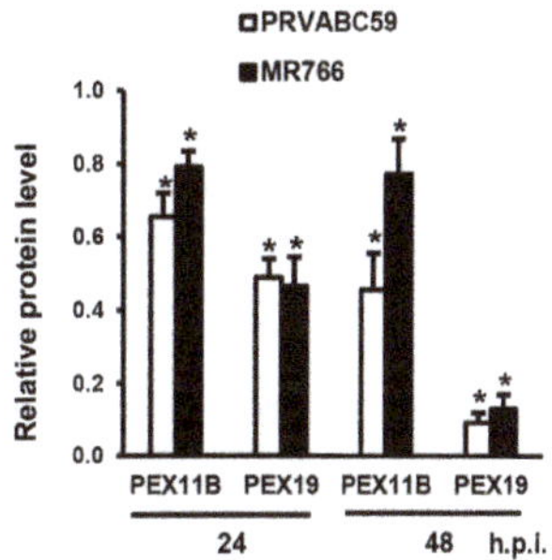

Figure 2. ZIKV infection decreases the number of peroxisomes in U251 cells. (**A**) U251 cells were infected with ZIKV PRVABC59 (PR) or MR766 (MR) using MOI = 3 for 48 h and then processed for confocal microscopy. Peroxisomes were detected using a mouse monoclonal to PMP70 and donkey anti-mouse IgG conjugated to Alexa Fluor 488. Infected cells were detected using a goat polyclonal antibody to ZIKV NS5 and chicken anti-goat IgG conjugated to Alexa Fluor 647. Nuclei were stained with DAPI. Images were obtained using a spinning disc confocal microscope. The number of peroxisomes in mock- and ZIKV-infected cells were determined using Volocity image analysis software. Averages were calculated from three independent experiments in which a minimum of 20 cells for each sample were analyzed. The average number of peroxisomes in mock-treated cells was normalized to 1.0. Bars represent standard error of the mean. ** $p < 0.01$, * $p < 0.05$. (**B**) U251 cells were infected with ZIKV PRVABC59 or MR766 (MOI = 5) for the indicated time periods, after which total cell lysates were processed for immunoblot analyses with antibodies to PEX11B, PEX19, ZIKV NS5 and actin. The relative levels of peroxisomal proteins (compared to actin) from three independent experiments were averaged and plotted. The average levels of peroxisomal proteins in mock-infected cells were normalized to 1.0. Error bars represent standard error of the mean. * $p < 0.05$. h.p.i.=hours post-infection.

3.2. ZIKV Capsid Protein Binds PEX19 and Induces Loss of Peroxisomes

Similar to WNV and DENV capsid proteins, the ZIKV capsid protein formed a stable complex with PEX19 (Figure 3A). Of note, interaction between ZIKV capsid and PEX19 was reported in a recent interactome study [34]. However, this was not verified by reciprocal co-immunoprecipitation or assays. To determine if the ZIKV capsid protein behaved similarly to the analogous proteins from WNV and DENV [20], U251 were transfected with a plasmid encoding FLAG-tagged ZIKV capsid or vector alone. Forty-eight hours post-transfection, cells were processed for immunoblotting or confocal microscopy.

Data in Figure 3B show that the expression of ZIKV capsid protein reduced PEX19 levels by 50%, indicating that ZIKV capsid protein may be the major viral determinant that impairs peroxisome biogenesis. We next investigated peroxisome numbers in ZIKV capsid expressing cells by confocal microscopy. The expression of capsid protein was detected using anti-FLAG, and peroxisomes were identified using an antibody to the tripeptide Ser-Lys-Leu (SKL), a targeting motif found at the carboxyl termini of many peroxisomal matrix proteins [35]. The quantification of SKL-positive structures showed that transient expression of ZIKV capsid protein reduced the number of peroxisomes by 30% versus the control (Figure 3C).

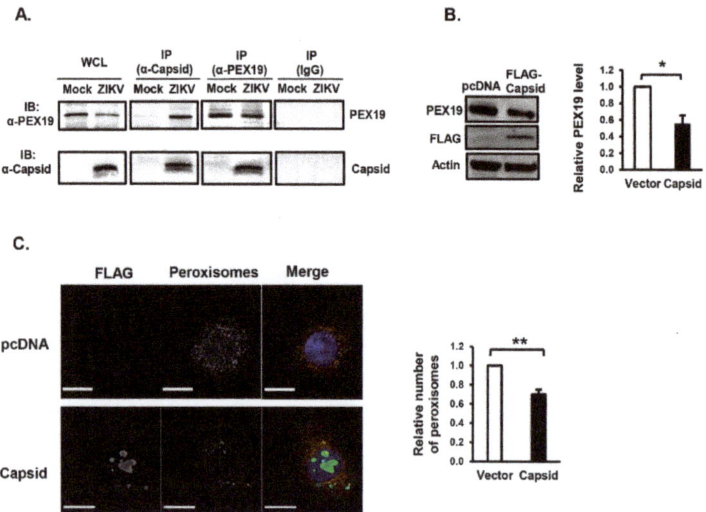

Figure 3. Expression of ZIKV capsid protein reduces peroxisome numbers. (**A**) U251 cells were infected with ZIKV PRVABC59 (MOI = 1). Forty-eight hours later, cell lysates were subjected to immunoprecipitation (IP) with rabbit anti-ZIKV capsid, rabbit anti-PEX19, or rabbit IgG followed by SDS-PAGE and immunoblotting (IB) with antibodies to PEX19 or ZIKV-capsid. WCL, whole-cell lysate. (**B**) HEK293T cells were transfected with a plasmid encoding FLAG-tagged ZIKV capsid or empty vector (pcDNA3.1) for 48 h. Cell lysates were subjected to SDS-PAGE and immunoblotting with antibodies to PEX19, ZIKV-capsid and actin. The relative levels of PEX19 (compared to actin) from three independent experiments were averaged and plotted. Error bars represent standard error of the mean. * $p < 0.05$. (**C**) U251 cells were transfected with a plasmid encoding FLAG-tagged ZIKV capsid or empty vector (pcDNA3.1) for 48 h and then processed for confocal microscopy. Peroxisomes were detected with a rabbit polyclonal antibody to the tri-peptide SKL and donkey anti-rabbit IgG conjugated to Alexa Fluor 546. Transfected cells expressing capsid were detected with a mouse anti-FLAG epitope antibody and donkey anti-mouse IgG conjugated to Alexa Fluor 488. Nuclei were stained using DAPI. Images were obtained using spinning disc confocal microscopy. The relative numbers of peroxisomes (SKL-positive structures) in cells transfected with or without ZIKV capsid plasmid were determined using Volocity image analysis software. Averages were calculated from three independent experiments, in which a minimum of 20 cells for each sample were analyzed. The average number of peroxisomes in mock-treated cells was normalized to 1.0. Bars represent standard error of the mean. ** $p < 0.01$.

3.3. Over-Expression of PEX11B Inhibits ZIKV Replication

Since flavivirus infection depletes peroxisomes, likely as a mechanism to impair the innate immune response, we questioned whether expanding the cellular pool of peroxisomes would restrict ZIKV replication. PEX11B is a cellular protein that induces peroxisome proliferation by stimulating the division of these organelles, and its over-expression results in increased numbers of peroxisomes [36,37].

U251 cells and Vero cells were transduced with lentiviruses encoding the reporter protein AcGFP alone as a control or AcGFP plus myc-tagged PEX11B for 48 h. Lentiviral-mediated over-expression of PEX11B resulted in a 20% increase in the number of peroxisomes in U251 cells (Figure 4A).

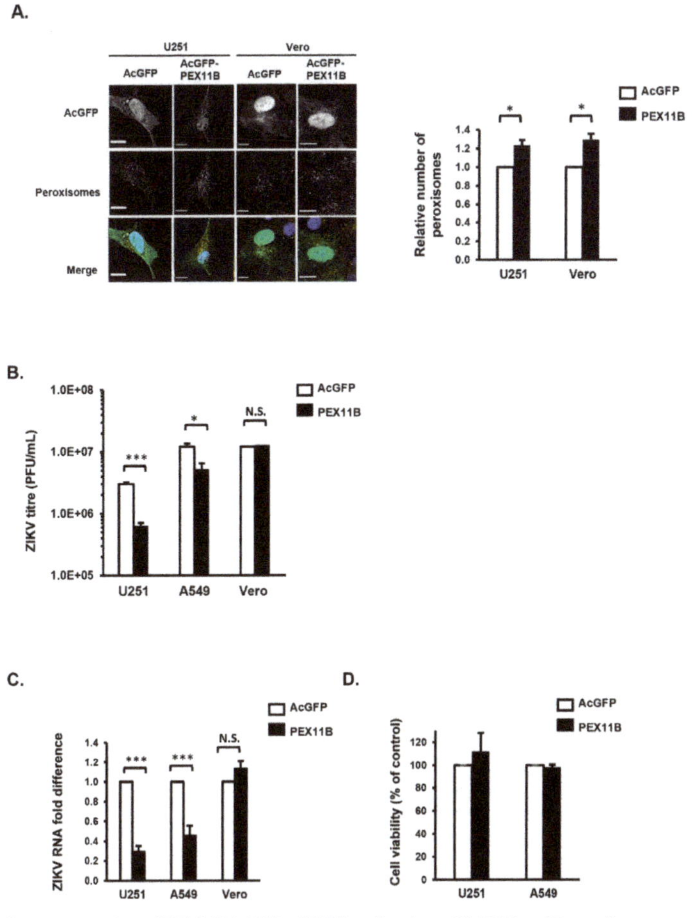

Figure 4. Over-expression of PEX11B inhibits ZIKV replication. (**A**) U251 or Vero cells were transduced with lentiviruses encoding AcGFP alone or AcGFP plus myc-tagged PEX11B for 48 h and then processed for confocal microscopy. Peroxisomes were detected with a rabbit polyclonal antibody to the tri-peptide SKL and donkey anti-rabbit IgG conjugated to Alexa Fluor 546. Nuclei were stained using DAPI. Images were obtained using spinning disc confocal microscopy. The relative numbers of peroxisomes (SKL-positive structures) in cells were determined using Volocity image analysis software. Averages were calculated from three independent experiments, in which a minimum of 20 cells for each sample were analyzed. The average number of peroxisomes in control cells was normalized to 1.0. Bars represent standard error of the mean. * $p < 0.05$. (**B**) U251, A549 or Vero cells were transduced with lentiviruses encoding the reporter protein AcGFP alone as a control or AcGFP plus myc-tagged PEX11B proteins for 48 h, after which the cells were infected with ZIKV PRVABC59 (MOI = 1) for another 48 h. Cell media were processed by plaque assay to determine viral titers. In parallel, cell lysates were also processed for RNA extraction and subsequent qRT-PCR to determine viral RNA level (**C**). U251 and Vero cell lysates were processed to determine cell viability (**D**). The data are averaged from the results of three independent experiments. Bars represent standard error of the mean. *** $p < 0.001$, * $p < 0.05$. N.S. = not significant.

While the effect of PEX11B expression on peroxisome proliferation was modest, the effect on ZIKV replication was dramatic. Specifically, ZIKV titers were reduced by more than 80% in U251 cells over-expressing PEX11B (Figure 4B). This effect was not limited to U251 cells as over-expression of PEX11B in A549 cells also resulted in a significant inhibition of ZIKV replication and reduction in viral titers (Figure 4B,C). To determine if the antiviral effect increasing peroxisome numbers was related to the ability of cells to mount an IFN response, PEX11B was over-expressed in Vero cells, a monkey kidney cell line which does not secrete type I IFN in response to viral infection [38]. While the over-expression of PEX11B increased the number of peroxisome in Vero cells by an average of 35% (Figure 4A), unlike in U251 or A549 cells, this peroxisome biogenesis factor did not reduce ZIKV replication or viral titers (Figure 4B,C). Data in Figure 4D confirm that the antiviral effects of PEX11B over-expression were not due to cytotoxicity in U251 or A549 cells.

3.4. Over-Expression of PEX11B Enhances the Innate Immune Response

One possible scenario to account for the antiviral effect of PEX11B is a more robust innate immune response due to an increase in the surface area of peroxisomes, which have been termed antiviral signaling platforms [19]. However, the expansion of the peroxisome pool without a corresponding increase in the level of MAVS may not enhance antiviral signaling in response to the detection of viral genomes by RIG-I and MDA5. As such, we first investigated how the over-expression of PEX11B affected MAVS protein levels. Immunoblotting data in Figure 5 show that in cells transduced with lentiviruses encoding PEX11B, levels of MAVS protein were increased two-fold. PEX13 protein levels were also increased in response to PEX11B over-expression, suggesting that structural components of this organelle were induced by PEX11B expression. Whether the additional MAVS protein induced by PEX11B over-expression localizes exclusively to peroxisomes, mitochondria, or both is not known at this point.

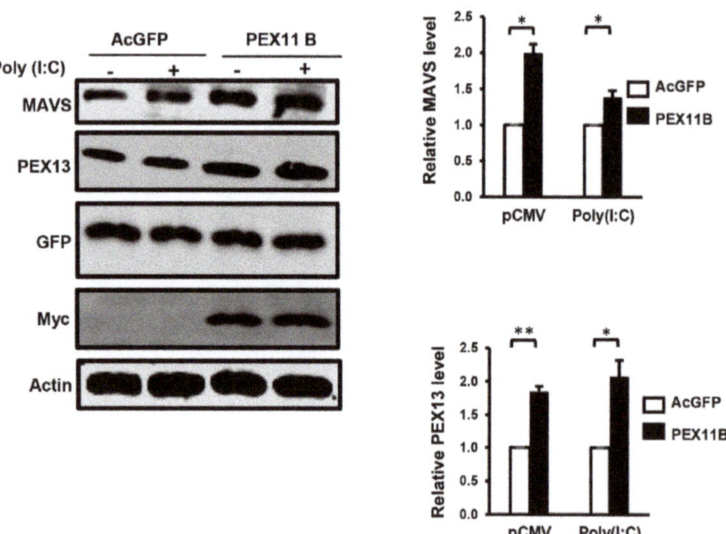

Figure 5. Over-expression of PEX11B increases the expression of MAVS protein. U251 cells were transduced with lentiviruses encoding AcGFP alone or AcGFP plus myc-tagged PEX11B for 48 h and then transfected with poly(I:C) (+) or an empty plasmid vector (−) for 12 h. The cell lysates were processed for immunoblot analyses with a mouse monoclonal antibody to MAVS, rabbit polyclonal PEX13, goat polyclonal antibody to GFP, and a mouse monoclonal antibody to the myc epitope. The relative levels of MAVS and PEX13 (compared to actin) from three independent experiments were averaged and plotted. Bars represent standard error of the mean. ** $p < 0.01$, * $p < 0.05$.

To determine if peroxisome proliferation enhanced the IFN response following the detection of viral RNA mimics such as poly(I:C), levels of type I and III IFN and ISGs were assessed in cells transduced with lentiviruses encoding PEX11B or AcGFP alone. Data in Figure 6 show that the induction of type I (*IFNβ*) and III (*IFNλ2*) IFNs was markedly elevated in cells over-expressing PEX11B. Similarly, the transcription of ISGs such as *Viperin, Mx2, IFIT1, RIG-I* and *MDA5* was significantly increased by PEX11B over-expression (Figure 6).

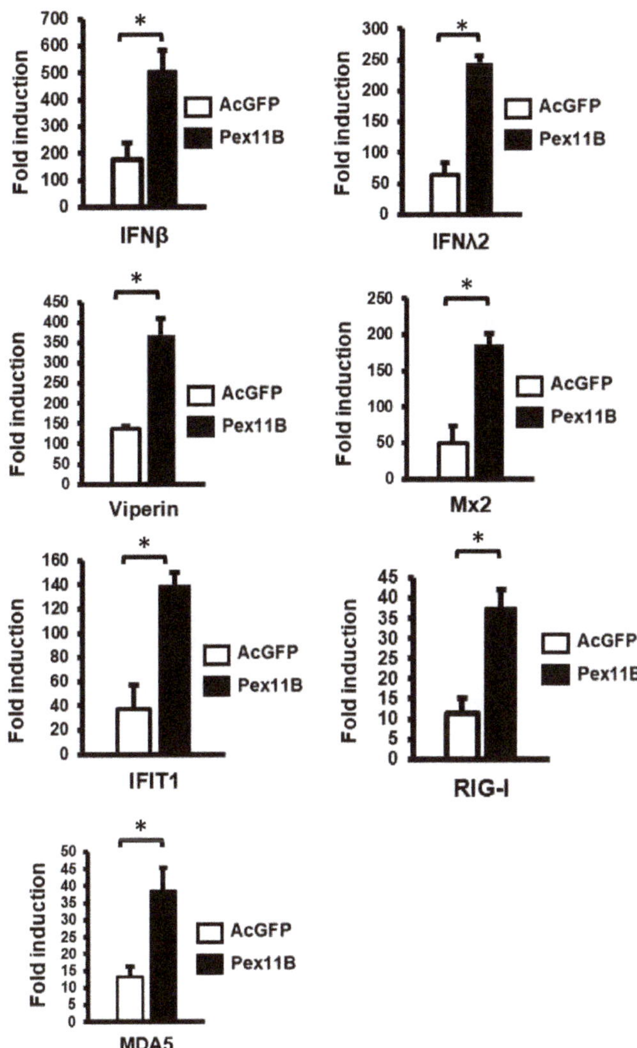

Figure 6. Over-expression of PEX11B enhances the innate immune response. U251 cells were transduced with lentiviruses encoding AcGFP alone or AcGFP plus myc-tagged PEX11B for 48 h and then transfected with poly(I:C) for 12 h. Cell lysates were processed for RNA extraction and subsequent qRT-PCR. Fold induction of selected ISG transcripts in response to poly(I:C) was determined. The mRNA levels of ISGs were normalized to *ACT-B* mRNA levels. The data represent the average from the results of three independent experiments. Bars represent standard error of the mean. * $p < 0.05$.

4. Discussion

The use of mouse models has been very useful in understanding how the IFN response affects ZIKV replication and pathogenesis. For example, whereas outbred mouse strains do not exhibit severe symptoms when infected with ZIKV, mice lacking type I IFN receptors are more susceptible to virus-induced neurological disease and death [5–7]. The ISG viperin has been shown to restrict the replication of many viruses in the Flaviviridae family, including Dengue, tickborne encephalitis, West Nile and hepatitis C viruses [39–43]. Similarly, genetic ablation of the viperin gene in mouse cells results in a more robust replication of ZIKV [44]. Of course, flaviviruses and hepaciviruses have evolved multiple strategies that are effective at blocking IFN induction and downstream signaling. Relevant examples include DENV NS2A and NS4B proteins that suppress IFN induction by inhibiting the phosphorylation of IRF3 [45,46]. During HCV infection, IFN induction can also be blocked by NS3/4A-mediated cleavage of MAVS [21]. Further downstream, the NS5 protein of DENV as well as ZIKV inhibits IFN signaling by inducing the degradation of the antiviral transcription factor STAT2 [28,47,48].

As well as canonical IFN induction pathways which involve RIG-I or MDA5 sensing of viral RNA followed by signaling through mitochondria-associated MAVS protein, peroxisomes are now known to play a role in IFN-based antiviral signaling [17–19,21,22]. A number of important pathogenic viruses have recently been shown to target peroxisomes during infection (reviewed in [49]). For instance, DENV and WNV infections result in the degradation of PEX19, which in turn leads to loss of peroxisomes and dampened induction of type III IFN [20]. Conversely, HIV infection induces the expression of miRNAs that suppress expression of multiple peroxisome biogenesis factors [23]. More recently, nsp1 protein of porcine epidemic diarrhea virus was shown to block IRF1-dependent type III IFN production by decreasing peroxisome pools, but the mechanism has yet to be elucidated [22].

Here, we show that ZIKV infection results in a dramatic loss of peroxisomes in primary human fetal astrocytes, a brain cell type that is highly permissive to the virus [25,50–52]. While we cannot rule out the potential effects of other ZIKV proteins on peroxisome biogenesis, the capsid protein appears to be the main viral determinant that causes the depletion of this organelle. The mechanism is not known but given that flavivirus capsid proteins have no enzymatic activity, they must act in concert with host cell proteins to interfere with peroxisome biogenesis and/or stability.

As intimated above, mounting evidence suggests that peroxisome depletion may be a common facet of RNA virus infection. A consensus among multiple studies [17–23] is that this phenomenon is yet another strategy used by viruses to interfere with the innate immune system. In contrast, Coyaud et al. [24] reported that these organelles are important for ZIKV infection, implying that peroxisome loss is the result of being consumed or used up during the replication process. Their conclusion was based on the observation that the replication of ZIKV in fibroblast lines derived from patients with peroxisomal biogenesis disorders is lower than in fibroblasts from control patients. This must be interpreted with caution because, for ZIKV at least, we found that the permissiveness of HFAs varied significantly depending upon the individual donor [25]. This indicates that differences within the host genetic background affects ZIKV infection. Given that the genetic background of the control patients were likely different than the peroxisome biogenesis disorder patients, it cannot be ruled out that the minor differences in ZIKV replication observed were independent of peroxisomes. Moreover, at early time points and during the peak of ZIKV replication (48 h), there were no differences in titers from normal and peroxisome-deficient cells [24]. Only at 96-h post-infection was a slight decrease in viral titers observed during the infection of peroxisome-deficient cells in the study by Coyaud et al.

While peroxisome depletion has been observed during the replication of multiple viruses, until the present study, the effects of peroxisome proliferation on viral replication had not been investigated. Our data indicate that even a modest increase in the number of peroxisomes through PEX11B over-expression results in a significant inhibition of ZIKV replication. The observation that over-expression of PEX11B in IFN-deficient Vero cells did not inhibit ZIKV replication, indicates that the enhanced antiviral response caused by peroxisome proliferation is IFN-dependent. Indeed, the induction of type I and III IFN, as well as ISGs in response to poly (I:C), was increased as much as five-fold in PEX11B-over-expressing cells. As well, levels of MAVS protein levels were two-fold higher in these cells, suggesting that the upregulation of peroxisomes is coordinated with the increased expression of antiviral signal transducing proteins.

Together, our findings further solidify the importance of peroxisomes in antiviral defense. Moreover, the fact that peroxisome activity and abundance can be pharmacologically modulated provides compelling rationale for the investigation of peroxisome-based antiviral strategies. However, as exciting as this prospect may be, it is important to point out that in some cases, viral infection seems to increase peroxisome numbers. Specifically, it was recently reported that human cytomegalovirus upregulates the biogenesis of these organelles during infection, a scenario that enhances synthesis of plasmalogen, a peroxisome-specific phospholipid that is important for the production of nascent virions [53]. Understanding how other viruses affect these organelles will be essential as we consider the prospect of antiviral therapeutics that affect the activity or abundance of peroxisomes.

Supplementary Materials: The following are available online at http://www.mdpi.com/2073-4409/8/7/725/s1, Table S1: Primer sets for PCR used in this study. Table S2: Primer sets for qRT-PCR used in this study.

Author Contributions: Conceptualization, C.P.W., S.H., A.K. and T.C.H.; methodology, C.P.W., S.H. and Z.X.; software, C.P.W. and S.H.; validation, T.C.H.; formal analysis, C.P.W., Z.X., S.H. and T.C.H.; investigation, C.P.W. and Z.X.; resources, D.L., C.P.; data curation, C.P.W. and Z.X.; writing—original draft preparation, C.P.W. and Z.X.; writing—review and editing, C.P.W., Z.X., S.H., D.L, A.K., C.P. and T.C.H.; project administration, C.P.W., Z.X., S.H., D.L, A.K., C.P. and T.C.H.; funding acquisition, C.P.W., A.K. and T.C.H.

Funding: This research was funded by Canadian Institutes of Health Research (PJT-148699 and ZV1-149782). C.P.W. is supported by a graduate studentship award from Alberta Innovates—Technology Futures, and Alberta Advanced Education.

Acknowledgments: The authors thank Eileen Reklow and Valeria Mancinelli for excellent technical assistance. We acknowledge use of The Cell Imaging Centre core facility in the Faculty of Medicine & Dentistry at the University of Alberta.

Conflicts of Interest: The authors declare no conflict of interest. The funders had no role in the design of the study; in the collection, analyses, or interpretation of data; in the writing of the manuscript, or in the decision to publish the results.

References

1. Kuno, G.; Chang, G.J. Full-length sequencing and genomic characterization of Bagaza, Kedougou, and Zika viruses. *Arch. Virol.* **2007**, *152*, 687–696. [CrossRef] [PubMed]
2. Gould, E.A.; Solomon, T. Pathogenic flaviviruses. *Lancet* **2008**, *371*, 500–509. [CrossRef]
3. Rubin, E.J.; Greene, M.F.; Baden, L.R. Zika Virus and Microcephaly. *N. Engl. J. Med.* **2016**, *374*, 984–985. [CrossRef] [PubMed]
4. Rodrigues, L.C. Microcephaly and Zika virus infection. *Lancet* **2016**, *387*, 2070–2072. [CrossRef]
5. Yockey, L.J.; Varela, L.; Rakib, T.; Khoury-Hanold, W.; Fink, S.L.; Stutz, B.; Szigeti-Buck, K.; Van den Pol, A.; Lindenbach, B.D.; Horvath, T.L.; et al. Vaginal Exposure to Zika Virus during Pregnancy Leads to Fetal Brain Infection. *Cell* **2016**, *166*, 1247–1256. [CrossRef] [PubMed]
6. Aliota, M.T.; Caine, E.A.; Walker, E.C.; Larkin, K.E.; Camacho, E.; Osorio, J.E. Characterization of Lethal Zika Virus Infection in AG129 Mice. *PLoS Negl. Trop. Dis.* **2016**, *10*, e0004682. [CrossRef]
7. Lazear, H.M.; Govero, J.; Smith, A.M.; Platt, D.J.; Fernandez, E.; Miner, J.J.; Diamond, M.S. A Mouse Model of Zika Virus Pathogenesis. *Cell Host Microbe* **2016**, *19*, 720–730. [CrossRef]
8. Said, E.A.; Tremblay, N.; Al-Balushi, M.S.; Al-Jabri, A.A.; Lamarre, D. Viruses Seen by Our Cells: The Role of Viral RNA Sensors. *J. Immunol. Res.* **2018**, *2018*, 9480497. [CrossRef]
9. Kell, A.M.; Gale, M. RIG-I in RNA virus recognition. *Virology* **2015**, *479*, 110–121. [CrossRef]
10. Pichlmair, A.; e Sousa, C.R. Innate recognition of viruses. *Immunity* **2007**, *27*, 370–383. [CrossRef]
11. Platanias, L.C. Mechanisms of type-I- and type-II-interferon-mediated signalling. *Nat. Rev. Immunol.* **2005**, *5*, 375–386. [CrossRef] [PubMed]

12. Levy, D.E.; Marié, I.J.; Durbin, J.E. Induction and function of type I and III interferon in response to viral infection. *Curr. Opin. Virol.* **2011**, *1*, 476–486. [CrossRef] [PubMed]

13. Beachboard, D.C.; Horner, S.M. Innate immune evasion strategies of DNA and RNA viruses. *Curr. Opin. Microbiol.* **2016**, *32*, 113–119. [CrossRef]

14. Hughes, R.; Towers, G.; Noursadeghi, M. Innate immune interferon responses to human immunodeficiency virus-1 infection. *Rev. Med. Virol.* **2012**, *22*, 257–266. [CrossRef] [PubMed]

15. Islinger, M.; Voelkl, A.; Fahimi, H.D.; Schrader, M. The peroxisome: An update on mysteries 2.0. *Histochem. Cell Biol.* **2018**, *150*, 1–29. [CrossRef] [PubMed]

16. Deori, N.M.; Kale, A.; Maurya, P.K.; Nagotu, S. Peroxisomes: Role in cellular ageing and age related disorders. *Biogerontology* **2018**, *19*, 303–324. [CrossRef] [PubMed]

17. Odendall, C.; Dixit, E.; Stavru, F.; Bierne, H.; Franz, K.M.; Durbin, A.F.; Boulant, S.; Gehrke, L.; Cossart, P.; Kagan, J.C. Diverse intracellular pathogens activate type III interferon expression from peroxisomes. *Nat. Immunol.* **2014**, *15*, 717–726. [CrossRef] [PubMed]

18. Bender, S.; Reuter, A.; Eberle, F.; Einhorn, E.; Binder, M.; Bartenschlager, R. Activation of Type I and III Interferon Response by Mitochondrial and Peroxisomal MAVS and Inhibition by Hepatitis C Virus. *PLoS Pathog.* **2015**, *11*, e1005264. [CrossRef]

19. Dixit, E.; Boulant, S.; Zhang, Y.; Lee, A.S.; Odendall, C.; Shum, B.; Hacohen, N.; Chen, Z.J.; Whelan, S.P.; Fransen, M.; et al. Peroxisomes are signaling platforms for antiviral innate immunity. *Cell* **2010**, *141*, 668–681. [CrossRef]

20. You, J.; Hou, S.; Malik-Soni, N.; Xu, Z.; Kumar, A.; Rachubinski, R.A.; Frappier, L.; Hobman, T.C. Flavivirus Infection Impairs Peroxisome Biogenesis and Early Antiviral Signaling. *J. Virol.* **2015**, *89*, 12349–12361. [CrossRef]

21. Ferreira, A.R.; Magalhães, A.C.; Camões, F.; Gouveia, A.; Vieira, M.; Kagan, J.C.; Ribeiro, D. Hepatitis C virus NS3-4A inhibits the peroxisomal MAVS-dependent antiviral signalling response. *J. Cell Mol. Med.* **2016**, *20*, 750–757. [CrossRef] [PubMed]

22. Zhang, Q.; Ke, H.; Blikslager, A.; Fujita, T.; Yoo, D. Type III interferon restriction by porcine epidemic diarrhea virus and the role of viral protein nsp1 in IRF1 signaling. *J. Virol.* **2018**, *92*, e01677-17. [CrossRef] [PubMed]

23. Xu, Z.; Asahchop, E.L.; Branton, W.G.; Gelman, B.B.; Power, C.; Hobman, T.C. MicroRNAs upregulated during HIV infection target peroxisome biogenesis factors: Implications for virus biology, disease mechanisms and neuropathology. *PLoS Pathog.* **2017**, *13*, e1006360. [CrossRef] [PubMed]

24. Coyaud, E.; Ranadheera, C.; Cheng, D.; Gonçalves, J.; Dyakov, B.J.A.; Laurent, E.M.N.; St-Germain, J.; Pelletier, L.; Gingras, A.C.; Brumell, J.H.; et al. Global Interactomics Uncovers Extensive Organellar Targeting by Zika Virus. *Mol. Cell Proteomics* **2018**, *17*, 2242–2255. [CrossRef] [PubMed]

25. Limonta, D.; Jovel, J.; Kumar, A.; Airo, A.; Hou, S.; Saito, L.; Branton, W.; Ka-Shu Wong, G.; Mason, A.; Power, C. Human Fetal Astrocytes Infected with Zika Virus Exhibit Delayed Apoptosis and Resistance to Interferon: Implications for Persistence. *Viruses* **2018**, *10*, 646. [CrossRef] [PubMed]

26. Kassmann, C.M.; Lappe-Siefke, C.; Baes, M.; Brügger, B.; Mildner, A.; Werner, H.B.; Natt, O.; Michaelis, T.; Prinz, M.; Frahm, J.; et al. Axonal loss and neuroinflammation caused by peroxisome-deficient oligodendrocytes. *Nat. Genet.* **2007**, *39*, 969–976. [CrossRef] [PubMed]

27. Bottelbergs, A.; Verheijden, S.; Hulshagen, L.; Gutmann, D.H.; Goebbels, S.; Nave, K.A.; Kassmann, C.; Baes, M. Axonal integrity in the absence of functional peroxisomes from projection neurons and astrocytes. *Glia* **2010**, *58*, 1532–1543. [CrossRef]

28. Kumar, A.; Hou, S.; Airo, A.M.; Limonta, D.; Mancinelli, V.; Branton, W.; Power, C.; Hobman, T.C. Zika virus inhibits type-I interferon production and downstream signaling. *EMBO Rep.* **2016**, *17*, 1766–1775. [CrossRef]

29. Schwarz, M.C.; Sourisseau, M.; Espino, M.M.; Gray, E.S.; Chambers, M.T.; Tortorella, D.; Evans, M.J. Rescue of the 1947 Zika virus prototype strain with a cytomegalovirus promoter-driven cDNA clone. *MSphere* **2016**, *1*, e00246-16. [CrossRef]

30. Urbanowski, M.D.; Hobman, T.C. The West Nile virus capsid protein blocks apoptosis through a phosphatidylinositol 3-kinase-dependent mechanism. *J. Virol.* **2013**, *87*, 872–881. [CrossRef]

31. Hou, S.; Kumar, A.; Xu, Z.; Airo, A.M.; Stryapunina, I.; Wong, C.P.; Branton, W.; Tchesnokov, E.; Götte, M.; Power, C. Zika virus hijacks stress granule proteins and modulates the host stress response. *J. Virol.* **2017**, *91*, e00474-17. [CrossRef] [PubMed]

32. Livak, K.J.; Schmittgen, T.D. Analysis of relative gene expression data using real-time quantitative PCR and the 2(-Delta Delta C(T)) Method. *Methods* **2001**, *25*, 402–408. [CrossRef] [PubMed]

33. Shepard, D.S.; Halasa, Y.A.; Tyagi, B.K.; Adhish, S.V.; Nandan, D.; Karthiga, K.S.; Chellaswamy, V.; Gaba, M.; Arora, N.K.; Group, I.S. Economic and disease burden of dengue illness in India. *Am. J. Trop. Med. Hyg.* **2014**, *91*, 1235–1242. [CrossRef] [PubMed]

34. Scaturro, P.; Stukalov, A.; Haas, D.A.; Cortese, M.; Draganova, K.; Płaszczyca, A.; Bartenschlager, R.; Götz, M.; Pichlmair, A. An orthogonal proteomic survey uncovers novel Zika virus host factors. *Nature* **2018**, *561*, 253–257. [CrossRef] [PubMed]

35. Gould, S.J.; Keller, G.A.; Hosken, N.; Wilkinson, J.; Subramani, S. A conserved tripeptide sorts proteins to peroxisomes. *J. Cell Biol.* **1989**, *108*, 1657–1664. [CrossRef] [PubMed]

36. Schrader, M.; Reuber, B.E.; Morrell, J.C.; Jimenez-Sanchez, G.; Obie, C.; Stroh, T.A.; Valle, D.; Schroer, T.A.; Gould, S.J. Expression of PEX11beta mediates peroxisome proliferation in the absence of extracellular stimuli. *J. Biol. Chem.* **1998**, *273*, 29607–29614. [CrossRef] [PubMed]

37. Opaliński, Ł.; Kiel, J.A.; Williams, C.; Veenhuis, M.; van der Klei, I.J. Membrane curvature during peroxisome fission requires Pex11. *EMBO J.* **2011**, *30*, 5–16. [CrossRef] [PubMed]

38. Desmyter, J.; Melnick, J.L.; Rawls, W.E. Defectiveness of interferon production and of rubella virus interference in a line of African green monkey kidney cells (Vero). *J. Virol.* **1968**, *2*, 955–961.

39. Helbig, K.J.; Lau, D.T.; Semendric, L.; Harley, H.A.; Beard, M.R. Analysis of ISG expression in chronic hepatitis C identifies viperin as a potential antiviral effector. *Hepatology* **2005**, *42*, 702–710. [CrossRef]

40. Helbig, K.J.; Eyre, N.S.; Yip, E.; Narayana, S.; Li, K.; Fiches, G.; McCartney, E.M.; Jangra, R.K.; Lemon, S.M.; Beard, M.R. The antiviral protein viperin inhibits hepatitis C virus replication via interaction with nonstructural protein 5A. *Hepatology* **2011**, *54*, 1506–1517. [CrossRef]

41. Szretter, K.J.; Brien, J.D.; Thackray, L.B.; Virgin, H.W.; Cresswell, P.; Diamond, M.S. The interferon-inducible gene viperin restricts West Nile virus pathogenesis. *J. Virol.* **2011**, *85*, 11557–11566. [CrossRef] [PubMed]

42. Helbig, K.J.; Carr, J.M.; Calvert, J.K.; Wati, S.; Clarke, J.N.; Eyre, N.S.; Narayana, S.K.; Fiches, G.N.; McCartney, E.M.; Beard, M.R. Viperin is induced following dengue virus type-2 (DENV-2) infection and has anti-viral actions requiring the C-terminal end of viperin. *PLoS Negl. Trop. Dis.* **2013**, *7*, e2178. [CrossRef] [PubMed]

43. Upadhyay, A.S.; Vonderstein, K.; Pichlmair, A.; Stehling, O.; Bennett, K.L.; Dobler, G.; Guo, J.T.; Superti-Furga, G.; Lill, R.; Överby, A.K.; et al. Viperin is an iron-sulfur protein that inhibits genome synthesis of tick-borne encephalitis virus via radical SAM domain activity. *Cell Microbiol.* **2014**, *16*, 834–848. [CrossRef] [PubMed]

44. Van der Hoek, K.H.; Eyre, N.S.; Shue, B.; Khantisitthiporn, O.; Glab-Ampi, K.; Carr, J.M.; Gartner, M.J.; Jolly, L.A.; Thomas, P.Q.; Adikusuma, F.; et al. Viperin is an important host restriction factor in control of Zika virus infection. *Sci. Rep.* **2017**, *7*, 4475. [CrossRef] [PubMed]

45. Dalrymple, N.A.; Cimica, V.; Mackow, E.R. Dengue Virus NS Proteins Inhibit RIG-I/MAVS Signaling by Blocking TBK1/IRF3 Phosphorylation: Dengue Virus Serotype 1 NS4A Is a Unique Interferon-Regulating Virulence Determinant. *MBio* **2015**, *6*, e00553-15. [CrossRef] [PubMed]

46. Xia, H.; Luo, H.; Shan, C.; Muruato, A.E.; Nunes, B.T.D.; Medeiros, D.B.A.; Zou, J.; Xie, X.; Giraldo, M.I.; Vasconcelos, P.F.C.; et al. An evolutionary NS1 mutation enhances Zika virus evasion of host interferon induction. *Nat. Commun.* **2018**, *9*, 414. [CrossRef]

47. Grant, A.; Ponia, S.S.; Tripathi, S.; Balasubramaniam, V.; Miorin, L.; Sourisseau, M.; Schwarz, M.C.; Sánchez-Seco, M.P.; Evans, M.J.; Best, S.M.; et al. Zika Virus Targets Human STAT2 to Inhibit Type I Interferon Signaling. *Cell Host Microbe* **2016**, *19*, 882–890. [CrossRef]

48. Ashour, J.; Laurent-Rolle, M.; Shi, P.Y.; García-Sastre, A. NS5 of dengue virus mediates STAT2 binding and degradation. *J. Virol.* **2009**, *83*, 5408–5418. [CrossRef]

49. Wong, C.P.; Xu, Z.; Power, C.; Hobman, T.C. Targeted Elimination of Peroxisomes During Viral Infection: Lessons from HIV and Other Viruses. *DNA Cell Biol.* **2018**, *37*, 417–421. [CrossRef]

50. Retallack, H.; Di Lullo, E.; Arias, C.; Knopp, K.A.; Laurie, M.T.; Sandoval-Espinosa, C.; Mancia Leon, W.R.; Krencik, R.; Ullian, E.M.; Spatazza, J.; et al. Zika virus cell tropism in the developing human brain and inhibition by azithromycin. *Proc. Natl. Acad. Sci. USA* **2016**, *113*, 14408–14413. [CrossRef]

51. Huang, Y.; Li, Y.; Zhang, H.; Zhao, R.; Jing, R.; Xu, Y.; He, M.; Peer, J.; Kim, Y.C.; Luo, J.; et al. Zika virus propagation and release in human fetal astrocytes can be suppressed by neutral sphingomyelinase-2 inhibitor GW4869. *Cell Discov.* **2018**, *4*, 19. [CrossRef] [PubMed]

52. Limonta, D.; Jovel, J.; Kumar, A.; Lu, J.; Hou, S.; Airo, A.M.; Lopez-Orozco, J.; Wong, C.P.; Saito, L.; Branton, W.; et al. Fibroblast growth factor 2 enhances Zika virus infection in human fetal brain. *J. Infect. Dis.* **2019**. [CrossRef] [PubMed]

53. Beltran, P.M.J.; Cook, K.C.; Hashimoto, Y.; Galitzine, C.; Murray, L.A.; Vitek, O.; Cristea, I.M. Infection-Induced Peroxisome Biogenesis Is a Metabolic Strategy for Herpesvirus Replication. *Cell Host Microbe* **2018**, *24*, 526–541. [CrossRef] [PubMed]

Review

The Cellular Impact of the ZIKA Virus on Male Reproductive Tract Immunology and Physiology

Raquel das Neves Almeida [1], Heloisa Antoniella Braz-de-Melo [1], Igor de Oliveira Santos [1], Rafael Corrêa [1], Gary P. Kobinger [2,3] and Kelly Grace Magalhaes [1,*]

[1] Laboratory of Immunology and Inflammation, Department of Cell Biology, University of Brasilia, 70910–900 Brasilia, Distrito Federal, Brazil; raquel.unb.bio@gmail.com (R.d.N.A.); heloisa.antoniella@gmail.com (H.A.B.-d.-M.); igorosantosbio@gmail.com (I.d.O.S.); rafaelcorrea31@gmail.com (R.C.)

[2] Département de Microbiologie-Infectiologie et d'Immunologie, Université Laval, Quebec City, QC G1V 4G2, Canada; Gary.Kobinger@crchudequebec.ulaval.ca

[3] Centre de Recherche en Infectiologie du CHU de Québec-Université Laval, Quebec City, QC G1V 4G2, Canada

* Correspondence: kellymagalhaes@unb.br; Tel.: +55-61-3107-3099-3103

Received: 11 March 2020; Accepted: 15 April 2020; Published: 18 April 2020

Abstract: Zika virus (ZIKV) has been reported by several groups as an important virus causing pathological damage in the male reproductive tract. ZIKV can infect and persist in testicular somatic and germ cells, as well as spermatozoa, leading to cell death and testicular atrophy. ZIKV has also been detected in semen samples from ZIKV-infected patients. This has huge implications for human reproduction. Global scientific efforts are being applied to understand the mechanisms related to arboviruses persistency, pathogenesis, and host cellular response to suggest a potential target to develop robust antiviral therapeutics and vaccines. Here, we discuss the cellular modulation of the immunologic and physiologic properties of the male reproductive tract environment caused by arboviruses infection, focusing on ZIKV. We also present an overview of the current vaccine effects and therapeutic targets against ZIKV infection that may impact the testis and male fertility.

Keywords: Sertoli cells; Leydig cells; ZIKA virus; arboviruses; infertility

---

1. Introduction

The testis is a reproductive gland that is part of the internal structures of the male reproductive tract (MRT) and is involved in spermatogenesis and steroidogenesis. Each testis is composed of a tangle of tubes, the seminiferous ducts. These ducts are formed by Sertoli cells (SCs) and the germinal epithelium, which is responsible for ensuring protection and nutrition to accurate spermatogenesis. Leydig cells (LCs) are found in the testis interstitium, adjacent to the seminiferous tubules. LCs promote steroidogenesis through the secretion of male sex hormones, especially testosterone, responsible for the development of male genital organs and secondary sexual characters [1,2].

The testis is considered an immune-privileged organ [3]. This is essential to ensure the immunogenic germ cell protection against immune system activation during spermatogenesis. This is mainly provided by the combination of a local immunosuppressive environment and systemic immune tolerance [4–6]. It has long been assumed that the blood–testis barrier (BTB) constitutes the main mechanism of the immune-privileged status of the testis [7]. In addition to BTB and anatomical impairment of external cells' and molecules' entrance to testis, SCs also provide anti-inflammatory mediator secretion aiming to maintain the tolerogenic microenvironment [8]. However, many local immune modulators, including macrophages, dendritic cells (DCs), natural killer cells (NKs), mast cells and T-lymphocytes, contribute to the intercommunication among testicular components [9–12].

The testis is commonly exposed to pathogens derived from blood, trauma, or through the genitourinary tract. To protect itself against all these pathogens, the testis also needs the ability to overpower immune privilege. This is achieved by inducing local innate immune responses [3]. Even counting this frontline protection, some pathogens have an immune scape mechanism that leads to infection and persistence in the MRT. Reproductive tract infections (RTI) can be caused by bacterial, parasitic, and viral pathogens [13]. RTI promoted by viral infections are notorious, as shown by the World Health Organization (WHO) in 2006, which estimated that 500 million people live with genital herpes, 300 million women have human papillomavirus (HPV), and approximately 240 million people suffer from chronic hepatitis B [14]. In 2016, the WHO also estimated that over 17 million people are living with HIV on antiretroviral therapy. However, the number of HIV-positive cases is increasing worldwide [15].

Some diseases can persist a long time in human semen. Ebola [16], Zika virus (ZIKV) [17], HIV [18], and 27 other types of viruses that contaminate humans have been found in semen and testis for differing periods [19]. Despite the knowledge that various types of viruses can be found in semen, their sexual transmission capacity is still poorly understood. Some of these are not considered sexually transmitted diseases because this route is not the main form of contagion. However, ZIKV has already been confirmed by the WHO to have sexual transmission (World Health Organization, 2016) and considered to be the first arbovirus reported to be associated with sexual transmission [20,21]. Due to this fact, attention is being turned to the possibility that other arboviruses may be present in the MRT. Compared to ZIKV, the literature regarding this effect is scarce, and the available data suggests that arbovirus sexual transmission is a relevant point of concern. The presence of ZIKV in the male genital tract and its ability of sexual transmission leads to unanswered questions such as (1) has the ZIKV a tropism for any specific cell in the male reproductive system?, (2) what features may favor the ZIKV persistence in testicles when compared to other arboviruses?, (3) can the spermatozoa harbor the virus?, (4) how long does the virus remain viable in the male genital tract?, (5) how can the prolonged presence of ZIKV in the male genital tract cause infertility?, (6) is this ZIKV-induced testicular damage reversible? Based on these questions, it is clear the importance of continuing to investigate the role of ZIKV in the male reproductive system. In addition, a vaccine against ZIKV may be the best way to protect the population from infection, and control the disease and its consequences. The vaccination should protect against future and possible damage to the male genital tract, avoiding fertility-related problems. Therefore, in this review, we will address recent findings of ZIKV infections in the MRT, focusing on cellular mechanisms, immune and physiological responses, and the ability to other arboviruses to remain in the testicle.

2. Male Reproductive Tract (MRT) and Cellular Composition of Testis

The MRT is composed of sexual organs that play a major role in the male germ cells (or sperm) production. It has mainly consisted of a pair of testicles that are specialized for androgen hormones and gamete production, an intromittent organ that is responsible for depositing sperm on the female reproductive tract and finally a couple of sexual accessories ducts and glands vital for sperm maturation, nutrition, and storage [22,23]. The different cell types that compound these tissues of MRT maintain crosstalk that allows the production of viable sperm in the testis (Figure 1). Once the homeostasis of the system is broken, this process is impaired and the fertility capacity is altered [24].

The testis is composed of interstitial LCs located between blood vessels and the seminiferous tubules, where sperm is produced [25]. LCs secrete androgens that participate in conjunction with pituitary hormones (gonadotropin) in germ cell development [26]. On the other hand, seminiferous tubules include the germ cells, which give rise to spermatozoa through a series of differentiation steps and the somatic SCs [23]. Somatic SCs are essential not just for testes formation but are one of the major conductors of gametogenesis [27]. The immunological infiltrate in the interstitial compartment of the normal testis, especially resident macrophages, is also important to directly influence testicular microenvironment [28].

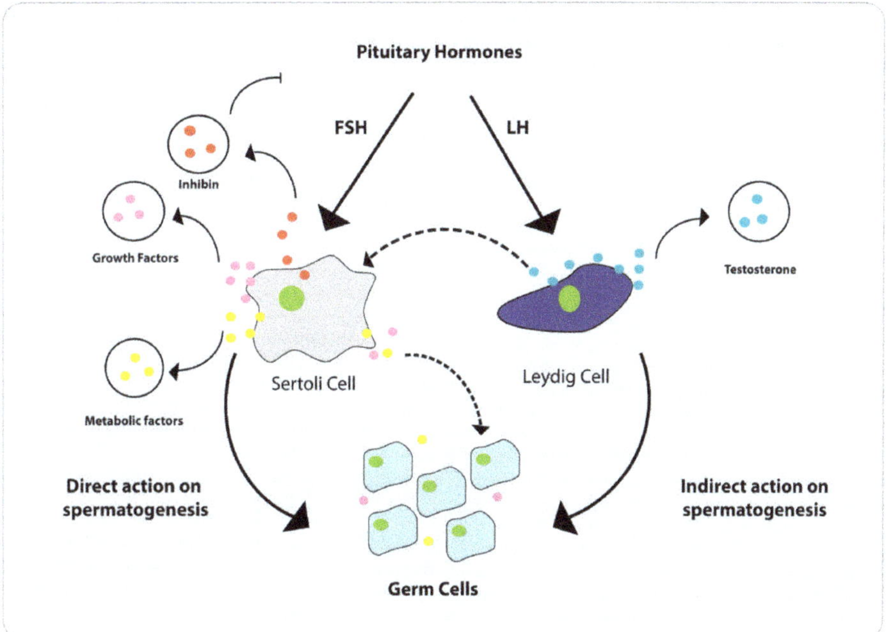

Figure 1. Cellular crosstalk during normal spermatogenesis. Pituitary hormones follicle-stimulating hormone (FSH) and Luteinizing Hormone (LH) have an important role in spermatogenesis. FSH leads to Sertoli cell proliferation stimulating the release of inhibin. LH triggers the production of the testosterone by Leydig cells, which can stimulate the release of metabolic and growth factors by Sertoli cells and indirectly trigger spermatogenesis in germ cells. Metabolic factors, such as lactate and growth factors, can directly drive the spermatogenesis in germ cells. Oppositely, inhibin produced by Sertoli cells can inhibit FSH release by pituitary gland acting as a negative feedback regulation.

The seminiferous tubules present an anatomical barrier that impairs the blood-derived factor input to the testicular microenvironment without any regulation [29]. BTB is the main factor responsible for regulating the paracellular transit of molecules. The BTB is the result of tightly cellular junctions of adjacent SCs in addition to epithelial and myeloid cell interaction [27]. The presence of this barrier creates separated compartments and protects against immunological infiltrate that could lead to testicular inflammation [30]. The unbalanced inflammatory response can disrupt BTB integrity, causing non-specific entry of harmful molecules that impair sperm cell maturation. Nevertheless, cytokine release is a regulatory factor during spermatogenesis in controlled levels [31]. It is important to emphasize that the transit of immune cells is not fully blocked once leukocytes have been reported in normal testicular surroundings, especially close to spermatozoa. Macrophages are the most abundant immunological cells that reside in seminiferous tubules environment and present an important role of immune-surveillance of the germ cell development process.

In the testis, macrophage characterization demonstrated novel functions associated with germ cell development, androgen hormone production, and maintenance of a homeostatic microenvironment [28]. Studies have shown that there are two distinct macrophages populations in testicular surroundings: the CD163$^-$ newly arrived macrophages and CD163$^+$ resident testicular macrophages. The CD163$^+$ macrophages are polarized to the type 2 macrophage (M2) profile that constantly secretes anti-inflammatory molecules, such as interleukin-10, in the seminiferous tubules acting as a protective component against sperm cell damage [32]. On the other hand, newly arrived CD163$^-$ macrophages are related to the inflammation maintained in the seminiferous tubules. These cells secrete higher levels of pro-inflammatory cytokines, such as interleukin-1β and tumoral necrosis factor-α, and present a higher expression of nitric oxide synthase (iNOS), demonstrating a pro-inflammatory profile, a key characteristic of type 1 macrophage (M1) [32]. The communication of these cells with LCs, SCs, and germ cells seems to be important in the development process that leads to sperm production. Macrophages are being called as true sentinels of testis function [28]. In addition, Matusali and colleagues have found evidence of ZIKV infection of the testicular CD163⁺ resident macrophages [33]. ZIKV-induced cell death of CD163⁺ resident macrophages could also contribute to the inflammation in testis.

In addition to macrophages, other immunological cells are found in testicular surroundings. DC are antigen-presenting cells found in testicular interstitial spaces and represent a minor population of leukocytes in the testis. DCs induce activation and differentiation of lymphocytes in response to allo-antigens and minimize autoimmune response by tolerating T-cells to auto-antigens under physiological conditions [12]. Other immune cells, including NKs, T-cells, and CD4⁺CD25⁺ regulatory T-cells (Tregs), are also found [10]. Besides, mast cells are present in a great number regarding immune cell populations in the testis during puberty [34]. However, the functions of these cells in the maintenance of testicular immune-privileged sites remain unclear [35].

The process of male mature gamete production is called spermatogenesis and consists of the intense proliferation and subsequent differentiation of spermatogonial stem cells to spermatozoa [25]. The crosstalk between constituent cells of the testis is essential in this process [36,37], once the energy source of gametes during differentiation depends on the lactate that is provided by SC. On the other hand, glucose capitation depends on androgen hormone signalization provided by LC, as well as pituitary hormones, insulin sensibility, and paracrine communication [36,37]. This is one of the central reasons that explains why altering testicular cell metabolism impairs the production of viable sperm [38]. Important factors as epigenetics (including miRNA regulatory activity), growth factors, and cytokine release also influence the process in the quantity and quality of the sperm [39,40].

Spermatogenesis starts in puberty, long after the perinatal self-tolerance process. For this reason, sperm cells contain a new repertory of proteins that present a great potential of activating an immune response, leading to autoimmunity [3]. Studies have shown that activation of T-lymphocytes and the production of specific antibodies against sperm cells are related to the infertility process. It was also reported that the production of intense pro-inflammatory cytokines is related to loss of BTB integrity and loss of viable sperm, leading to infertility [30,41]. Avoiding this massive activation, the testis presents a unique tolerogenic microenvironment, making the organ immune-privileged, and protecting mature gametes against the immune-cell-induced death and inflammation.

The immune-privileged microenvironment is essential for the viability of sperm cells and maintenance of testis function while, at the same time, serving as a site for the persistence of infections due to the tolerogenic surrounding. Microorganisms coming from blood or urogenital infections enter into a testicular environment and disrupt tissue homeostasis, leading to activation of local immune system [3]. This process triggers testicular inflammation and may alter tissue metabolism, signalization, cellular function, and leading to impaired spermatogenesis and spermiogenesis [42,43]. Many pathogens have been shown to cause male infertility by many mechanisms, induced inflammation being the key for most of them.

3. Flavivirus and ZIKV Features

The *Flavivirus* genus is composed by viruses of small single-stranded RNA. The flaviviruses can cause mild symptoms, such as fever, pain, and cutaneous rash but also covers severe disturbances, such as encephalitis, neurological complications, and hemorrhagic fever [44]. Flaviviruses are arthropod-borne pathogens typically transmitted by mosquitoes or tick vectors and are related to significant mortality and morbidity worldwide [45]. Members with clinical relevance of this genus include Dengue virus (DENV), Yellow Fever virus (YFV), Japanese Encephalitis virus (JEV), West Nile virus (WNV) and ZIKV. The geographic distribution of flaviviruses and the diversity of arthropod vectors make them of great interest for epidemiological surveillance. Moreover, the easy entry and adaptation of these viruses in new environments make this genus relevant to extensive research and experimental studies [44].

ZIKV is a vector-borne flavivirus belonging to the *Flaviviridae* family, with two main lineages: the African and the Asian lineage [46]. It is an enveloped virus measuring about 50 nm in diameter with a non-segmented, positive single-stranded ribonucleic acid (RNA) genome (Figure 2). The genome is made up around of 11 kb with a single open reading frame that codes structural proteins: Capsid (C), Envelope (E), precursor membrane (prM); and non-structural proteins (NS1, NS2A, NS2B, NS3, NS4A, NS4B, and NS5) [47] (Figure 2).

The first ZIKV isolate was identified in primates in 1947 in Uganda Protectorate in a program for surveillance of yellow fever in primates [48]. The first human infection was reported in 1954 in Nigeria; for decades, ZIKV cases were restricted to Africa and Asia [49]. Since 1954, several outbreaks with increasing number cases have been reported worldwide [50,51]. The last outbreak was documented in 2015 in America, which was the largest epidemic ever described of ZIKV affecting more than 20 countries [52,53]. In 2016, WHO considered ZIKV a public health emergency of international concern [20].

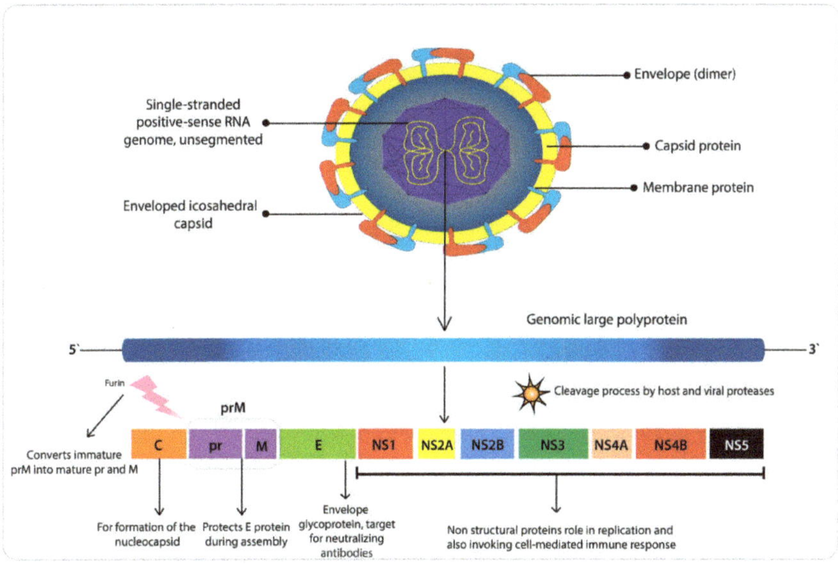

Figure 2. Zika virus (ZIKV) structure and features. ZIKV is an enveloped positive-sense single-stranded RNA virus composed by envelope, capsid, membrane protein, and single-stranded positive-sense RNA. The lower part represents the polyprotein which is cleaved by viral and cellular proteases four structural proteins: capsid (C), envelope (E), precursor membrane (prM), and membrane (M) and seven non-structural proteins (NS1, NS2A, NS2B, NS3, NS4A, NS4B, and NS5). During infection, the ZIKV E proteins bind to host cell receptors and the viral particle is endocytosed. The E proteins enable the fusion of the virus with the endosomal membrane, leading the release of the genomic RNA into the host cell cytoplasm. The translation of the RNA genome occurs in the endoplasmic reticulum. The RNA is translated as a single polypeptide chain encompassing all the viral proteins: C-prM-E-NS1-NS2A-NS2B-NS3-NS4A-NS4B-NS5.

ZIKV has different pathways of transmission. The ZIKV transmission in humans was firstly reported through bites of infected *Aedes aegypti* or *Aedes albopictus* mosquito [54]. However, the virus was identified and isolated from seventeen different *Aedes mosquitos* species, *Culex quinquefasciatus*, *Culex perfuscus*, *Mansonia uniformis*, *Anopheles coustani*, and *Anopheles gambiae* mosquitoes [55–59].

Another important fact about ZIKV transmission became apparent during the 2015 outbreak, when several cases of ZIKV vertical transmission were identified from an infected mother through the placenta to the fetus and sexual transmission (male-to-female; female-to-male; male-to-male) [60]. This novel mode of ZIKV transmission in humans had never been reported before in flavivirus infection [60–62]. ZIKV was the first arbovirus detected in human semen [63]. While needing more consistent evidence about the ZIKV transmission, these findings suggest the complexity of ZIKV dynamics transmission [64,65].

4. ZIKV on Male Reproductive Tract

The male reproductive system includes the penis, scrotum, testicles, epididymis, vas deferens, prostate and seminal vesicles (Figure 3). Recent studies have demonstrated the presence of ZIKV RNA in semen, as well as in male and female reproductive tracts, indicating the occurrence of the sexual transmission [66]. The first sexual transmission became evident in 2011, and many cases have supported the idea of one potential transmission pathway [62]. Moreover, ZIKV could be detected in semen six months after infection in negative ZIKV serum from a patient [67]. Similarly, ZIKV RNA was detected in semen in symptomatic and asymptomatic-infected patients [68–70]. A case report showed ZIKV RNA presence in total semen and also in the sperm fraction used in assisted reproductive technology up to 112 days after infection [71]. Taken together, all these data indicate that infected men can be a potential reservoir for sexual transmission, even a long time after the infection [72].

In a mouse model, ZIKV sexual transmission was recently characterized, showing that epididymal epithelial cells and leukocytes should be the main source of ZIKV RNA shedding [73]. ZIKV can persist and replicate in MRT [74]. In cases of ZIKV infection, is it known that SCs can support a high level of ZIKV replication [75,76]. In the early stages of infection, ZIKV suppresses cell growth, cell proliferation, and dysregulation of germ cell–SC junction signaling [77]. ZIKV downregulated the secretion of inhibin B, a hormone mostly produced by SCs [78]. Strange and colleagues demonstrated a unique cross-talk between ZIKV infection and SC immune response, which in the course of infection, the viral persistence was associated with activation of canonical pro-inflammatory pathways. That includes the upregulation of genes of the human leukocyte antigen (HLA) class I, pro-inflammatory genes such as interleukin-23 subunit alpha (IL23A) and lymphotoxin beta (LTB), NF-kappa-B-epsilon (NFKBIE), IL6, STAT1, STAT2, and IFN [77].

The IFN response is a strong key in the innate immune response against virus dissemination in testicles. Two animal models of *Mus musculus* species, susceptible to ZIKV infection, are important for understanding the pathogenesis of this virus. These models are A129 and AG129 mice, both immunocompromised mice. A129 mice do not have the receptor for interferon type I (IFN α/β). AG129 mice do not have the receptor for interferon type I and II (IFN $\alpha/\beta/\gamma$) [79]. IFNAR$^{-/-}$ mice are one of the best mice models for ZIKV susceptibility studies [80]. Siemann and colleagues have shown that in the first hours of infection, ZIKV does not induce IFN-α in SC, but it presents a modest induction after 48 and 72 h of infection. However, high levels of pro-inflammatory cytokines such as interleukin-1α (IL-1α), IL-1β, IL-6, IL-8, and TNF-α were found in the supernatant of infected SC, and in the chemokines such as RANTES (CCL5), fractalkine (CX3CL1), and IP-10 (CXCL10). These levels increased significantly 72 h after infection. Although SCs generate a strong immune response against ZIKV, the virus can persist in the male reproductive tract for a long time [81].

The TAM receptor, AXL, promotes the ZIKV entrance in SCs and contributes negatively to the antiviral states of SCs [82]. SCs are one type of cell that expresses high levels of TAM receptors, TGF-β expression, and activin-A to maintain the immune regulation in the seminiferous tubules. SCs play an important role in testicular physiology, creating a BTB and contributing to the nourishment of the spermatozoa. This cellular physiology and ZIKV modulation can develop an important factor that may lead to the establishment of viruses in this organ. Other cell types in the testicle can support the ZIKV infection, such as testicular fibroblast, germ cells, and spermatocyte [43,83].

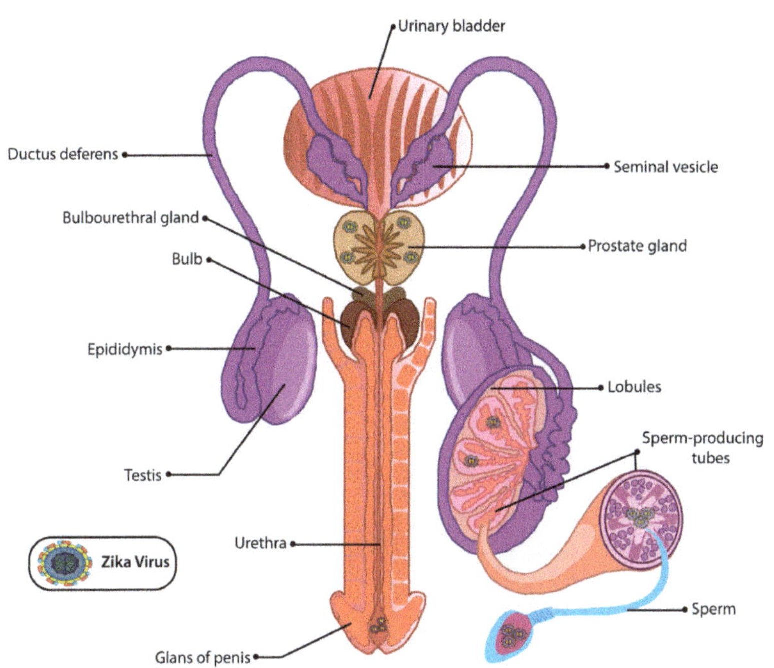

Figure 3. ZIKV reservoir in the male reproductive tract. ZIKV has been found in several portions of the male reproductive tract, including the prostate gland, testicle, epididymis, and seminiferous tubules. ZIKV-infected men have presented prostatitis, hematospermia, and microhematospermia. ZIKV RNA has been detected in semen from ZIKV-infected men and sexual transmission is an important route of contagious ZIKV. Some testicular cells are susceptible to ZIKV infection, such as spermatogonia, primary spermatocytes, Sertoli cells, and spermatozoa. Moreover, ZIKV can infect and replicate in mature sperm, leading to male infertility.

LCs and testicular macrophages are part of the first line of defense in the seminiferous tubules [84]. LCs are not highly susceptible to ZIKV infection in mice models, but more studies in humans are necessary. However, LCs are the main source of testosterone in testis, and during ZIKV infection, the levels of testosterone are significantly modulated [78]. Testicular macrophages are infected by ZIKV [33], and the infection promotes an increase of mRNA transcript levels of the IFN-α and IFIT1 genes, inducing the secretion of pro-inflammatory cytokine TNF-α, IL-1α, and IL-8 and chemokines, such as GRO, IP-10, and monocyte chemoattractant protein 1 (MCP-1). These inflammatory mediators are correlated with the possibility that ZIKV infection can compromise SC barrier integrity [81]. ZIKV does not modulate the expression of tight junction proteins (TJPs). The virus can cross BTB efficiently and persist in abluminal side seminiferous tubules by the induction of adhesion molecules expression such as VCAM-1, which facilitates the adhesion of immune cells, compromising BTB permeability [81].

In spermatogonia, the infection can promote cell death, leading to the destruction of seminiferous tubules and triggering male infertility by damaging the male reproductive system [75]. Low sperm counts are observed in patients infected with ZIKV [69,85]. Several studies have shown the effect of the ZIKV infection promoting genital damage, modulation of testicular immunity leading to orchitic and viral replication, promoting a long infection establishment. ZIKV does not affect only the testes. In mice and monkey models, ZIKV infection causes acute and chronic prostatitis [86]. Male rats infected with the Mexican ZIKV strain presented a significant decrease in testicle size compared to uninfected rats. Testicle atrophy may have occurred due to decreased testosterone levels in cells infected with this virus [87].

Several studies have shown alterations in mature sperm infected by ZIKV [85,88]. Such findings may also be an additional indication that ZIKV reduces male fertility. Furthermore, it is important to evaluate sperm banks regarding the presence of ZIKV-infection in donors due to the implications for assisted reproduction.

Therefore, ZIKV is capable of entering the testicular microenvironment, disrupting cellular metabolism, altering testicular physiology, and activating an intense immune response, which can result in severe testicular damage and infertility. A better understanding of how ZIKV affects the regulation of cell survival pathways and the testicle physiology can help evaluate pathogenesis and may be used for vaccine studies to identify intervention strategies (Figure 4).

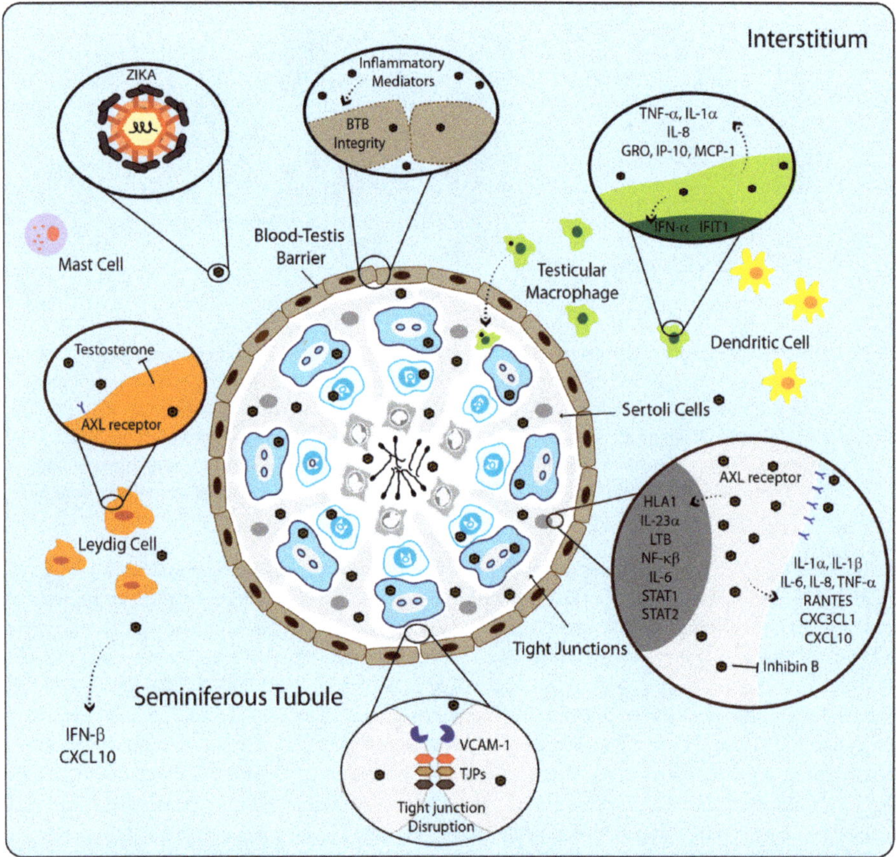

Figure 4. Testicular cells infection by ZIKV. ZIKV infection can cause serious physiological, immunological, and endocrine damage in the testes, impairing spermatogenesis. ZIKV can infect several cells in the male reproductive tract. Leydig cells are less susceptible to the infection when compared to other cells in the male reproductive tract. Testosterone, the main hormone produced by Leydig cells, is modulated by ZIKV, impairing the endocrinological function. Testicular macrophage is infected by ZIKV, triggering upregulation of IFN-α, IFIT1, TNF-α, IL-1a and IL-8, GRO, IP-10, and MCP-1. Inside the seminiferous tubule, Sertoli cells have high expression levels of AXL receptors,

which is used by ZIKV to invade cells. Sertoli cells support high levels of ZIKV replication, and the infection promotes the upregulation of genes related to antigen presentation (HLA-1), proinflammatory cytokines (lymphotoxin-beta LTB, IL-6, IL-23a) and transcription factor related to inflammation (NF-kb, STAT1, and STAT2). The release of proinflammatory cytokines such as IL-1α, IL-1β, IL-6, IL-8, TNF-α, and chemokines such as RANTES, CXC3CL1, and CXCL10 in SCs is also promoted by the infection. These molecules can promote the chemoattraction of more immunological cells and lead to an inflammatory profile, impairing efficient spermatogenesis. Inhibin-B, produced predominantly by SCs, can control follicle-stimulating hormone (FSH) secretion and is downregulated by ZIKV infection. ZIKV increases the expression of VCAM-1 in SCs which can facilitate the immune cells adhesion. Inside the seminiferous tubules, ZIKV can infect spermatogonia, primary spermatocytes, and mature spermatozoa.

5. The Immune System of Testis during Viral Infection

MRT requires a homeostatic microenvironment for viable germ cell production and nutrition. The crosstalk between SCs and LCs is fundamental to spermatozoa development [89,90].

In the testicular surroundings, an important immunological component maintains a proper environment for spermatogenesis, turning the testis into an immune-privileged organ [91]. Once MRT homeostasis is broken, spermatogenesis key steps are impaired and inflammation can be triggered. Many pathogens have shown to infect and persist in the MRT [3,26,84]. Testicular abnormalities, infertility, or sexual transmission are some of the major consequences of pathogen persistence in the MRT. Considering the important findings regarding ZIKV RNA detection in the semen, the scientific community has turned their attention to the possibility that other flaviviruses promote similar effects [92]. Once their detection becomes proven, the possibility of sexual transmission or impaired spermatogenesis is another important factor to be explored. Preliminary studies about this have provided us with information on a possible threat derived from different flaviviruses in the MRT. Nevertheless, this question is far from clear and molecular mechanisms still under investigation.

Some studies have been reported flavivirus infection in the MRT [19]. The viral load could be found for some of them, and the presence of leukocytes in the semen suggests an inflammatory process caused by the infection. Salam and colleagues found viruses from several families in the semen, including *Adenoviridae*, *Filoviridae*, *Flaviviridae*, *Herpesviridae*, and *Retroviridae* [19].

DENV is a considerably more common flavivirus than ZIKV, and the knowledge about DENV effects in the testis is scarce. The first case report linking DENV infection to MRT modulation was published in 2011 [93]. In this report, scrotal and penile edema was a rare complication associated with DENV infection. However, the mechanism by which this edema was formed was not evaluated; neither could DENV be detected in penile fluids. Currently, there is no data reporting if testicular abnormalities could be triggered by DENV-associated inflammation in MRT. In 2018, two controversial publications raised questions about the possible impact of DENV in the MRT. The first one demonstrated that DENV RNA was not detected in the semen of five confirmed patients during the acute infection [94]. The second one is a case report released a few days later, demonstrating that DENV was detected in the semen of an infected man 37 days after the related symptoms. The report demonstrated DENV RNA in the cellular fraction, suggesting the possibility of sexual transmission [95]. New evidence of DENV sexual transmission was published in 2019, where a case report from Spain detected the viral RNA in the semen of two men who were partners [96]. Only one of the men had contact with a DENV endemic area and his partner presented the symptoms a few days after the first one. This is the first evidence of DENV sexual transmission. Nevertheless, clinical trials aiming to concisely respond to this question are underway and may be published soon (Clinical Trial Identifier: NCT03612609).

In 2018, a case report was published regarding YFV RNA detection in the semen and urine of a Brazilian man in the convalescent phase of the disease [97]. The integrity and infectivity of the viral particles were accessed and confirmed in the report. This strongly suggests that this virus can be sexually transmitted once it is capable of maintaining infective parameters, although no data are available confirming the capability of YFV persistence and impact in the MRT, or sexual transmission associated with the infection.

Zheng and colleagues showed that the JEV infection induces inflammation of pig testicles by activating RIG-I/NF-kB pathway signaling [98]. This also leads to orchitis, which is a type of chronic inflammation in the testes caused by viral or bacterial infections, associated with pain, swelling, along with blood and swelling in prostate ejaculate [81,99]. Testes infection with JEV showed a differential production of pro-inflammatory cytokines, such as IL-1β, IL-6, IL-8, chemokine RANTES, and TNF-α, as well as an increased presence of NS5 (non-structural protein of the virus), RIG-I, TLR3 and -7 [98]. Smith and colleagues showed that a 43-year-old patient presented signs of encephalitis and orchitis caused by WNV [100]. In this report, histological sections showed lymphocyte, SCs, and interstitial multinucleated cells infiltrate, as well as marked thickening of the basement tubular membranes and absence of spermatogenesis, an indication of atrophy. Numerous foci of dense chronic interstitial inflammatory infiltrate and necrotic cell death was observed in the seminiferous tubules [100].

DENV, YFV, and JEV are classified both as arbovirus and flavivirus and present major clinical relevance within these groups. Nevertheless, another important virus that compounds arbovirus group but is a member of a distinct family, presents important findings regarding MRT infection. For this reason, an analysis of the available data for this arbovirus is relevant and will also be explored in this section.

The Chikungunya virus (CHIKV) is a small, enveloped, single-stranded positive-sense RNA virus that belongs to *Alphavirus* genus and *Togaviridae* family. Chikungunya is a vector-borne disease, also transmitted by the bites of mosquitoes from the *Aedes* genus, mainly *Ae. aegypti* and *Ae. albopictus*, causing arthritis or arthralgia, which is accompanied by fever and rash [101]. CHIKV RNA has been detected in semen and urine, as reported in a case published in 2016. This study showed a patient presenting CHIKV and DENV (type 3) dual infection, in which only CHIKV was detected [102] in both the acute and convalescence phases of the disease, within 30 days after symptoms. Thereby, it is important to emphasize that CHIKV presents tropism and cytotoxic effects on monocyte-derived macrophages [103], which can be later recruited to testicular microenvironment [32]. In this context, macrophages are being identified as a possible source of CHIKV RNA in the testis, acting as a testicular trojan horse. However, more studies are necessary to verify this hypothesis [102].

Numerous questions related to viruses infection in MRT remain to be answered. The long-term effects of persistent infection for several flaviviruses in male reproductive function, as well as production and fertility of spermatozoa need to be investigated. Importantly, in the case of ZIKV, cryptorchidism, hypospadias and micropenis have been reported in newborn infants of infected mothers [104], although its prevalence is unknown. An effect of arboviruses infection in male fertility will only be fully understood in long-term epidemiological studies and suitable animal model experiment design.

6. ZIKV Vaccines and Treatment to Improve the Host Response in the MRT

Sexual transmission of ZIKV and the viral persistence in the MRT are the strongest challenges for outbreaks control, vaccines, and antiviral drug development [105–107]. The impact of ZIKV infection in the population leads to a significant global efforts to develop vaccines. Spectacular progress has been made in ZIKV vaccine development, and several strategies have been proposed to increase vaccine protection in immune-privileged organs [105,108–112].

Antibody usage has shown a promising strategy to protect ZIKV in the testicle. Some subclasses of immunoglobulin (IgG) can cross the BTB [113]. The administration of human antibodies to DENV E-dimer epitope (EDE1-B10) 3 days after infection was able to reduce the viral load in testis, reducing the inflammation and preserving sperm count. The protection is not effective for the long duration [114]. Further studies in this area have explored the pathogenesis pathways and the host cellular response, suggesting potential targets to develop vaccines, including DNA-based vaccines. DNA-based vaccines, and live attenuated ZIKV have shown testicular protection against infection, avoiding atrophy, damage, and male infertility [74,115].

The combined strategies of DNA-based vaccines and live attenuated ZIKV vaccines demonstrated efficacy when used in a single-dose in A129 mice. This vaccination promotes the complete prevention of testicle infection, injury, and oligospermia [116]. Another live-attenuated ZIKV vaccine, which presents one deletion in the 3′ untranslated region of the ZIKV genome (ZIKV-3′UTR-LAV), presented protection after a single vaccination in mice and non-human primates. This protection was evaluated for preventing mother-child vertical transmission and the prevention of testicle damages [117].

DNA-based vaccination of recombinant chimpanzee adenovirus type 7 (AdC7) expressing ZIKV M/E glycoproteins presents high efficacy in a single vaccination. AdC7-M/E induced a potent neutralizing antibody in immunocompetent and immunodeficient mice and full protection against ZIKV-induced testicular damage [118]. Another DNA-based vaccine, encoding ZIKV pre-membrane and envelope (prME) in pVAX vector, protected mice completely against ZIKV, promoting protection in testes and sperm and decreasing viral persistence in MRT [115]. Moreover, this vaccine was also effective in reversing mouse infertility [119].

A few drugs against ZIKV have also been proposed and may have an impact on testicles [106,120,121]. Recently, Z2 an amphipathic peptide derived from the stem region of ZIKV envelope protein was reported to inhibit vertical ZIKV transmission in a mouse model and reduce viral load in the testicle and epididymis. This was also reported to reduce pathological damage while improving sperm quality [122]. Simanjuntak and colleagues demonstrated that ZIKV-infected testicles presented progressive damage with a significant oxidative microenvironment, with high levels of reactive oxygen species, nitric oxide, glutathione peroxidase 4 and pro-inflammatory cytokines as IL-1β, IL-6, and G-CSF. They proposed the use of the antioxidant ebselen (EBS) to prevent the sexual transmission of the virus and to improve host testicular immune response [123].

7. Conclusions

ZIKV can infect and persist in testicular somatic and germ cells, as well as, spermatozoa, leading to cell death and testicular atrophy. ZIKV has also been detected in semen samples from ZIKV-infected patients. This has huge implications for human reproduction. DNA-based vaccination and/or live attenuated ZIKV vaccines showed high efficacy against MRT damage induced by ZIKV and are a very prominent therapeutic tool to prevent male infertility caused by ZIKV.

It is important to note that, often, no evident testicular inflammatory response is usually observed against ZIKV infection in testes, with normal testicular morphology and hormone production remaining unaffected after ZIKV infection. This indicates that ZIKV can remain quiescent in the testes, acting as a trojan horse, and maintaining asymptomatic ZIKV sexual transmission. The better understanding of the mechanisms that mediate the cellular impact of the ZIKV on MRT, regulating testicular immune and physiological responses, is the key factor to the correct design of efficient anti-ZIKV therapeutic strategies to prevent male infertility caused by ZIKV.

Author Contributions: R.d.N.A., H.A.B.-d.-M., I.d.O.S., R.C. and K.G.M. wrote the manuscript. R.d.N.A., I.d.O.S. and H.A.B.-d.-M. generated original figures. G.P.K. edited the manuscript. All authors have read and agreed to the published version of the manuscript.

Funding: The authors declare no funding for this manuscript.

Conflicts of Interest: The authors declare no conflict of interest.

References

1. Gurung, P.; Jialal, I. *Physiology, Male Reproductive System*; StatPearls Publishing [Internet]: Treasure Island, FL, USA, 2019.
2. Griswold, M.D. The central role of Sertoli cells in spermatogenesis. *Semin. Cell Dev. Biol.* **1998**, *9*, 411–416. [CrossRef]

3. Zhao, S.; Zhu, W.; Xue, S.; Han, D. Testicular defense systems: Immune privilege and innate immunity. *Cell. Mol. Immunol.* **2014**, 428–437. [CrossRef] [PubMed]

4. Filippini, A.; Riccioli, A.; Padula, F.; Lauretti, P.; D'Alessio, A.; De Cesaris, P.; Gandini, L.; Lenzi, A.; Ziparo, E. Control and impairment of immune privilege in the testis and in semen. *Hum. Reprod. Update* **2001**, *7*, 444–449. [CrossRef] [PubMed]

5. Jiang, X.H.; Bukhari, I.; Zheng, W.; Yin, S.; Wang, Z.; Cooke, H.J.; Shi, Q.H. Blood—Testis barrier and spermatogenesis: Lessons from genetically-modified mice. *Asian J. Androl.* **2014**, *16*, 572–580. [PubMed]

6. Wang, M.; Fijak, M.; Hossain, H.; Nüsing, R.M.; Lochnit, G.; Michaela, F.; Wudy, S.A.; Zhang, L.; Gu, H.; Konrad, L.; et al. Characterization of the Micro-Environment of the Testis that Shapes the Phenotype and Function of Testicular Macrophages. *J. Immunol.* **2017**, *198*, 4327–4340. [CrossRef]

7. Mital, P.; Hinton, B.T.; Dufour, J.M. The Blood-Testis and Blood-Epididymis Barriers Are More than Just Their Tight Junctions. *Biol. Reprod.* **2011**, *84*, 851–858. [CrossRef]

8. Luca, G.; Baroni, T.; Arato, I.; Hansen, B.C.; Cameron, D.F.; Calafiore, R. Role of Sertoli Cell Proteins in Immunomodulation. *Protein Pept. Lett.* **2018**, *25*, 440–445. [CrossRef]

9. Keller, N.M.; Gentek, R.; Gimenez, G.; Bigot, S.; Mailfert, S.; Sieweke, M.H. Developmental origin and maintenance of distinct testicular macrophage populations. *J. Exp. Med.* **2017**, 2829–2841. [CrossRef]

10. Hedger, M.P.; Meinhardt, A. Local regulation of T cell numbers and lymphocyte-inhibiting activity in the interstitial tissue of the adult rat testis. *J. Reprod. Immunol.* **2000**, *48*, 69–80. [CrossRef]

11. Windsch, S. Are testicular mast cells involved in the regulation of germ cells in man? *Andrology* **2014**, *2*, 615–622. [CrossRef]

12. Rival, C.; Lustig, L.; Iosub, R.; Guazzone, V.A.; Schneider, E.; Meinhardt, A.; Fijak, M. Identification of a dendritic cell population in normal testis and in chronically inflamed testis of rats with autoimmune orchitis. *Cell Tissue Res.* **2006**, *324*, 311–318. [CrossRef] [PubMed]

13. Patel, D.A.; Burnett, N.M.; Curtis, K.M. *Reproductive Tract Infections. Reproductive Health Epidemiology Series Module 3*; CDC, Department of Health and Human Services: Atlanta, GA, USA, 2003; pp. 3–4.

14. World Health Organization, W. Sexually transmitted infections (STIs). Available online: https://www.who. int/en/news-room/fact-sheets/detail/sexually-transmitted-infections-(stis) (accessed on 15 April 2020).

15. World Health Organization (WHO). *Global Health Sector Strategy on HIV 2016–2021 towards Ending AIDS*; WHO: Geneva, Switzerland, 2018; ISBN 9789241501651.

16. Mackay, I.M.; Arden, K.E. Ebola virus in the semen of convalescent men. *Lancet Infect. Dis.* **2015**, *15*, 149–150. [CrossRef]

17. Oliveira Souto, I.; Alejo-Cancho, I.; Gascón Brustenga, J.; Peiró Mestres, A.; Muñoz Gutiérrez, J.; Martínez Yoldi, M.J. Persistence of Zika virus in semen 93 days after the onset of symptoms. *Enferm. Infec. Micr. Cl.* **2018**, *36*, 21–23. [CrossRef] [PubMed]

18. Pudney, J.; Anderson, D. Orchitis and human immunodeficiency virus type 1 infected cells in reproductive tissues from men with the acquired immune deficiency syndrome. *Am. J. Pathol.* **1991**, *139*, 149–160.

19. Salam, A.P.; Horby, P.W. The breadth of viruses in human semen. *Emerg. Infect. Dis.* **2017**, *23*, 1922–1924. [CrossRef]

20. World Health Organization (WHO). WHO statement on the first meeting of the International Health Regulations (IHR 2005). In Proceedings of the Emergency Committee on Zika Virus and Observed Increase in Neurological Disorders and Neonatal Malformations, Washington, WA, USA, 18 November 2016.

21. Wang, J.; Counotte, M.J.; Kim, C.R.; Bernstein, K.; Broutet, N.J.N.; Deal, C.D.; Low, N. Sexual transmission of Zika virus and other flaviviruses: A living systematic review. *PLoS Med.* **2018**, *15*, 30–49.

22. Kelly, D.A.; Moore, B.C. Integrative and Comparative Biology The Morphological Diversity of Intromittent Organs. *Integr. Comp. Biol.* **2016**, *56*, 630–634. [CrossRef]

23. Jones, R.E.; Lopez, K.H. *Human Reproductive Biology*, 4th ed.; Academic Press: Cambridge, MA, USA, 2013.

24. Fujisawa, M. Cell-to-cell cross talk in the testis. *Urol. Res.* **2001**, *29*, 144–151. [CrossRef]

25. Picut, C.A.; De Rijk, E.P.C.T.; Dixon, D. Immunopathology of the Male Reproductive Tract. In *Immunopathology in Toxicology and Drug Development. Molecular and Integrative Toxicology*; Parker, G., Ed.; Humana Press: Cham, Switzerland, 2017.

26. Ramaswamy, S.; Weinbauer, G.F. Endocrine control of spermatogenesis: Role of FSH and LH/testosterone. *Spermatogenesis* **2014**, *4*, 1–15. [CrossRef]

27. Kaur, G.; Vadala, S.; Dufour, J.M. An overview of a Sertoli cell transplantation model to study testis morphogenesis and the role of the Sertoli cells in immune privilege. *Environ. Epigenet.* **2017**, *3*, 1–10. [CrossRef]

28. Mossadegh-Keller, N.; Sieweke, M.H. Testicular macrophages: Guardians of fertility. *Cell. Immunol.* **2018**, *330*, 120–125. [CrossRef] [PubMed]

29. Cheng, C.Y.; Mruk, D.D. The Blood-Testis Barrier and Its Implications for Male Contaception. *Pharmacol. Rev.* **2012**, *64*, 16–64. [CrossRef] [PubMed]

30. Syriou, V.; Papanikolaou, D.; Kozyraki, A.; Goulis, D.G. Cytokines and male infertility. *Eur. Cytokine Netw.* **2018**, *29*, 73–82. [CrossRef] [PubMed]

31. Li, M.W.M.; Mruk, D.D.; Lee, W.M.; Cheng, C.Y. Cytokines and junction restructuring events during spermatogenesis in the testis: An emerging concept of regulation. *Cytokine Growth Factor Rev.* **2009**, *20*, 329–338. [CrossRef]

32. Winnall, W.R.; Hedger, M.P. Phenotypic and functional heterogeneity of the testicular macrophage population: A new regulatory model. *J. Reprod. Immunol.* **2013**, *97*, 147–158. [CrossRef]

33. Matusali, G.; Houzet, L.; Satie, A.-P.; Mahé, D.; Aubry, F.; Couderc, T.; Frouard, J.; Bourgeau, S.; Bensalah, K.; Lavoué, S.; et al. Zika virus infects human testicular tissue and germ cells. *J. Clin. Invest.* **2018**, *128*, 4697–4710. [CrossRef]

34. Nistal, M.; Santamaria, L.; Paniagua, R. Mast Cells in the Human Testis and Epididymis from Birth to Adulthood. *Acta Anat.* **1984**, *119*, 155–160. [CrossRef]

35. Hussein, M.R.; Abou-Deif, E.S.; Bedaiwy, M.A.; Said, T.M.; Mustafa, M.G.; Nada, E.; Ezat, A.; Agarwal, A. Phenotypic characterization of the immune and mast cell infiltrates in the human testis shows normal and abnormal spermatogenesis. *Fertil. Steril.* **2005**, *83*, 1447–1453. [CrossRef]

36. Schrade, A.; Akinrinade, O.; Pihlajoki, M.; Fischer, S.; Rodriguez, V.M.; Otte, K.; Velagapudi, V.; Toppari, J.; Wilson, D.B.; Heikinheimo, M. GATA4 regulates blood-testis barrier function and lactate metabolism in mouse Sertoli cells. *Endocrinology* **2016**, *157*, 1–17. [CrossRef]

37. Alves, M.G.; Rato, L.; Carvalho, R.A.; Moreira, P.I.; Socorro, S.; Oliveira, P.F. Hormonal control of Sertoli cell metabolism regulates spermatogenesis. *Cell. Mol. Life Sci.* **2013**, *70*, 777–793. [CrossRef]

38. Alves, M.G.; Martins, A.D.; Cavaco, J.E.; Socorro, S.; Oliveira, P.F. Diabetes, insulin-mediated glucose metabolism and Sertoli/blood-testis barrier function. *Tissue Barriers* **2013**, *1*, 1–10. [CrossRef] [PubMed]

39. Jenkins, T.G.; Yeste, M. The role of miRNAs in male human reproduction: A systematic review. *Andrology* **2020**, *8*, 7–26.

40. Tenorio, F.; Neto, L.; Bach, P.V.; Najari, B.B.; Li, P.S.; Goldstein, M. Spermatogenesis in Humans and Its affecting factors. *Semin. Cell Dev. Biol.* **2016**, *59*, 10–26.

41. Raghupathy, R.; Shaha, C.; Gupta, S.K. Autoimmunity to sperm antigens. *Curr. Opin. Immunol.* **1990**, *2*, 757–760. [CrossRef]

42. Roulet, V.; Satie, A.P.; Ruffault, A.; Le Tortorec, A.; Denis, H.; Patard, J.J.; Rioux-Leclerq, N.; Gicquel, J.; Jégou, B.; Dejucq-Rainsford, N. Susceptibility of Human Testis to Human Immunodeficiency Virus-1 Infection in Situ and in Vitro. *Am. J. Pathol.* **2006**, *169*, 2094–2103. [CrossRef]

43. Robinson, C.L.; Chong, A.C.N.; Ashbrook, A.W.; Jeng, G.; Jin, J.; Chen, H.; Tang, E.I.; Martin, L.A.; Kim, R.S.; Kenyon, R.M.; et al. Male germ cells support long-term propagation of Zika virus. *Nat. Commun.* **2018**, *9*, 1–11. [CrossRef]

44. Holbrook, M.R. Historical perspectives on flavivirus research. *Viruses* **2017**, *9*, 97. [CrossRef]

45. Laureti, M.; Narayanan, D.; Rodriguez-Andres, J.; Fazakerley, J.K.; Kedzierski, L. Flavivirus Receptors: Diversity, Identity, and Cell Entry. *Front. Immunol.* **2018**, *9*, 2180. [CrossRef]

46. Lanciotti, R.S.; Lambert, A.J.; Holodniy, M.; Saavedra, S.; del Carmen Castillo Signor, L. Phylogeny of zika virus in western Hemisphere. *Emerg. Infect. Dis.* **2016**, *22*, 933–935. [CrossRef]

47. Hussain, A.; Ali, F.; Latiwesh, O.B.; Hussain, S. A Comprehensive Review of the Manifestations and Pathogenesis of Zika Virus in Neonates and Adults. *Cureus* **2018**, *10*, 1–10. [CrossRef]

48. Dick, G.W.A.; Kitchen, S.F.; Haddow, A.J. Zika virus (I). Isolations and Serological Specificity. *Trans. R. Soc. Trop. Med. Hyg.* **1952**, *46*, 509–520. [CrossRef]

49. MacNamara, F.N. Zika virus: A report on three cases of human infection during an epidemic of jaundice in Nigeria. *Trans. R. Soc. Trop. Med. Hyg.* **1954**, *48*, 139–145. [CrossRef]

50. Duffy, M.R.; Chen, T.-H.; Hancock, W.T.; Powers, A.M.; Kool, J.L.; Lanciotti, R.S.; Pretrick, M.; Marfel, M.; Holzbauer, S.; Dubray, C.; et al. Zika Virus Outbreak on Yap Island, Federated States of Micronesia. *N. Engl. J. Med.* **2009**, *360*, 2536–2543. [CrossRef] [PubMed]

51. Cao-Lormeau, V.M. RE: Zika virus, French Polynesia, South Pacific, 2013. *Emerg. Infect. Dis.* **2014**, *20*, 1085–1086. [CrossRef] [PubMed]

52. Zanluca, C.; De Melo, V.C.A.; Mosimann, A.L.P.; Dos Santos, G.I.V.; dos Santos, C.N.D.; Luz, K. First report of autochthonous transmission of Zika virus in Brazil. *Mem. Inst. Oswaldo Cruz* **2015**, *110*, 569–572. [CrossRef] [PubMed]

53. Enfissi, A.; Codrington, J.; Roosblad, J.; Kazanji, M.; Rousset, D. Zika virus genome from the Americas. *Lancet* **2016**, *387*, 227–228. [CrossRef]

54. Sharma, A.; Lal, S.K. Zika Virus: Transmission, Detection, Control, and Prevention. *Front. Microbiol.* **2017**, *8*, 1–14. [CrossRef]

55. Slavov, S.N.; Otaguiri, K.K.; Kashima, S.; Covas, D.T. Overview of Zika virus (ZIKV) infection in regards to the Brazilian epidemic. *Braz. J. Med. Biol. Res.* **2016**, *49*, 1–11. [CrossRef]

56. Ayres, C.F.J. Identification of Zika virus vectors and implications for control. *Lancet Infect. Dis.* **2016**, *16*, 278–279. [CrossRef]

57. Villar, L.; Dayan, H.G.; García-Arredondo, J.L.; Rivera, M.D.; Cunha, R.; Deseda, C.; Reynales, H.; Costa, S.M.; Ramírez-Morales, O.J.; Carraquilha, G.; et al. Efficacy of a Tetravalent Dengue Vaccine in Children in Latin America. *N. Engl. J. Med.* **2015**, *372*, 113–123. [CrossRef]

58. Zhao, B.; Yi, G.; Du, F.; Chuang, Y.; Vaughan, R.C.; Sankaran, B.; Kao, C.C.; Li, P. Structure and function of the Zika virus full-length NS5 protein. *Nat. Commun.* **2017**, *8*, 1–9. [CrossRef] [PubMed]

59. Saiz, J.; Vázquez-Calvo, Á.; Blázquez, A.B.; Merino-Ramos, T.; Escribano-Romero, E.; Martín-Acebes, M.A. Zika Virus: The Latest Newcomer. *Front. Microbiol.* **2016**, *7*, 1–19. [CrossRef] [PubMed]

60. Center for Disease Control and Prevention, CDC. First female-to-male sexual transmission of Zika virus infection reported in New York City. Available online: https://www.cdc.gov/media/releases/2016/s0715-zika-female-to-male.html (accessed on 15 April 2020).

61. Imperato, P.J. The Convergence of a Virus, Mosquitoes, and Human Travel in Globalizing the Zika Epidemic. *J. Community Health* **2016**, *41*, 674–679. [CrossRef] [PubMed]

62. Foy, B.D.; Kobylinski, K.C.; Foy, J.L.C.; Blitvich, B.J.; da Rosa, A.T.; Haddow, A.D.; Lanciotti, R.S.; Tesh, R.B. Probable Non–Vector-borne Transmission of Zika Virus, Colorado, USA. *Emerg. Infect. Dis.* **2011**, *17*, 880–882. [CrossRef] [PubMed]

63. Musso, D.; Roche, C.; Robin, E.; Nhan, T.; Teissier, A.; Lormeau-Cao, V. Potential Sexual Transmission of Zika Virus. *Emerg. Infect. Dis.* **2015**, *21*, 359–361. [CrossRef]

64. Guedes, D.R.D.; Paiva, M.H.S.; Donato, M.M.A.; Barbosa, P.P.; Krokovsky, L.; Rocha, S.W.S.; Saraiva, K.L.A.; Crespo, M.M.; Rezende, T.M.T.; Wallau, G.L.; et al. Zika virus replication in the mosquito Culex quinquefasciatus in Brazil. *Emerg. Microbes Infect.* **2017**, *6*, 1–11. [CrossRef]

65. Roundy, C.M.; Azar, S.R.; Brault, A.C.; Ebel, G.D.; Failloux, A.; Fernandez-Salas, I.; Kitron, U.; Kramer, L.D.; Lourenço-de-Oliveira, R.; Osorio, J.E.; et al. Lack of evidence for Zika virus transmission by Culex mosquitoes. *Emerg. Microbes Infect.* **2017**, *6*, 1–2. [CrossRef]

66. D'Ortenzio, E.; Matheron, S.; Yazdanpanah, Y. Evidence of Sexual Transmission of Zika Virus. *N. Engl. J. Med.* **2016**, *374*, 2195–2198. [CrossRef]

67. Nicastri, E.; Castilletti, C.; Liuzzi, G.; Iannetta, M.; Capobianchi, M.R.; Ippolito, G. Persistent detection of Zika virus RNA in semen for six months after symptom onset in a traveller returning from Haiti to Italy, February 2016. *Euro Surveill.* **2016**, *21*, 1–4. [CrossRef]

68. Matheron, S.; D'Ortenzio, E.; Leparc-Goffart, I.; Hubert, B.; de Lamballerie, X.; Yazdanpanah, Y. Long-Lasting Persistence of Zika Virus in Semen. *Clin. Infect. Dis.* **2016**, *63*, 1264. [CrossRef]

69. Martin-Blondel, G. Zika virus in semen and spermatozoa. *Lancet Infect. Dis.* **2016**, *16*, 1106–1107.

70. Musso, D.; Richard, V.; Teissier, A.; Stone, M.; Lanteri, M.C.; Latoni, G.; Alsina, J.; Reik, R.; Busch, M.P.; Epidemiology, R.; et al. Detection of Zika virus RNA in semen of asymptomatic blood donors. *Clin. Microbiol. Infect.* **2017**, *23*, 1001.e1–1001.e3. [CrossRef] [PubMed]

71. Cassuto, N.G.; Marras, G.; Jacomo, V.; Bouret, D. Persistence of Zika virus in gradient sperm preparation. *J. Gynecol. Obstet. Hum. Reprod.* **2018**, *47*, 211–212. [CrossRef] [PubMed]

72. Atkinson, B.; Hearn, P.; Afrough, B.; Lumley, S.; Carter, D.; Aarons, E.J.; Simpson, A.J.; Brooks, T.J. Detection of Zika Virus in Semen. *Emerg. Infect. Dis.* **2016**, *22*, 940. [CrossRef]

73. McDonald, E.M.; Duggal, N.K.; Ritter, J.M.; Brault, A.C. Infection of epididymal epithelial cells and leukocytes drives seminal shedding of Zika virus in a mouse model. *PLoS Negl. Trop. Dis.* **2018**, *12*, 1–22. [CrossRef]

74. Stassen, L.; Armitage, C.; van der Heide, D.; Beagley, K.; Frentiu, F. Zika Virus in the Male Reproductive Tract. *Viruses* **2018**, *10*, 198. [CrossRef]

75. Kumar, A.; Jovel, J.; Lopez-Orozco, J.; Limonta, D.; Airo, A.M.; Hou, S.; Stryapunina, I.; Fibke, C.; Moore, R.B.; Hobman, T.C. Human sertoli cells support high levels of zika virus replication and persistence. *Sci. Rep.* **2018**, *8*, 1–11. [CrossRef]

76. Sheng, Z.-Y.; Gao, N.; Wang, Z.-Y.; Cui, X.-Y.; Zhou, D.-S.; Fan, D.-Y.; Chen, H.; Wang, P.-G.; An, J. Sertoli Cells Are Susceptible to ZIKV Infection in Mouse Testis. *Front. Cell. Infect. Microbiol.* **2017**, *7*, 1–13. [CrossRef]

77. Strange, D.P.; Green, R.; Siemann, D.N.; Gale, M.; Verma, S. Immunoprofiles of human Sertoli cells infected with Zika virus reveals unique insights into host-pathogen crosstalk. *Sci. Rep.* **2018**, *8*, 1–15. [CrossRef]

78. Govero, J.; Richner, J.M.; Caine, E.A.; Salazar, V.; Diamond, M.S.; Esakky, P.; Scheaffer, S.M.; Drury, A.; Moley, K.H.; Fernandez, E.; et al. Zika virus infection damages the testes in mice. *Nature* **2016**, *540*, 438–442. [CrossRef]

79. Rossi, S.L.; Tesh, R.B.; Azar, S.R.; Muruato, A.E.; Hanley, K.A.; Auguste, A.J.; Langsjoen, R.M.; Paessler, S.; Vasilakis, N.; Weaver, S.C. Characterization of a novel murine model to study zika virus. *Am. J. Trop. Med. Hyg.* **2016**, *94*, 1362–1369. [CrossRef] [PubMed]

80. Dong, S.; Liang, Q. Recent Advances in Animal Models of Zika Virus Infection. *Virol. Sin.* **2018**, *33*, 125–130. [CrossRef] [PubMed]

81. Siemann, D.N.; Strange, D.P.; Maharaj, P.N.; Shi, P.-Y.; Verma, S. Zika Virus Infects Human Sertoli Cells and Modulates the Integrity of the In Vitro Blood-Testis Barrier Model. *J. Virol.* **2017**, *91*, 1–17. [CrossRef] [PubMed]

82. Strange, D.P.; Jiyarom, B.; Pourhabibi Zarandi, N.; Xie, X.; Baker, C.; Sadri-Ardekani, H.; Shi, P.-Y.; Verma, S. Axl Promotes Zika Virus Entry and Modulates the Antiviral State of Human Sertoli Cells. *MBio* **2019**, *10*, 1–16. [CrossRef]

83. Mlera, L.; Bloom, M.E. Differential zika virus infection of testicular cell lines. *Viruses* **2019**, *11*, 42. [CrossRef]

84. Bhushan, S.; Schuppe, H.C.; Fijak, M.; Meinhardt, A. Testicular infection: Microorganisms, clinical implications and host-pathogen interaction. *J. Reprod. Immunol.* **2009**, *83*, 164–167. [CrossRef]

85. Joguet, G.; Mansuy, J.; Matusali, G.; Hamdi, S.; Walschaerts, M.; Pavili, L.; Guyomard, S.; Prisant, N.; Lamarre, P.; Dejucq-rainsford, N.; et al. Effect of acute Zika virus infection on sperm and virus clearance in body fluids: A prospective observational study. *Lancet Infect. Dis.* **2017**, *17*, 1200–1208. [CrossRef]

86. Halabi, J.; Jagger, B.W.; Salazar, V.; Winkler, E.S.; White, J.P.; Humphrey, P.A.; Hirsch, A.J.; Streblow, D.N.; Diamond, M.S.; Moley, K. Zika virus causes acute and chronic prostatitis in mice and macaques. *J. Infect. Dis.* **2020**, *221*, 1506–1517. [CrossRef]

87. Uraki, R.; Hwang, J.; Jurado, K.A.; Householder, S.; Yockey, L.J.; Hastings, A.K.; Homer, R.J.; Iwasaki, A.; Fikrig, E. Zika virus causes testicular atrophy. *Sci. Adv.* **2017**, *3*, 1–6. [CrossRef]

88. Huits, R.; De Smet, B.; Ariën, K.K.; Van Esbroeck, M.; Bottieau, E.; Cnops, L. Zika virus in semen: A prospective cohort study of symptomatic travellers returning to Belgium. *Bull. World Health Organ.* **2017**, *95*, 802–809. [CrossRef]

89. Wu, S.; Yan, M.; Ge, R.; Cheng, C.Y. Crosstalk between Sertoli and Germ Cells in Male Fertility. *Trends Mol. Med.* **2020**, *26*, 215–231. [CrossRef] [PubMed]

90. Siu, M.K.Y.; Cheng, C.Y. Dynamic cross-talk between cells and the extracellular matrix in the testis. *Bioessays* **2004**, *26*, 978–992. [CrossRef] [PubMed]

91. McCarrey, J.R.; Berg, W.M.; Paragioudakis, S.J.; Zhang, P.L.; Dilworth, D.D.; Arnold, B.L.; Rossi, J.J. Differential Transcription of Pgk Genes during Spermatogenesis in the Mouse. *Dev. Biol.* **1992**, *168*, 160–168. [CrossRef]

92. Alvarado-Arnez, L.E. International Journal of Infectious Diseases. *Int. J. Infect. Dis.* **2018**, *76*, 126–127.

93. Shamim, M.; Zohaib, S.; Naqvi, G. Case Report Dengue fever associated with acute scrotal oedema: Two case reports. *J. Pakistan Med. Assoc.* **2011**, *61*, 601–603.

94. Molton, J.S.; Low, I.; Ming, M.; Choy, J.; Poh, P.; Aw, K.; Hibberd, M.L.; Tambyah, P.A.; Wilder-Smith, A. Dengue virus not detected in human semen. *J. Travel Med.* **2018**, *25*, 1–3. [CrossRef]

95. Lalle, E.; Colavita, F.; Iannetta, M.; Teklè, S.G.; Carletti, F.; Scorzolini, L.; Vincenti, D.; Castilletti, C.; Ippolito, G.; Capobianchi, M.R.; et al. Prolonged detection of dengue virus RNA in the semen of a man returning from Thailand to Italy, January 2018. *Euro Surveill.* **2018**, *23*, 2–4. [CrossRef]

96. European Centre for Disease Prevention and Control. Rapid Risk Assessment—Sexual Transmission of Dengue in Spain. Available online: https://www.ecdc.europa.eu/en/publications-data/rapid-risk-assessment-sexual-transmission-dengue-spain (accessed on 15 April 2020).

97. Barbosa, C.M.; Di Paola, N.; Cunha, M.P.; Rodrigues-Jesus, M.J.; Araujo, D.B.; Silveira, V.B.; Leal, F.B.; Mesquita, F.S.; Botosso, V.F.; Zanotto, P.M.A.; et al. Yellow Fever Virus RNA in Urine and Semen of Convalescent Patient, Brazil. *Emerg. Infect. Dis.* **2018**, *24*, 176–178. [CrossRef]

98. Zheng, B.; Wang, X.; Liu, Y.; Li, Y.; Long, S.; Gu, C.; Ye, J.; Xie, S.; Cao, S. Japanese Encephalitis Virus infection induces inflammation of swine testis through RIG-I—NF-kB signaling pathway. *Vet. Microbiol.* **2019**, *238*, 108430. [CrossRef]

99. Han, D.; Liu, Z.; Yan, K. Immunology of the Testis and Privileged Sites. *Encycl. Immunobiol.* **2016**, *5*, 46–53.

100. Smith, R.D.; Konoplev, S.; DeCourten-Myers, G.; Brown, T. West Nile Virus Encephalitis with Myositis and Orchitis. Case Stud. *Hum. Pathol.* **2004**, *35*, 254–258. [CrossRef] [PubMed]

101. Moizéis, R.N.C.; Fernandes, T.A.A.M.; Guedes, P.M.D.M.; Pereira, H.W.B.; Lanza, D.C.F.; Azevedo, J.W.V.; Galvão, J.M.A.; Fernandes, J.V. Chikungunya fever: A threat to global public health. *Pathog. Glob. Health* **2018**, *112*, 182–194. [CrossRef] [PubMed]

102. Carlos, A.; Soares, G.; França, V.; Rocha, D.; Solano, B.; Souza, D.F.; Botelho, M.; Soares, P.; Araujo, A.; Carvalho, Y.; et al. IDCases Prolonged shedding of Chikungunya virus in semen and urine: A new perspective for diagnosis and implications for transmission. *IDCases* **2016**, *6*, 100–103.

103. Sourisseau, M.; Schilte, C.; Casartelli, N.; Trouillet, C.; Guivel-Benhassine, F.; Rudnicka, D.; Sol-Foulon, N.; Le Roux, K.; Prevost, M.C.; Fsihi, H.; et al. Characterization of Reemerging Chikungunya Virus. *PLoS Pathog.* **2007**, *3*, 804–817. [CrossRef]

104. Vouga, M.; Baud, D. Imaging of congenital Zika virus infection: The route to identification of prognostic factors. *Prenat. Diagn.* **2016**, *36*, 799–811. [CrossRef] [PubMed]

105. Das Neves Almeida, R.; Racine, T.; Magalhães, K.; Kobinger, G. Zika Virus Vaccines: Challenges and Perspectives. *Vaccines* **2018**, *6*, 62. [CrossRef]

106. Brault, A.C.; Domi, A.; McDonald, E.M.; Talmi-Fran, D.; McCurley, N.; Basu, R.; Robinson, H.L.; Hellerstein, M.; Duggal, N.K.; Bowen, R.A.; et al. A Zika Vaccine Targeting NS1 Protein Protects Immunocompetent Adult Mice in a Lethal Challenge Model. *Sci. Rep.* **2017**, *7*, 1–11. [CrossRef]

107. Munjal, A.; Khandia, R.; Dhama, K.; Sachan, S.; Karthik, K.; Tiwari, R.; Malik, Y.S.; Kumar, D.; Singh, R.K.; Iqbal, H.M.N.; et al. Advances in developing therapies to combat zika virus: Current knowledge and future perspectives. *Front. Microbiol.* **2017**, *8*, 1–19. [CrossRef]

108. Morrison, C. DNA vaccines against Zika virus speed into clinical trials. *Nat. Publ. Gr.* **2016**, *15*, 521–522. [CrossRef]

109. Sun, H.; Chen, Q.; Lai, H. Development of Antibody Therapeutics against Flaviviruses. *Int. J. Mol. Sci.* **2018**, *19*, 1–26.

110. Beaver, J.T.; Lelutiu, N.; Habib, R.; Skountzou, I. Evolution of Two Major Zika Virus Lineages: Implications for Pathology, Immune Response, and Vaccine Development. *Front. Immunol.* **2018**, *9*, 1–17. [CrossRef] [PubMed]

111. Abbink, P.; Larocca, R.A.; De La Barrera, R.A.; Bricault, C.A.; Moseley, E.T.; Boyd, M.; Kirilova, M.; Li, Z.; Ng'ang'a, D.; Nanayakkara, O.; et al. Protective efficacy of multiple vaccine platforms against Zika virus challenge in rhesus monkeys. *Science* **2016**, *353*, 1129–1132. [CrossRef] [PubMed]

112. Larocca, R.A.; Abbink, P.; Peron, J.P.S.; Zanotto, P.M.; Iampietro, M.J.; Badamchi-Zadeh, A.; Boyd, M.; Ng'ang'a, D.; Kirilova, M.; Nityanandam, R.; et al. Vaccine protection against Zika virus from Brazil. *Nature* **2016**, *536*, 474–478. [CrossRef] [PubMed]

113. Bagasra, O.; Addanki, K.C.; Goodwin, G.R.; Hughes, B.W.; Pandey, P.; McLean, E. Cellular Targets and Receptor of Sexual Transmission of Zika Virus. *Appl. Immunohistochem. Mol. Morphol.* **2017**, *25*, 679–686. [CrossRef]

114. Fernandez, E.; Dejnirattisai, W.; Cao, B.; Scheaffer, S.M.; Supasa, P.; Wongwiwat, W.; Esakky, P.; Drury, A.; Mongkolsapaya, J.; Moley, K.H.; et al. Human antibodies to the dengue virus E-dimer epitope have therapeutic activity against Zika virus infection. *Nat. Immunol.* **2017**, *18*, 1261–1269. [CrossRef]

115. Griffin, B.D.; Muthumani, K.; Warner, B.M.; Majer, A.; Hagan, M.; Audet, J.; Stein, D.R.; Ranadheera, C.; Racine, T.; De LaVega, M.A.; et al. DNA vaccination protects mice against Zika virus-induced damage to the testes. *Nat. Commun.* **2017**, *8*, 1–8. [CrossRef]
116. Zou, J.; Xie, X.; Luo, H.; Shan, C.; Muruato, A.E.; Weaver, S.C.; Wang, T.; Shi, P.Y. A single-dose plasmid-launched live-attenuated Zika vaccine induces protective immunity. *EBioMedicine* **2018**, *36*, 92–102. [CrossRef]
117. Shan, C.; Muruato, A.E.; Jagger, B.W.; Richner, J.; Nunes, B.T.D.; Medeiros, D.B.A.; Xie, X.; Nunes, J.G.C.; Morabito, K.M.; Kong, W.; et al. A single-dose live-attenuated vaccine prevents Zika virus pregnancy transmission and testis damage. *Nat. Commun.* **2017**, *8*, 1–9. [CrossRef]
118. Xu, K.; Song, Y.; Dai, L.; Zhang, Y.; Lu, X.; Xie, Y.; Zhang, H.; Cheng, T.; Wang, Q.; Huang, Q.; et al. Recombinant Chimpanzee Adenovirus Vaccine AdC7-M/E Protects against Zika Virus Infection and Testis Damage. *J. Virol.* **2018**, *92*, 1–16. [CrossRef]
119. De La Vega, M.; Piret, J.; Griffin, B.D.; Rhéaume, C.; Venable, M.; Carbonneau, J.; Couture, C.; Almeida, N.; Tremblay, R.R.; Magalhães, K.G.; et al. Zika-Induced Male Infertility in Mice Is Potentially Reversible and Preventable by Deoxyribonucleic Acid Immunization. *J. Infect. Dis.* **2018**, *219*, 365–374. [CrossRef]
120. Abbink, P.; Stephenson, K.E.; Barouch, D.H. Zika virus vaccines. *Nat. Rev. Microbiol.* **2018**, *16*, 594–600. [CrossRef] [PubMed]
121. Garg, H.; Mehmetoglu-Gurbuz, T.; Joshi, A. Recent advances in Zika virus vaccines. *Viruses* **2018**, *10*, 631. [CrossRef] [PubMed]
122. Si, L.; Meng, Y.; Tian, F.; Li, W.; Zou, P.; Wang, Q.; Xu, W.; Wang, Y.; Xia, M.; Hu, J.; et al. A Peptide-Based Virus Inactivator Protects Male Mice Against Zika Virus-Induced Damage of Testicular Tissue. *Front. Microbiol.* **2019**, *10*, 1–13. [CrossRef] [PubMed]
123. Simanjuntak, Y.; Liang, J.J.; Chen, S.Y.; Li, J.K.; Lee, Y.L.; Wu, H.C.; Lin, Y.L. Ebselen alleviates testicular pathology in mice with Zika virus infection and prevents its sexual transmission. *PLoS Pathog.* **2018**, *14*, 1–23. [CrossRef]

 *cells*

Review

The Potential Role of the ZIKV NS5 Nuclear Spherical-Shell Structures in Cell Type-Specific Host Immune Modulation during ZIKV Infection

Min Jie Alvin Tan [1,†], Kitti Wing Ki Chan [1,†], Ivan H. W. Ng [1], Sean Yao Zu Kong [1], Chin Piaw Gwee [1], Satoru Watanabe [1] and Subhash G. Vasudevan [1,2,3,*]

1   Program in Emerging Infectious Diseases, Duke-NUS Medical School, Singapore 169857, Singapore;
    alvin.tan@duke-nus.edu.tabsg (M.J.A.T.); kitti.chan@duke-nus.edu.sg (K.W.K.C.);
    hwing10@gmail.com (I.H.W.N.); sean_kong@duke-nus.edu.sg (S.Y.Z.K.);
    chinpiaw.gwee@duke-nus.edu.sg (C.P.G.); satoru.watanabe@duke-nus.edu.sg (S.W.)
2   Department of Microbiology and Immunology, National University of Singapore, 5 Science Drive 2,
    Singapore 117545, Singapore
3   Institute for Glycomics, Griffith University, Gold Coast Campus, Queensland 4022, Australia
*   Correspondence: subhash.vasudevan@duke-nus.edu.sg; Tel.: +65-6516-6718
†   These authors contributed equally as first authors to this work.

Received: 3 October 2019; Accepted: 23 November 2019; Published: 26 November 2019

Abstract: The Zika virus (ZIKV) non-structural protein 5 (NS5) plays multiple viral and cellular roles during infection, with its primary role in virus RNA replication taking place in the cytoplasm. However, immunofluorescence assay studies have detected the presence of ZIKV NS5 in unique spherical shell-like structures in the nuclei of infected cells, suggesting potentially important cellular roles of ZIKV NS5 in the nucleus. Hence ZIKV NS5's subcellular distribution and localization must be tightly regulated during ZIKV infection. Both ZIKV NS5 expression or ZIKV infection antagonizes type I interferon signaling, and induces a pro-inflammatory transcriptional response in a cell type-specific manner, but the mechanisms involved and the role of nuclear ZIKV NS5 in these cellular functions has not been elucidated. Intriguingly, these cells originate from the brain and placenta, which are also organs that exhibit a pro-inflammatory signature and are known sites of pathogenesis during ZIKV infection in animal models and humans. Here, we discuss the regulation of the subcellular localization of the ZIKV NS5 protein, and its putative role in the induction of an inflammatory response and the occurrence of pathology in specific organs during ZIKV infection.

Keywords: flavivirus; Zika virus; NS5 protein; nuclear localization; inflammation; innate immunity

---

1. Introduction

Flaviviruses are small, enveloped RNA viruses that make up one genus in the *Flaviviridae* family of viruses. Many of these flaviviruses, including dengue virus (DENV), West Nile virus (WNV), yellow fever virus (YFV) and Zika virus (ZIKV), are medically important and cause human disease. There is currently no antiviral therapeutics available for clinical use against these viruses. ZIKV was initially identified as a rather innocuous member of the flavivirus genus in 1947 [1,2], and was only highlighted as a threat during the 2015 epidemic in the Americas [3–5] due to the correlation of ZIKV infection with severe neurological pathologies. Like many of the other flaviviruses, ZIKV is transmitted from human-to-human via the *Aedes* mosquitoes, although horizontal human-to-human transmission through sexual intercourse has also been identified as a unique mode of transmission of ZIKV.

The flavivirus genome is made up of a single-stranded positive sense RNA of approximately 11,000 nucleotides that encodes an ~3400 amino acid residue polyprotein precursor. This polyprotein is post-translationally cleaved into three structural proteins (capsid, pre-membrane/membrane, and envelope) and seven nonstructural (NS) proteins (NS1, NS2A, NS2B, NS3, NS4A, NS4B, and NS5). Flavivirus replication has been rigorously studied, and shown to require massive intracellular membrane redistribution to enable efficient replication within the cytoplasm of the host cell [6–8]. The membrane reorganization creates structures that are visualized by electron microscopy as vesicle packets within which all the viral NS proteins form the virus replication complex (RC) with host proteins that have not been fully characterized [9]. Flavivirus RNA replication occurs exclusively in these RCs within cytoplasmic membrane vesicles, where the virus genome can safely replicate out of reach of the host immune system [6,10–12].

The non-structural protein 5 (NS5) is the largest and most conserved protein encoded by the members of the flaviviruses [13,14], and plays crucial enzymatic roles during virus RNA replication as part of the virus RC. Apart from these functions, flavivirus NS5 has other viral and cellular roles during virus infection, including the modulation of the host immune response. Interestingly, the subcellular localization of NS5 proteins is different amongst the flaviviruses when examined by immunofluorescence assays, as they have been found in the nucleus and/or cytoplasm [15–24]. In particular, the observation of these flavivirus NS5 proteins in the nucleus suggests that NS5 plays important cellular and virus roles there, despite its major virus enzymatic role in virus RNA replication in the cytoplasm. In the case of ZIKV, most, if not all, of the detected NS5 protein form discrete spherical shell-like structures in the nucleus [21,25,26]. Taken together, the localization and distribution of ZIKV NS5 has to be tightly regulated for it to fulfill all its different roles during virus infection, but this is not well elucidated at the moment. As with many other flavivirus NS5 proteins, the significance of its localization to the nucleus on its cellular functions during infection is not known. In the first part of this review, we will use this recent demonstration of unique ZIKV NS5 nuclear localization as a springboard to discuss the regulation and function of the sub-cellular localization and, in particular, the nuclear role of the flavivirus NS5 protein.

During virus infection, the host responds by triggering a variety of signaling cascades, especially those involved in the immune response. The activation of these pathways such as the antiviral and pro-inflammatory signaling pathways leads to the induction of an antiviral state that helps the host repress virus replication and clear the virus [27]. In turn, viruses have developed mechanisms to modulate and impair these responses, and often utilize multiple strategies to target these signaling cascades. Flaviviruses do so by encoding multiple virus proteins that can antagonize or suppress the host immune response [28], as multiple DENV [29–31] and WNV NS proteins [32–34] have been described to repress antiviral pathways such as the type I interferon signaling pathway. ZIKV is no exception, as its NS2A, NS2B, NS4A and NS4B proteins have been described to antagonize the type I interferon pathway by targeting distinct components of the signaling cascade [35]. The flavivirus NS5 protein is another major player in this process [36,37], as it has been characterized to antagonize type I IFN signaling at multiple points [26,31,38,39]. Like its other flavivirus NS5 counterparts, ZIKV NS5 is able to inhibit the host antiviral response [25,26,39,40], although the significance and contribution of its nuclear localization to this process is not known.

Besides antagonizing the host antiviral response, ZIKV infection induces the transcriptional activation of pro-inflammatory genes [21,25]. Intriguingly, this activation is only observed in certain cell types such as neural and placental cells, but not in others such as liver cells. While the detailed mechanism of this cell type-specific activation is not known, ZIKV NS5, and its nuclear localization, has been shown to contribute to this transcriptional activation, as the expression of ZIKV NS5 alone in neural cells is able to trigger a pro-inflammatory transcriptional response. In the second part of this review, we will discuss the possible role of ZIKV NS5's subcellular localization in the modulation of the host immune response during ZIKV infection, and the pathological outcome in specific tissue during infection.

2. Subcellular Localization of Flavivirus NS5

The flavivirus NS5 protein bears two well-characterized enzymatic activities that are essential for virus replication: an N-terminal methyltransferase (MTase) function and a C-terminal RNA-dependent RNA polymerase (RdRp) activity (Figure 1). Additionally, it is also believed to contain a guanylyltransferase activity required for RNA cap formation [41,42]. While these flavivirus functions occur in the cytoplasm of infected cells, the NS5 protein from several flaviviruses are already present in the nucleus of the infected cells from earliest detectable time points post-infection [15–24] (Figure 1). Immunofluorescence assay studies have shown that >95% of fluorescence signal associated with NS5 can be located in the nucleus from around 24 h post-infection. Flavivirus NS5 subcellular localization has been best studied in the context of dengue virus (DENV) [15,17–20,22,23,43], trace amounts of DENV2 NS5 have been found co-localising with dsRNA, in addition to the NS5 detected in the nucleus [23]. Other flavivirus NS5 such as Zika virus (ZIKV), the main subject of this review, Japanese encephalitis virus (JEV), yellow fever virus (YFV) and West Nile virus (WNV) [16,21,24,44] have also been reported to accumulate in the nucleus. The regulation and function of the sub-cellular localization NS5 has to be tightly controlled in order for it to perform all its functions in the cell, but this is not well elucidated.

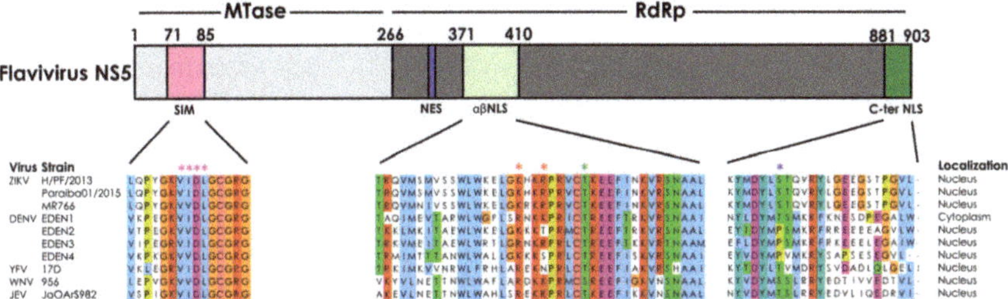

Figure 1. Sequences involved in the regulation of the subcellular localization of flavivirus NS5. (Top) The flavivirus NS5 consists of two functional domains, the methyltransferase (MTase) and the RNA-dependent RNA polymerase (RdRP). The regions involved in its subcellular localization (SIM, NES, αβNLS and C-ter NLS) are also denoted. (Bottom) Amino acid sequence alignment of the denoted regions of NS5 from various flaviviruses showing the sequence conservation using the Clustal X color scheme. The aligned NS5 regions contain the SIM (left), αβNLS (nuclear localization signal) (middle) and C-terminal NLS (right) sequences. The SIM motif is indicated with pink asterisks, two residues mutated in the Zika virus (ZIKV) NS5 NLS mutant are indicated with a red asterisk, CK2 phosphorylated threonine residue is indicated with a green asterisk, and the critical C-terminal NLS residue is indicated with a blue asterisk. The predominant subcellular localization of each NS5 protein is indicated to the right of the alignment. The numbering of amino acid residues is based on H/PF/2013. The virus strains in the alignment and GenBank accession numbers are as follows: H/PF/2013 (KJ776791), Paraiba01/2015 (KX280026), MR766 (LC002520), EDEN1 (EU081230), EDEN2 (EU081177), EDEN3 (EU081190), EDEN4 (GQ398256), YFV (NC 002031.1), WNV (NC_001563.2) and JEV (NC_001437.1).

The NS5 of flaviviruses is the largest virus-encoded protein of ~900 amino acids in size (Figure 1). At over 100kDa, the flavivirus NS5 protein can only enter the nucleus at a slow rate via passive diffusion [45], and hence will need to rely on active transport through the nuclear pore complex for efficient entry into the nucleus. In order for a protein to be transported in and out the nucleus, it needs to contain a nuclear targeting sequence that is recognized by its specific receptior from the karyopherin superfamily of proteins. In the case of nuclear import, the protein contains a specific sequence of amino acids (typically arginine and lysine) called the nuclear localization signal (NLS), and this is recognized by the importin family of proteins that mediate the movement of proteins from the cytoplasm into the nucleus [46,47].

Previous studies have revealed two highly conserved NLS sequences in the flavivirus NS5 proteins (light green in Figure 1) [15,22]. The αβNLS region was initially characterized to be the NLS of DENV serotype 2 (DENV2) NS5 [20,43] in studies involving the fusion of the NS5 NLS region to β-galactosidase and in vitro nuclear localization strategies. However the subsequent availability of a NS5-specific antibody [48] and a DENV2 infectious cDNA permitted an extensive structure-based assessment of the predominantly nuclear DENV2 NS5 and predominantly cytoplasmic DENV1 NS5, leading to the surprising finding that the previously characterized αβNLS sequence of DENV2 NS5 could not localize a truncated NS5 protein to the nucleus [23]. Instead, the NLS sequence within the C-terminal 18 residues region of DENV2 NS5 (dark green in Figure 1 and referred to as C-ter NLS) was sufficient to direct NS5 to the nucleus. Moreover the introduction of a single substitution at NS5 amino acid position 884 (blue asterisk in Figure 1) from Proline (as found in DENV2, 3 & 4) to Threonine (as found DENV1) into the DENV2 infectious clone produced a recombinant DENV mutant virus that replicated like the wild-type virus, yet showed a predominantly cytoplasmic NS5 somewhat similar to DENV1 NS5 sub-cellular localization [23]. This study showed that the sequence motif recognized by nuclear transport proteins is present in DENV1 NS5, but its conformation needs to be accessible in order for efficient transport to occur. Collectively this led to the hypothesis that NS5 nuclear localization is independent of viral RNA replication and its impact on pathogenesis must be linked to its non-enzymatic functions.

Nuclear export can also play a role in the subcellular localization of proteins as has been reported for several virus proteins [24,49–51], including flavivirus NS5 which has a conserved nuclear export sequence (NES) (Figure 1). But broader validation studies are needed to evaluate the functional role of nuclear export in flaviviruses, especially in the context of ZIKV NS5 which forms supramolecular spherical shell-like structures in the nucleus [21].

The role of nuclear NS5 has been investigated for several flaviviruses [20,22,49]. However the ZIKV NS5 protein was found to form discrete spherical shell-like nuclear structures, which is distinct from the diffuse distribution throughout the nucleus noted for other NS5 proteins [16,23]. But the significance of these spherical shell-like structures is not known, and further studies have to be done to elucidate this.

2.1. The Nuclear Localization Signal Sequence of ZIKV NS5

An examination of the NS5 sequences from a variety of flaviviruses reveal that the C-ter NLS sequence in DENV NS5 (dark green in Figure 1) is not conserved in ZIKV NS5. In contrast, the αβNLS region (light green in Figure 1) is highly conserved amongst several members of the flavivirus genus. Hence the functional NLS in the context of the full-length ZIKV NS5 protein appears to be within this αβNLS region involving residues 388–393, and this was confirmed by mutagenesis studies in which a K390A/R393A double mutation (red asterisks in Figure 1) was sufficient to alter the subcellular localization of ZIKV NS5 from being in the nucleus to being in the cytoplasm [21]. In addition, the spherical shell-like structures formed by wild type ZIKV NS5 in the nucleus is not observed, as the ZIKV NS5 NLS mutant is diffusely localized in the cytoplasm. This suggests that the formation of nuclear spherical shell-like structures that colocalize with host proteins such as STAT1 and importin-α [21] may require host nuclear proteins and/or be regulated by additional cellular processes.

2.2. Other Cellular Processes Regulating the Subcellular Localization of Flavivirus NS5

While the crucial role of NLS sequences in the nuclear transport process of flavivirus NS5 proteins has been demonstrated through site-specific mutagenesis, additional cellular processes such as post-translational modifications (PTMs) have also been shown to play a role in the regulation of the subcellular localization of cellular proteins [52]. In fact, correct localization of cellular proteins has also been shown to be important for proper function of these proteins [53,54]. In the case of flavivirus NS5 proteins, YFV [16] and DENV NS5 [43,55] are phosphorylated and localized into the nucleus, with a hyperphosphorylated form of DENV NS5 being associated with nuclear localization [19]. Interestingly the study by Kumar et al. showed that mutations of the residues adjacent to the highly conserved threonine residue in the putative CK2 phosphorylation site of DENV NS5 can alter its subcellular localization, presumably since the C-ter NLS of DENV2 NS5 is structurally part of the thumb subdomain close to the CK2 site [23]. Phosphorylation, in particular CK2 phosphorylation, of sites near NLS sequences has been utilized by multiple viruses to regulate localization of their virus proteins [56–60]. We note that this CK2 site is also conserved in ZIKV NS5 as well as other flaviviruses but the exact order or physiological setting that causes the phosphorylation of the CK2 site to influence sub-cellular localization is unclear. Nevertheless, the proximity to the CK2 residues and the potential phosphorylation of the conserved threonine residue (green asterisk in Figure 1) might add an additional layer of regulation to ZIKV NS5's subcellular localization and warrants a revisit in the context of DENV2 NS5 in the light of the discovery of its C-terminal NLS [23]. Interestingly, the DENV4 NS5 NLS at the C-terminal region might be regulated by phosphorylation, based on sequence prediction although this is yet to be evaluated experimentally [23,61].

Another PTM that regulates cellular nuclear localization involves the small ubiquitin-like modifier (SUMO) proteins that are a family of small proteins closely related to ubiquitin. Like ubiquitin, it can be reversibly attached to other cellular proteins via a series of enzymatic reactions. Modification of the SUMO target protein or SUMOlyation can functionally alter the protein by inducing a conformation change, or hindering or creating binding sites for interaction with other proteins [62]. SUMOlyation of a variety of proteins including CtBP1, CREB, Mdm2, KLF5 and Lipin-1α [63–66] have been associated with nuclear localization. The flavivirus NS5 protein has a highly conserved SUMO-interacting motif (SIM) within its methyltransferase (MTase) domain [67] (pink in Figure 1), and this SIM has been demonstrated to be necessary for the SUMOlyation of the DENV NS5 protein [68]. But it is not known what role SUMOlyation plays in the subcellular localization of the NS5 protein. Further studies need to be performed to understand the contribution of the SIM domain and SUMOlyation process to the regulation of ZIKV NS5 subcellular localization, and how that relates to ZIKV pathogenesis.

Overall it can be summarized that the regulation of the subcellular localization of flavivirus NS5 proteins is complex and multifaceted, and a thorough evaluation of this process is needed in future studies. Although there may be broad similarities in the mechanism of subcellular localization of flavivirus NS5, the consequences in the context of pathogenesis may be different in line with the proposed divergent evolution of flaviviral NS5 proteins based on their conformational flexibility [69]. The location of the NLS sequence in DENV2 NS5 appears to be clearly in the C-terminal part of the protein (residues 883–900) [23] whilst that for ZIKV NS5 is within the $\alpha\beta$NLS region [21]. However, both these sequences are structurally in the thumb sub-domain of NS5 [69–74] and will be located close to the conserved CK2 phosphorylation site. One possible scenario is that SUMOylation of NS5 could result in increased dynamics of the region around the thumb subdomain that may lead to its engagement with host proteins. In this context the structural studies of DENV NS5 using hydrogen-deuterium exchange (HDX) mass-spectrometry analysis [69] showed that the thumb subdomain appears to be more flexible and consistent with its identification as a "hot spot" for protein-protein interaction. It is likely that flavivirus NS5 proteins are separated into distinct sub-populations by their subcellular localization during various stages of flavivirus infection, and that these subpopulations are dynamic and fluid as the post-translational modifications alter the relative proportion of flavivirus NS5 in each subpopulation. Detailed investigation of this would require the reverse engineering of the virus to identify mutations that could partition the various functions without impacting viral RNA replication. This has not yet been achieved in the context ZIKV NS5 studies and suitably reverse engineered viruses would be needed to get a better understanding of these non-enzymatic functions of NS5. However the thumb sub-domain region is also critical for RNA replication function and involves direct interaction with RNA, so it might be challenging to examine the consequence of the site-specific mutations in the $\alpha\beta$NLS in the context of infectious cDNA clones. Nevertheless, our hypothesis, at least in the context of ZIKV NS5, is that the nuclear spheres that form in the infected cells can result in different innate immune modulation within different tissues extrapolated from the demonstration in liver Huh7 cells and neural LN229 cells [21]. It can be envisaged that the ZIKV NS5 nuclear bodies sequester different subsets of host proteins at different time points in different tissues, and this point would be further elaborated below.

3. Cell-Type-Specific Modulation of the Host Immune Response by ZIKV and ZIKV NS5

In response to virus infection, the host activates or suppresses various signaling pathways. Many of the pathways activated upon virus infection are part of the host immune response. The antiviral and pro-inflammatory signaling cascades are two major pathways of this host immune response, and work to suppress virus replication and clear the virus through the establishment of an antiviral state in the cell (Figure 2). Thus all viruses have also evolved to subvert the host immune response. The flaviviruses utilize a variety of strategies to evade, antagonize and suppress the host immune response to virus infection. One of the major pathways activated as part of the host antiviral response is the type I interferon (IFN) response. The production of type I IFN is triggered by host sensors that detect the presence of virus components such as nucleic acid. Type I IFN then activates a number of interferon-stimulated genes (ISGs) that mediate the antivirus response and establishment of an antiviral state [36,37]. Like its other flavivirus NS5 counterparts, ZIKV NS5 has been well characterized to inhibit the host antiviral response [25,26,39,40] (Table 1). But the mechanisms involved including whether ZIKV NS5's nuclear localization plays a role have not been fully elucidated.

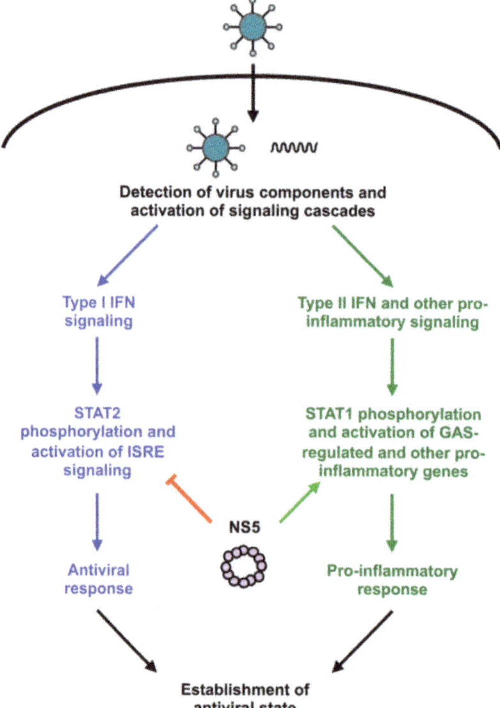

Figure 2. Modulation of host immune response by ZIKV NS5. Virus infection leads to the activation of host signaling cascades in response to the recognition of virus components such as RNA and protein. The type I IFN signaling pathway is one of the many pathways that mediate the host antiviral response. Type I IFN induces the phosphorylation of STAT2 and activation of interferon-stimulated response element (ISRE)-regulated genes, and this signaling cascade is targeted by ZIKV NS5 through the degradation of STAT2. Type II IFN and other pro-inflammatory molecules induces the pro-inflammatory responses through STAT1 phosphorylation and activation of interferon-gamma activated site (GAS)-regulated and other pro-inflammatory genes. ZIKV NS5 can activate the transcription of pro-inflammatory genes in a cell type-specific manner, but the mechanism is not known.

Table 1. Effects of ZIKV NS5 and infection on host immune responses.

NS5 Expression		
Cell Type	**Effect**	**Citation**
293/293T	Inhibits ISRE-luc signaling	[67]
	Inhibits ISG54-luc signaling	[26]
	Blocks type I IFN-induced STAT1 phosphorylation	[39]
	Antagonizes type I IFN production	[40]
A549	Antagonizes type I IFN production	[75]
LN229	Activates pro-inflammatory genes	[21]
ZIKV Infection		
Cell Type	**Effect**	**Citation**
DCs	Inhibits type I IFN signaling	[76]
293T	Inhibits type I IFN signaling	[26]
Dendritic cells	Inhibits type I IFN induced STAT1 and STAT2 phosphorylation	[75]
JEG3 SF268	Activates STAT1-mediated type II IFN and pro-inflammatory pathways	[25]
LN229	Induces STAT1 phosphorylation and activation of pro-inflammatory genes	[21]
Brain Microglia	Induces inflammation	[77]
Retina Pigment Epithelial	Induces inflammatory response	[76]

While the typical countermeasure of the virus to the host antiviral response is one of suppression, the relationship between the virus and the host inflammatory response is more complex. The host pro-inflammatory response is a double-edged sword. A controlled response can suppress virus replication and promote virus clearance. In contrast, an excessive response can actually lead to pathogenesis as in the case of DENV and influenza virus infections [78]. In addition, some viruses such as human cytomegalovirus (CMV) and human immunodeficiency virus (HIV) [79,80] have their replication enhanced through the induction of a host inflammatory response. ZIKV infection can induce a pro-inflammatory response in cell culture models in a variety of cell types from different organs, including the brain, testes, placenta and eyes [21,25,76,77,81] (Table 2). Interestingly, many of these organs are also sites of virus replication, persistence and pathology in multiple animal models [82–87]. Virus persistence has also been detected in human samples from these tissues, with the brain being an example of a known site of pathology for ZIKV infection [88–92]. Despite all these studies, the relationship between the host inflammatory response, and ZIKV replication, persistence and pathology is still unclear. Hence further and more comprehensive studies need to be performed to further our understanding of this complex relationship. In addition, it is unclear what aspects of ZIKV infection leads to the induction of this inflammatory response, although the expression of ZIKV NS5 alone appears to be sufficient to induce a pro-inflammatory transcriptional response in some of these cell types [21,25].

In the section below, we attempt to compile and reconcile the many studies that have examined the effects of ZIKV NS5 expression and ZIKV infection on various aspects of the host immune response in various cell types and tissue types. We then follow with a discussion on these topics and to explore our hypothesis that the spherical shell-like nuclear structures formed by NS5 in the infected cells can result in different immune modulation within different tissues.

Table 2. Summary of organs in which ZIKV infection leads to an inflammatory response and/or pathology in cell culture (grey), animal models (green) and human samples (blue). Y denotes a positive finding while N means it was not demonstrated in the cited paper.

Organ	Cell/Tissue Type	Inflammatory Response	Pathology	Citation
	SF268	Y	N	[25]
	LN229	Y	N	[21]
	Brain Microglia	Y	N	[77]
Brain	NPCs	N	Y	[93]
	NPCs	Y	N	[81]
	Neurospheres/Organoids	N	Y	[94]
	Brain	N	N	[86]
	Brain	N	N	[95]
	Retina Pigment Epithelial	Y	N	[76]
Eye	Eye	N	N	[82]
	Conjunctival fluid	N	N	[88]
	Sertoli	N	N	[96]
	Testis	Y	Y	[85]
Male Reproductive Tract	Testis	N	N	[86]
	Testis	N	Y	[84]
	Sperm/Semen	N	N	[91,92]
	JEG3	Y	N	[25]
	Uterine fibroblasts	N	N	[28]
Female Reproductive Tract	Ovary	Y	N	[97]
	Cervical mucus	N	N	[89]
	Vagina secretions	N	N	[90]
	Placenta	N	N	[95]

3.1. ZIKV NS5 Differentially Modulates the Various Arms of the Host Immune Response

The flavivirus NS5 protein is a well-studied antagonist of the type I IFN response, and ZIKV NS5 is no exception. Multiple studies using both reporter assays and by measuring cellular gene and protein levels, have been performed in different cell types, and they all demonstrate a repression of this pathway (Table 1). While ZIKV NS5 has been shown to degrade STAT2 and block TBK1 activation, all the mechanisms that contribute to this inhibition are still to be elucidated.

We have previously demonstrated that ZIKV NS5 expression alone can induce a cell type-specific transcriptional activation of pro-inflammatory genes in neural LN229 cells, but not in liver Huh7 cells [21]. Many of these pro-inflammatory genes are also upregulated in response to ZIKV infection, but the magnitude of transcriptional activation is much smaller in response to NS5 expression than it is to ZIKV infection [21]. This suggests that ZIKV NS5 can contribute partially to this gene signature. While the detailed mechanism is not known, there is evidence to suggest the involvement of the type II IFN and STAT1-mediated signaling pathways in this pro-inflammatory transcription activation [25]. Further experiments need to be performed to identify the molecular mechanisms involved in this process.

This stark difference between LN229 and Huh7 transcriptional activation profile is also highly fascinating, as ZIKV NS5 subcellular localization is identical in both cell lines [21] (Figure 3). This suggests that the spherical shell-like nuclear structures of ZIKV NS5 could potentially be interacting with different sets of host proteins during ZIKV infection at different sites of the host.

Further experiments to identify the host determinants of the differential transcriptional profiles observed in these different cell types will contribute towards the elucidation of this process.

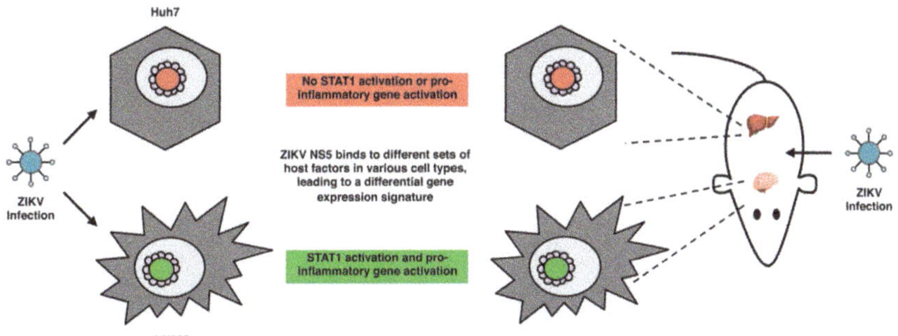

Figure 3. Cell type-specific activation of pro-inflammatory genes by ZIKV NS5. ZIKV infection of cells derived from the brain (LN229) leads to STAT1 activation and pro-inflammatory gene expression. In contrast, ZIKV infection of cells derived from the liver (Huh7) does not trigger any pro-inflammatory response. In an animal model of ZIKV infection, the same pattern of pro-inflammatory gene expression is observed. As NS5's subcellular localization is the same in both types of cells, it is likely to be interacting with different sets of host factors (green and red circles) in these cells, resulting in the differential gene expression signature observed (Image of organs: Freepik.com).

3.2. Role of ZIKV NS5's Subcellular Localization in Its Modulation of the Host Immune Response

The mechanism by which ZIKV NS5 is able to modulate various arms of the host immune response is not fully elucidated. As discussed in the previous section, ZIKV NS5 has been detected to localize in spherical shell-like structures in the nucleus, despite its known role in virus RNA replication that takes place in RCs in the cytoplasm. But it is not known whether the NS5 proteins that localize to these structures in the nucleus play a role in this cellular function. The ZIKV NS5 NLS mutant protein has a clear and distinct difference in the subcellular localization from the wild-type NS5 protein [21]. When these proteins were separately expressed in neural LN229 cells, the wild-type NS5 protein strongly upregulated expression of many pro-inflammatory genes, while the NS5 NLS mutant induced a much weaker upregulation [21].

This data suggest that ZIKV NS5's ability to modulate the host immune response can be altered by mutating the NLS sequence and preventing the ZIKV NS5 proteins from entering the nucleus. Hence the ZIKV NS5 proteins that localize in the nucleus appear to be playing a role in the modulation of the host immune response. It is also possible that this observed phenotype is due to the NLS mutation having a direct impact on an interaction with an important host factor or the formation of the spherical shell-like structures. The identification of additional NS5 mutations that alter the subcellular localization of ZIKV NS5 will provide additional tools to study its relationship with its immune modulation function. Interestingly, ZIKV NS5 forms spherical shell-like structures in the nuclei of all known cells, suggesting that the formation of structures alone does not direct the transcriptional response. Hence it is likely that ZIKV NS5 is interacting with separate sets of host factors in the different cell types, and these distinct NS5-host protein interactomes are responsible. The identification of these host proteins will go a long way towards our understanding of these observations. One major caveat of this data at the moment is that these differences are being observed in systems utilizing the overexpression of the ZIKV NS5 protein. Further studies using such a recombinant ZIKV containing these NS5 mutations are necessary to rigorously interrogate this hypothesis in the context of ZIKV infection.

3.3. Cell-Type-Specific Pro-Inflammatory Response and Virus Persistence during ZIKV Infection

The host immune response to ZIKV infection has been extensively studied in numerous cell types, and ZIKV infection has been well characterized to antagonize type I IFN signaling (Table 1). On the other hand, we and others have observed that ZIKV infection induces type II IFN signaling and pro-inflammatory gene expression in multiple cell types, including neural and placental cells. In our case, we observed a cell type-specific activation of pro-inflammatory genes when comparing ZIKV infection of liver Huh7 and neural LN229 cells [21] (Figure 3).

As with our studies in Huh7 and LN229 cells, many of the studies showing a pro-inflammatory response to ZIKV infection have been performed in individual cell lines (Table 2). The origin of these cell lines includes the brain, eye, and both male and female reproductive organs. In line with these studies in cell lines, this differential pro-inflammatory response has also been observed in animal models of ZIKV infection [21] (Figure 3), as a pro-inflammatory gene expression signature was only observed in the brain but not the liver of ZIKV infected mice. Early studies in our laboratory also suggest a pro-inflammatory gene signature in the testes. Intriguingly these organs are also sites of persistence and maintenance of viremia in animal models of ZIKV infection as well as studies of humans infected with ZIKV [21,25,98]. This correlation between the persistence of the virus and the activation of pro-inflammatory genes at these sites is notable, and suggests that the transcriptional activation of pro-inflammatory genes is not leading to the clearance of the virus from these sites.

Interestingly, many of these tissues where a pro-inflammatory gene expression signature is observed are also known sites of immune privilege [91,92,99], which allows ZIKV to establish and replicate unimpeded while protected from the peripheral immune system. In addition, the induction of a pro-inflammatory response did not lead to successful clearance of the virus from these sites, and hence it is possible that the pro-inflammatory response might actually be assisting in virus replication and persistence, as has been described for other viruses such as human cytomegalovirus (CMV) and human immunodeficiency virus (HIV) [79,80]. During this process, the virus might be inadvertently causing damage to these sites that become manifested as pathogenesis of virus infection.

Based on the current knowledge in the field, we propose a model in which ZIKV infection induces a pro-inflammatory response in a cell-type-specific matter, and this results in the persistence of the virus and pathological damage at these sites. ZIKV NS5, and its localization into spherical shell-like structures in the nucleus, plays a role in this, possibly by recruiting a cell-type-specific set of host factors to mediate a portion of this pro-inflammatory response (Figure 3). Further studies need to be performed to test this hypothesis, and they will help elucidate how ZIKV infection and the ZIKV NS5 is inducing this organ-specific pro-inflammatory response, and the role of this pro-inflammatory response during ZIKV infection.

In conclusion, the regulation of the subcellular localization of ZIKV NS5, and the role of ZIKV NS5 and infection on the cell-type-specific activation of pro-inflammatory genes are highly complex topics that require detailed analysis in conjunction with reverse genetic and appropriate in vivo model systems. Detailed studies on these topics will help further our understanding of ZIKV infection and perhaps contribute towards the development of antiviral therapeutics and other countermeasures.

Funding: Our research is funded by National Medical Research Council grants NMRC/OFYIRG/0006/2016 to M.J.A.T., NMRC/OFYIRG/0003/2016 to S.W. and NMRC/CBRG/0103/2016 to S.G.V., a National Research Foundation grant NRF2016NRF-CRP001-063 as co-PI to S.G.V., and a Ministry of Health grant MOH-000086 (MOH-OFIRG18may-0006/2019) to S.G.V.

Conflicts of Interest: The authors declare no conflict of interest. The funders had no role in the design or in the writing of the manuscript, and in the decision to publish the manuscript.

References

1. Dick, G.W. Zika virus. II. Pathogenicity and physical properties. *Trans. R. Soc. Trop. Med. Hyg.* **1952**, *46*, 521–534. [CrossRef]
2. Dick, G.W.; Kitchen, S.F.; Haddow, A.J. Zika virus. I. Isolations and serological specificity. *Trans. R. Soc. Trop. Med. Hyg.* **1952**, *46*, 509–520. [CrossRef]
3. Campos, G.S.; Bandeira, A.C.; Sardi, S.I. Zika Virus Outbreak, Bahia, Brazil. *Emerg. Infect. Dis.* **2015**, *21*, 1885–1886. [CrossRef]
4. Enfissi, A.; Codrington, J.; Roosblad, J.; Kazanji, M.; Rousset, D. Zika virus genome from the Americas. *Lancet* **2016**, *387*, 227–228. [CrossRef]
5. Hennessey, M.; Fischer, M.; Staples, J.E. Zika Virus Spreads to New Areas—Region of the Americas, May 2015–January 2016. *Mmwr. Morb. Mortal. Wkly. Rep.* **2016**, *65*, 55–58. [CrossRef]
6. Welsch, S.; Miller, S.; Romero-Brey, I.; Merz, A.; Bleck, C.K.; Walther, P.; Fuller, S.D.; Antony, C.; Krijnse-Locker, J.; Bartenschlager, R. Composition and three-dimensional architecture of the dengue virus replication and assembly sites. *Cell Host Microbe* **2009**, *5*, 365–375. [CrossRef]
7. Cortese, M.; Goellner, S.; Acosta, E.G.; Neufeldt, C.J.; Oleksiuk, O.; Lampe, M.; Haselmann, U.; Funaya, C.; Schieber, N.; Ronchi, P.; et al. Ultrastructural Characterization of Zika Virus Replication Factories. *Cell Rep.* **2017**, *18*, 2113–2123. [CrossRef]
8. Offerdahl, D.K.; Dorward, D.W.; Hansen, B.T.; Bloom, M.E. Cytoarchitecture of Zika virus infection in human neuroblastoma and Aedes albopictus cell lines. *Virology* **2017**, *501*, 54–62. [CrossRef]
9. Klema, V.J.; Padmanabhan, R.; Choi, K.H. Flaviviral Replication Complex: Coordination between RNA Synthesis and 5′-RNA Capping. *Viruses* **2015**, *7*, 4640–4656. [CrossRef] [PubMed]
10. Gillespie, L.K.; Hoenen, A.; Morgan, G.; Mackenzie, J.M. The endoplasmic reticulum provides the membrane platform for biogenesis of the flavivirus replication complex. *J. Virol.* **2010**, *84*, 10438–10447. [CrossRef] [PubMed]
11. Davidson, A.D. Chapter 2. New insights into flavivirus nonstructural protein 5. *Adv. Virus Res.* **2009**, *74*, 41–101. [CrossRef]
12. Uchida, L.; Espada-Murao, L.A.; Takamatsu, Y.; Okamoto, K.; Hayasaka, D.; Yu, F.; Nabeshima, T.; Buerano, C.C.; Morita, K. The dengue virus conceals double-stranded RNA in the intracellular membrane to escape from an interferon response. *Sci. Rep.* **2014**, *4*, 7395. [CrossRef]
13. Zhu, Z.; Chan, J.F.; Tee, K.M.; Choi, G.K.; Lau, S.K.; Woo, P.C.; Tse, H.; Yuen, K.Y. Comparative genomic analysis of pre-epidemic and epidemic Zika virus strains for virological factors potentially associated with the rapidly expanding epidemic. *Emerg. Microbes Infect.* **2016**, *5*, e22. [CrossRef]
14. Bollati, M.; Alvarez, K.; Assenberg, R.; Baronti, C.; Canard, B.; Cook, S.; Coutard, B.; Decroly, E.; de Lamballerie, X.; Gould, E.A.; et al. Structure and functionality in flavivirus NS-proteins: perspectives for drug design. *Antivir. Res.* **2010**, *87*, 125–148. [CrossRef]
15. Brooks, A.J.; Johansson, M.; John, A.V.; Xu, Y.; Jans, D.A.; Vasudevan, S.G. The interdomain region of dengue NS5 protein that binds to the viral helicase NS3 contains independently functional importin beta 1 and importin alpha/beta-recognized nuclear localization signals. *J. Biol. Chem.* **2002**, *277*, 36399–36407. [CrossRef]
16. Buckley, A.; Gaidamovich, S.; Turchinskaya, A.; Gould, E.A. Monoclonal antibodies identify the NS5 yellow fever virus non-structural protein in the nuclei of infected cells. *J. Gen. Virol.* **1992**, *73 Pt 5*, 1125–1130. [CrossRef]
17. Jans, D.A.; Martin, A.J. Nucleocytoplasmic Trafficking of Dengue Non-structural Protein 5 as a Target for Antivirals. *Adv. Exp. Med. Biol.* **2018**, *1062*, 199–213. [CrossRef]
18. Johansson, M.; Brooks, A.J.; Jans, D.A.; Vasudevan, S.G. A small region of the dengue virus-encoded RNA-dependent RNA polymerase, NS5, confers interaction with both the nuclear transport receptor importin-beta and the viral helicase, NS3. *J. Gen. Virol.* **2001**, *82*, 735–745. [CrossRef]
19. Kapoor, M.; Zhang, L.; Ramachandra, M.; Kusukawa, J.; Ebner, K.E.; Padmanabhan, R. Association between NS3 and NS5 proteins of dengue virus type 2 in the putative RNA replicase is linked to differential phosphorylation of NS5. *J. Biol. Chem.* **1995**, *270*, 19100–19106. [CrossRef]
20. Kumar, A.; Buhler, S.; Selisko, B.; Davidson, A.; Mulder, K.; Canard, B.; Miller, S.; Bartenschlager, R. Nuclear localization of dengue virus nonstructural protein 5 does not strictly correlate with efficient viral RNA replication and inhibition of type I interferon signaling. *J. Virol.* **2013**, *87*, 4545–4557. [CrossRef]

21. Ng, I.H.W.; Chan, K.W.; Tan, M.J.A.; Gwee, C.P.; Smith, K.M.; Jeffress, S.J.; Saw, W.G.; Swarbrick, C.M.D.; Watanabe, S.; Jans, D.A.; et al. Zika Virus NS5 Forms Supramolecular Nuclear Bodies That Sequester Importin-alpha and Modulate the Host Immune and Pro-Inflammatory Response in Neuronal Cells. *ACS Infect. Dis.* **2019**. [CrossRef]

22. Pryor, M.J.; Rawlinson, S.M.; Butcher, R.E.; Barton, C.L.; Waterhouse, T.A.; Vasudevan, S.G.; Bardin, P.G.; Wright, P.J.; Jans, D.A.; Davidson, A.D. Nuclear localization of dengue virus nonstructural protein 5 through its importin alpha/beta-recognized nuclear localization sequences is integral to viral infection. *Traffic (Cph. Den.)* **2007**, *8*, 795–807. [CrossRef]

23. Tay, M.Y.; Smith, K.; Ng, I.H.; Chan, K.W.; Zhao, Y.; Ooi, E.E.; Lescar, J.; Luo, D.; Jans, D.A.; Forwood, J.K.; et al. The C-terminal 18 Amino Acid Region of Dengue Virus NS5 Regulates its Subcellular Localization and Contains a Conserved Arginine Residue Essential for Infectious Virus Production. *PLoS Pathog.* **2016**, *12*, e1005886. [CrossRef]

24. Lopez-Denman, A.J.; Russo, A.; Wagstaff, K.M.; White, P.A.; Jans, D.A.; Mackenzie, J.M. Nucleocytoplasmic shuttling of the West Nile virus RNA-dependent RNA polymerase NS5 is critical to infection. *Cell. Microbiol.* **2018**, *20*, e12848. [CrossRef]

25. Chaudhary, V.; Yuen, K.S.; Chan, J.F.; Chan, C.P.; Wang, P.H.; Cai, J.P.; Zhang, S.; Liang, M.; Kok, K.H.; Chan, C.P.; et al. Selective Activation of Type II Interferon Signaling by Zika Virus NS5 Protein. *J. Virol.* **2017**, *91*. [CrossRef] [PubMed]

26. Grant, A.; Ponia, S.S.; Tripathi, S.; Balasubramaniam, V.; Miorin, L.; Sourisseau, M.; Schwarz, M.C.; Sanchez-Seco, M.P.; Evans, M.J.; Best, S.M.; et al. Zika Virus Targets Human STAT2 to Inhibit Type I Interferon Signaling. *Cell Host Microbe* **2016**, *19*, 882–890. [CrossRef]

27. Crosse, K.M.; Monson, E.A.; Beard, M.R.; Helbig, K.J. Interferon-Stimulated Genes as Enhancers of Antiviral Innate Immune Signaling. *J. Innate Immun.* **2018**, *10*, 85–93. [CrossRef]

28. Chen, S.; Wu, Z.; Wang, M.; Cheng, A. Innate Immune Evasion Mediated by Flaviviridae Non-Structural Proteins. *Viruses* **2017**, *9*, 291. [CrossRef] [PubMed]

29. Dalrymple, N.A.; Cimica, V.; Mackow, E.R. Dengue Virus NS Proteins Inhibit RIG-I/MAVS Signaling by Blocking TBK1/IRF3 Phosphorylation: Dengue Virus Serotype 1 NS4A Is a Unique Interferon-Regulating Virulence Determinant. *MBIO* **2015**, *6*. [CrossRef]

30. Munoz-Jordan, J.L.; Sanchez-Burgos, G.G.; Laurent-Rolle, M.; Garcia-Sastre, A. Inhibition of interferon signaling by dengue virus. *Proc. Natl. Acad. Sci. USA* **2003**, *100*, 14333–14338. [CrossRef] [PubMed]

31. Morrison, J.; Laurent-Rolle, M.; Maestre, A.M.; Rajsbaum, R.; Pisanelli, G.; Simon, V.; Mulder, L.C.; Fernandez-Sesma, A.; Garcia-Sastre, A. Dengue virus co-opts UBR4 to degrade STAT2 and antagonize type I interferon signaling. *PLoS Pathog.* **2013**, *9*, e1003265. [CrossRef] [PubMed]

32. Liu, W.J.; Chen, H.B.; Wang, X.J.; Huang, H.; Khromykh, A.A. Analysis of adaptive mutations in Kunjin virus replicon RNA reveals a novel role for the flavivirus nonstructural protein NS2A in inhibition of beta interferon promoter-driven transcription. *J. Virol.* **2004**, *78*, 12225–12235. [CrossRef] [PubMed]

33. Munoz-Jordan, J.L.; Laurent-Rolle, M.; Ashour, J.; Martinez-Sobrido, L.; Ashok, M.; Lipkin, W.I.; Garcia-Sastre, A. Inhibition of alpha/beta interferon signaling by the NS4B protein of flaviviruses. *J. Virol.* **2005**, *79*, 8004–8013. [CrossRef] [PubMed]

34. Liu, W.J.; Wang, X.J.; Mokhonov, V.V.; Shi, P.Y.; Randall, R.; Khromykh, A.A. Inhibition of interferon signaling by the New York 99 strain and Kunjin subtype of West Nile virus involves blockage of STAT1 and STAT2 activation by nonstructural proteins. *J. Virol.* **2005**, *79*, 1934–1942. [CrossRef]

35. Xia, H.; Luo, H.; Shan, C.; Muruato, A.E.; Nunes, B.T.D.; Medeiros, D.B.A.; Zou, J.; Xie, X.; Giraldo, M.I.; Vasconcelos, P.F.C.; et al. An evolutionary NS1 mutation enhances Zika virus evasion of host interferon induction. *Nat. Commun.* **2018**, *9*, 414. [CrossRef]

36. Thurmond, S.; Wang, B.; Song, J.; Hai, R. Suppression of Type I Interferon Signaling by Flavivirus NS5. *Viruses* **2018**, *10*, 712. [CrossRef]

37. Best, S.M. The Many Faces of the Flavivirus NS5 Protein in Antagonism of Type I Interferon Signaling. *J. Virol.* **2017**, *91*. [CrossRef]

38. Laurent-Rolle, M.; Boer, E.F.; Lubick, K.J.; Wolfinbarger, J.B.; Carmody, A.B.; Rockx, B.; Liu, W.; Ashour, J.; Shupert, W.L.; Holbrook, M.R.; et al. The NS5 protein of the virulent West Nile virus NY99 strain is a potent antagonist of type I interferon-mediated JAK-STAT signaling. *J. Virol.* **2010**, *84*, 3503–3515. [CrossRef]

39. Hertzog, J.; Dias Junior, A.G.; Rigby, R.E.; Donald, C.L.; Mayer, A.; Sezgin, E.; Song, C.; Jin, B.; Hublitz, P.; Eggeling, C.; et al. Infection with a Brazilian isolate of Zika virus generates RIG-I stimulatory RNA and the viral NS5 protein blocks type I IFN induction and signaling. *Eur. J. Immunol.* **2018**, *48*, 1120–1136. [CrossRef]

40. Lin, S.; Yang, S.; He, J.; Guest, J.D.; Ma, Z.; Yang, L.; Pierce, B.G.; Tang, Q.; Zhang, Y.J. Zika virus NS5 protein antagonizes type I interferon production via blocking TBK1 activation. *Virology* **2019**, *527*, 180–187. [CrossRef]

41. Issur, M.; Geiss, B.J.; Bougie, I.; Picard-Jean, F.; Despins, S.; Mayette, J.; Hobdey, S.E.; Bisaillon, M. The flavivirus NS5 protein is a true RNA guanylyltransferase that catalyzes a two-step reaction to form the RNA cap structure. *RNA (New York)* **2009**, *15*, 2340–2350. [CrossRef] [PubMed]

42. Ray, D.; Shah, A.; Tilgner, M.; Guo, Y.; Zhao, Y.; Dong, H.; Deas, T.S.; Zhou, Y.; Li, H.; Shi, P.Y. West Nile virus 5′-cap structure is formed by sequential guanine N-7 and ribose 2′-O methylations by nonstructural protein 5. *J. Virol.* **2006**, *80*, 8362–8370. [CrossRef]

43. Forwood, J.K.; Brooks, A.; Briggs, L.J.; Xiao, C.Y.; Jans, D.A.; Vasudevan, S.G. The 37-amino-acid interdomain of dengue virus NS5 protein contains a functional NLS and inhibitory CK2 site. *Biochem. Biophys. Res. Commun.* **1999**, *257*, 731–737. [CrossRef]

44. Edward, Z.; Takegami, T. Localization and functions of Japanese encephalitis virus nonstructural proteins NS3 and NS5 for viral RNA synthesis in the infected cells. *Microbiol. Immunol.* **1993**, *37*, 239–243. [CrossRef]

45. Timney, B.L.; Raveh, B.; Mironska, R.; Trivedi, J.M.; Kim, S.J.; Russel, D.; Wente, S.R.; Sali, A.; Rout, M.P. Simple rules for passive diffusion through the nuclear pore complex. *J. Cell Biol.* **2016**, *215*, 57–76. [CrossRef]

46. Marfori, M.; Mynott, A.; Ellis, J.J.; Mehdi, A.M.; Saunders, N.F.; Curmi, P.M.; Forwood, J.K.; Boden, M.; Kobe, B. Molecular basis for specificity of nuclear import and prediction of nuclear localization. *Biochim. Biophys. Acta* **2011**, *1813*, 1562–1577. [CrossRef]

47. Lopez-Denman, A.J.; Mackenzie, J.M. The IMPORTance of the Nucleus during Flavivirus Replication. *Viruses* **2017**, *9*, 14. [CrossRef]

48. Zhao, Y.; Moreland, N.J.; Tay, M.Y.; Lee, C.C.; Swaminathan, K.; Vasudevan, S.G. Identification and molecular characterization of human antibody fragments specific for dengue NS5 protein. *Virus Res.* **2014**, *179*, 225–230. [CrossRef]

49. Rawlinson, S.M.; Pryor, M.J.; Wright, P.J.; Jans, D.A. CRM1-mediated nuclear export of dengue virus RNA polymerase NS5 modulates interleukin-8 induction and virus production. *J. Biol. Chem.* **2009**, *284*, 15589–15597. [CrossRef]

50. Ringel, M.; Behner, L.; Heiner, A.; Sauerhering, L.; Maisner, A. Replication of a Nipah Virus Encoding a Nuclear-Retained Matrix Protein. *J. Infect. Dis.* **2019**. [CrossRef]

51. Wang, Y.E.; Park, A.; Lake, M.; Pentecost, M.; Torres, B.; Yun, T.E.; Wolf, M.C.; Holbrook, M.R.; Freiberg, A.N.; Lee, B. Ubiquitin-regulated nuclear-cytoplasmic trafficking of the Nipah virus matrix protein is important for viral budding. *PLoS Pathog.* **2010**, *6*, e1001186. [CrossRef] [PubMed]

52. Bauer, N.C.; Doetsch, P.W.; Corbett, A.H. Mechanisms Regulating Protein Localization. *Traffic (Cph. Den.)* **2015**, *16*, 1039–1061. [CrossRef] [PubMed]

53. Guardia, C.M.; De Pace, R.; Mattera, R.; Bonifacino, J.S. Neuronal functions of adaptor complexes involved in protein sorting. *Curr. Opin. Neurobiol.* **2018**, *51*, 103–110. [CrossRef]

54. Bridges, R.J.; Bradbury, N.A. Cystic Fibrosis, Cystic Fibrosis Transmembrane Conductance Regulator and Drugs: Insights from Cellular Trafficking. *Handb. Exp. Pharmacol.* **2018**, *245*, 385–425. [CrossRef] [PubMed]

55. Bhattacharya, D.; Mayuri; Best, S.M.; Perera, R.; Kuhn, R.J.; Striker, R. Protein kinase G phosphorylates mosquito-borne flavivirus NS5. *J. Virol.* **2009**, *83*, 9195–9205. [CrossRef]

56. Alvisi, G.; Marin, O.; Pari, G.; Mancini, M.; Avanzi, S.; Loregian, A.; Jans, D.A.; Ripalti, A. Multiple phosphorylation sites at the C-terminus regulate nuclear import of HCMV DNA polymerase processivity factor ppUL44. *Virology* **2011**, *417*, 259–267. [CrossRef]

57. Alvisi, G.; Jans, D.A.; Guo, J.; Pinna, L.A.; Ripalti, A. A protein kinase CK2 site flanking the nuclear targeting signal enhances nuclear transport of human cytomegalovirus ppUL44. *Traffic (Cph. Den.)* **2005**, *6*, 1002–1013. [CrossRef] [PubMed]

58. Hubner, S.; Xiao, C.Y.; Jans, D.A. The protein kinase CK2 site (Ser111/112) enhances recognition of the simian virus 40 large T-antigen nuclear localization sequence by importin. *J. Biol. Chem.* **1997**, *272*, 17191–17195. [CrossRef]

59. Xiao, C.Y.; Jans, P.; Jans, D.A. Negative charge at the protein kinase CK2 site enhances recognition of the SV40 large T-antigen NLS by importin: effect of conformation. *FEBS Lett.* **1998**, *440*, 297–301. [CrossRef]

60. Nardozzi, J.D.; Lott, K.; Cingolani, G. Phosphorylation meets nuclear import: a review. *Cell Commun. Signal.* **2010**, *8*, 32. [CrossRef]

61. Schwartz, D.; Church, G.M. Collection and motif-based prediction of phosphorylation sites in human viruses. *Sci. Signal.* **2010**, *3*, rs2. [CrossRef] [PubMed]

62. Kerscher, O. SUMO junction-what's your function? New insights through SUMO-interacting motifs. *EMBO Rep.* **2007**, *8*, 550–555. [CrossRef] [PubMed]

63. Liu, G.H.; Gerace, L. Sumoylation regulates nuclear localization of lipin-1alpha in neuronal cells. *PLoS ONE* **2009**, *4*, e7031. [CrossRef] [PubMed]

64. Eifler, K.; Vertegaal, A.C. Mapping the SUMOylated landscape. *FEBS J.* **2015**, *282*, 3669–3680. [CrossRef]

65. Maarifi, G.; Fernandez, J.; Portilho, D.M.; Boulay, A.; Dutrieux, J.; Oddos, S.; Butler-Browne, G.; Nisole, S.; Arhel, N.J. RanBP2 regulates the anti-retroviral activity of TRIM5alpha by SUMOylation at a predicted phosphorylated SUMOylation motif. *Commun. Biol.* **2018**, *1*, 193. [CrossRef]

66. Geiss-Friedlander, R.; Melchior, F. Concepts in sumoylation: A decade on. *Nat. Reviews. Mol. Cell Biol.* **2007**, *8*, 947–956. [CrossRef]

67. Zhu, Z.; Chu, H.; Wen, L.; Yuan, S.; Chik, K.K.; Yuen, T.T.; Yip, C.C.; Wang, D.; Zhou, J.; Yin, F.; et al. Targeting SUMO Modification of the Non-Structural Protein 5 of Zika Virus as a Host-Targeting Antiviral Strategy. *Int. J. Mol. Sci.* **2019**, *20*, 392. [CrossRef]

68. Su, C.I.; Tseng, C.H.; Yu, C.Y.; Lai, M.M.C. SUMO Modification Stabilizes Dengue Virus Nonstructural Protein 5 To Support Virus Replication. *J. Virol.* **2016**, *90*, 4308–4319. [CrossRef]

69. Zhao, Y.; Soh, T.S.; Zheng, J.; Chan, K.W.; Phoo, W.W.; Lee, C.C.; Tay, M.Y.; Swaminathan, K.; Cornvik, T.C.; Lim, S.P.; et al. A crystal structure of the Dengue virus NS5 protein reveals a novel inter-domain interface essential for protein flexibility and virus replication. *PLoS Pathog.* **2015**, *11*, e1004682. [CrossRef]

70. Zhao, B.; Yi, G.; Du, F.; Chuang, Y.C.; Vaughan, R.C.; Sankaran, B.; Kao, C.C.; Li, P. Structure and function of the Zika virus full-length NS5 protein. *Nat. Commun.* **2017**, *8*, 14762. [CrossRef]

71. Godoy, A.S.; Lima, G.M.; Oliveira, K.I.; Torres, N.U.; Maluf, F.V.; Guido, R.V.; Oliva, G. Crystal structure of Zika virus NS5 RNA-dependent RNA polymerase. *Nat. Commun.* **2017**, *8*, 14764. [CrossRef] [PubMed]

72. Lu, G.; Gong, P. Crystal Structure of the full-length Japanese encephalitis virus NS5 reveals a conserved methyltransferase-polymerase interface. *PLoS Pathog.* **2013**, *9*, e1003549. [CrossRef] [PubMed]

73. Upadhyay, A.K.; Cyr, M.; Longenecker, K.; Tripathi, R.; Sun, C.; Kempf, D.J. Crystal structure of full-length Zika virus NS5 protein reveals a conformation similar to Japanese encephalitis virus NS5. *Acta Crystallographica. Sect. Fstructural Biol. Commun.* **2017**, *73*, 116–122. [CrossRef]

74. Wang, B.; Tan, X.F.; Thurmond, S.; Zhang, Z.M.; Lin, A.; Hai, R.; Song, J. The structure of Zika virus NS5 reveals a conserved domain conformation. *Nat. Commun.* **2017**, *8*, 14763. [CrossRef] [PubMed]

75. Bowen, J.R.; Quicke, K.M.; Maddur, M.S.; O'Neal, J.T.; McDonald, C.E.; Fedorova, N.B.; Puri, V.; Shabman, R.S.; Pulendran, B.; Suthar, M.S. Zika Virus Antagonizes Type I Interferon Responses during Infection of Human Dendritic Cells. *PLoS Pathog.* **2017**, *13*, e1006164. [CrossRef] [PubMed]

76. Simonin, Y.; Erkilic, N.; Damodar, K.; Cle, M.; Desmetz, C.; Bollore, K.; Taleb, M.; Torriano, S.; Barthelemy, J.; Dubois, G.; et al. Zika virus induces strong inflammatory responses and impairs homeostasis and function of the human retinal pigment epithelium. *EBIO Med.* **2019**, *39*, 315–331. [CrossRef] [PubMed]

77. Lum, F.M.; Low, D.K.; Fan, Y.; Tan, J.J.; Lee, B.; Chan, J.K.; Renia, L.; Ginhoux, F.; Ng, L.F. Zika Virus Infects Human Fetal Brain Microglia and Induces Inflammation. *Clin. Infect. Dis.* **2017**, *64*, 914–920. [CrossRef]

78. Forrester, J.V.; McMenamin, P.G.; Dando, S.J. CNS infection and immune privilege. *Nat. Reviews. Neurosci.* **2018**, *19*, 655–671. [CrossRef]

79. Teijaro, J.R.; Walsh, K.B.; Rice, S.; Rosen, H.; Oldstone, M.B. Mapping the innate signaling cascade essential for cytokine storm during influenza virus infection. *Proc. Natl. Acad. Sci. USA* **2014**, *111*, 3799–3804. [CrossRef]

80. Srikiatkhachorn, A.; Mathew, A.; Rothman, A.L. Immune-mediated cytokine storm and its role in severe dengue. *Semin. Immunopathol.* **2017**, *39*, 563–574. [CrossRef]

81. Lima, M.C.; de Mendonca, L.R.; Rezende, A.M.; Carrera, R.M.; Anibal-Silva, C.E.; Demers, M.; D'Aiuto, L.; Wood, J.; Chowdari, K.V.; Griffiths, M.; et al. The Transcriptional and Protein Profile From Human Infected Neuroprogenitor Cells Is Strongly Correlated to Zika Virus Microcephaly Cytokines Phenotype Evidencing a Persistent Inflammation in the CNS. *Front. Immunol.* **2019**, *10*, 1928. [CrossRef] [PubMed]

82. Miner, J.J.; Sene, A.; Richner, J.M.; Smith, A.M.; Santeford, A.; Ban, N.; Weger-Lucarelli, J.; Manzella, F.; Ruckert, C.; Govero, J.; et al. Zika Virus Infection in Mice Causes Panuveitis with Shedding of Virus in Tears. *Cell Rep.* **2016**, *16*, 3208–3218. [CrossRef] [PubMed]

83. Miner, J.J.; Cao, B.; Govero, J.; Smith, A.M.; Fernandez, E.; Cabrera, O.H.; Garber, C.; Noll, M.; Klein, R.S.; Noguchi, K.K.; et al. Zika Virus Infection during Pregnancy in Mice Causes Placental Damage and Fetal Demise. *Cell* **2016**, *165*, 1081–1091. [CrossRef] [PubMed]

84. Govero, J.; Esakky, P.; Scheaffer, S.M.; Fernandez, E.; Drury, A.; Platt, D.J.; Gorman, M.J.; Richner, J.M.; Caine, E.A.; Salazar, V.; et al. Zika virus infection damages the testes in mice. *Nature* **2016**, *540*, 438–442. [CrossRef] [PubMed]

85. Ma, W.; Li, S.; Ma, S.; Jia, L.; Zhang, F.; Zhang, Y.; Zhang, J.; Wong, G.; Zhang, S.; Lu, X.; et al. Zika Virus Causes Testis Damage and Leads to Male Infertility in Mice. *Cell* **2016**, *167*, 1511–1524.e1510. [CrossRef] [PubMed]

86. Lazear, H.M.; Govero, J.; Smith, A.M.; Platt, D.J.; Fernandez, E.; Miner, J.J.; Diamond, M.S. A Mouse Model of Zika Virus Pathogenesis. *Cell Host Microbe* **2016**, *19*, 720–730. [CrossRef] [PubMed]

87. Hanners, N.W.; Eitson, J.L.; Usui, N.; Richardson, R.B.; Wexler, E.M.; Konopka, G.; Schoggins, J.W. Western Zika Virus in Human Fetal Neural Progenitors Persists Long Term with Partial Cytopathic and Limited Immunogenic Effects. *Cell Rep.* **2016**, *15*, 2315–2322. [CrossRef]

88. Sun, J.; Wu; Zhong, H.; Guan, D.; Zhang, H.; Tan, Q.; Ke, C. Presence of Zika Virus in Conjunctival Fluid. *Jama Ophthalmol.* **2016**, *134*, 1330–1332. [CrossRef]

89. Prisant, N.; Bujan, L.; Benichou, H.; Hayot, P.H.; Pavili, L.; Lurel, S.; Herrmann, C.; Janky, E.; Joguet, G. Zika virus in the female genital tract. *Lancet. Infect. Dis.* **2016**, *16*, 1000–1001. [CrossRef]

90. Murray, K.O.; Gorchakov, R.; Carlson, A.R.; Berry, R.; Lai, L.; Natrajan, M.; Garcia, M.N.; Correa, A.; Patel, S.M.; Aagaard, K.; et al. Prolonged Detection of Zika Virus in Vaginal Secretions and Whole Blood. *Emerg. Infect. Dis.* **2017**, *23*, 99–101. [CrossRef]

91. Barzon, L.; Pacenti, M.; Franchin, E.; Lavezzo, E.; Trevisan, M.; Sgarabotto, D.; Palu, G. Infection dynamics in a traveller with persistent shedding of Zika virus RNA in semen for six months after returning from Haiti to Italy, January 2016. *Euro Surveill.* **2016**, *21*. [CrossRef] [PubMed]

92. Mansuy, J.M.; Dutertre, M.; Mengelle, C.; Fourcade, C.; Marchou, B.; Delobel, P.; Izopet, J.; Martin-Blondel, G. Zika virus: High infectious viral load in semen, a new sexually transmitted pathogen? *Lancet. Infect. Dis.* **2016**, *16*, 405. [CrossRef]

93. Tang, H.; Hammack, C.; Ogden, S.C.; Wen, Z.; Qian, X.; Li, Y.; Yao, B.; Shin, J.; Zhang, F.; Lee, E.M.; et al. Zika Virus Infects Human Cortical Neural Progenitors and Attenuates Their Growth. *Cell Stem Cell* **2016**, *18*, 587–590. [CrossRef] [PubMed]

94. Garcez, P.P.; Loiola, E.C.; Madeiro da Costa, R.; Higa, L.M.; Trindade, P.; Delvecchio, R.; Nascimento, J.M.; Brindeiro, R.; Tanuri, A.; Rehen, S.K. Zika virus impairs growth in human neurospheres and brain organoids. *Sci. (New York)* **2016**, *352*, 816–818. [CrossRef] [PubMed]

95. Bhatnagar, J.; Rabeneck, D.B.; Martines, R.B.; Reagan-Steiner, S.; Ermias, Y.; Estetter, L.B.; Suzuki, T.; Ritter, J.; Keating, M.K.; Hale, G.; et al. Zika Virus RNA Replication and Persistence in Brain and Placental Tissue. *Emerg. Infect. Dis.* **2017**, *23*, 405–414. [CrossRef]

96. Kumar, A.; Jovel, J.; Lopez-Orozco, J.; Limonta, D.; Airo, A.M.; Hou, S.; Stryapunina, I.; Fibke, C.; Moore, R.B.; Hobman, T.C. Human Sertoli cells support high levels of Zika virus replication and persistence. *Sci. Rep.* **2018**, *8*, 5477. [CrossRef]

97. Caine, E.A.; Scheaffer, S.M.; Broughton, D.E.; Salazar, V.; Govero, J.; Poddar, S.; Osula, A.; Halabi, J.; Skaznik-Wikiel, M.E.; Diamond, M.S.; et al. Zika virus causes acute infection and inflammation in the ovary of mice without apparent defects in fertility. *J. Infect. Dis.* **2019**. [CrossRef]

98. Miner, J.J.; Diamond, M.S. Zika Virus Pathogenesis and Tissue Tropism. *Cell Host Microbe* **2017**, *21*, 134–142. [CrossRef]

99. Aid, M.; Abbink, P.; Larocca, R.A.; Boyd, M.; Nityanandam, R.; Nanayakkara, O.; Martinot, A.J.; Moseley, E.T.; Blass, E.; Borducchi, E.N.; et al. Zika Virus Persistence in the Central Nervous System and Lymph Nodes of Rhesus Monkeys. *Cell* **2017**, *169*, 610–620.e614. [CrossRef]

Review

Protective to a T: The Role of T Cells during Zika Virus Infection

Ryan D. Pardy [1,2] and Martin J. Richer [1,2,*]

[1] Department of Microbiology & Immunology, McGill University, Montreal, QC H3A 2B4, Canada
[2] Rosalind & Morris Goodman Cancer Research Centre, McGill University, Montreal, QC H3G 1Y6, Canada
* Correspondence: martin.j.richer@mcgill.ca; Tel.: +1-514-398-4400 (ext. 00538)

Received: 4 July 2019; Accepted: 2 August 2019; Published: 3 August 2019

Abstract: CD4 and CD8 T cells are an important part of the host's capacity to defend itself against viral infections. During *flavivirus* infections, T cells have been implicated in both protective and pathogenic responses. Given the recent emergence of Zika virus (ZIKV) as a prominent global health threat, the question remains as to how T cells contribute to anti-ZIKV immunity. Furthermore, high homology between ZIKV and other, co-circulating *flaviviruses* opens the possibility of positive or negative effects of cross-reactivity due to pre-existing immunity. In this review, we will discuss the CD4 and CD8 T cell responses to ZIKV, and the lessons we have learned from both mouse and human infections. In addition, we will consider the possibility of whether T cells, in the context of *flavivirus*-naïve and *flavivirus*-immune subjects, play a role in promoting ZIKV pathogenesis during infection.

Keywords: zika virus; flaviviruses; T cells; host-pathogen interactions

---

1. Introduction

T cell responses represent a crucial aspect of the adaptive immune response to infection. In the context of viral infections, both CD4 and CD8 T cells play important roles in controlling and clearing the pathogen. CD4 T cells (or helper T cells) support the immune response through the production of effector cytokines such as interferon (IFN)-γ and tumor necrosis factor (TNF)-α, licensing dendritic cells (DCs) to promote activation of CD8 T cell responses, and activating humoral immunity [1,2]. Meanwhile, CD8 T cells (or cytotoxic T cells) are capable of directly killing infected cells in addition to producing effector cytokines, which makes them critical for controlling viral infections [3]. In addition, both T cell pools are capable of generating long-lived memory populations in order to rapidly respond to re-infection and provide greater protection [3]. While T cell responses during *flavivirus* infections have been shown to be protective, they have also been implicated in pathogenic responses [4]. For example, in mice lacking B cells, CD8 T cells were shown to be critical for controlling yellow fever virus (YFV) infection [5]. In contrast, CD8 T cell infiltration has been associated with increased tissue damage and neurological symptoms in mouse models of Japanese encephalitis virus (JEV) and West Nile virus (WNV) infection [6]. This illustrates that, as is often the case for T cells, T cell responses to *flavivirus* infections must strike a balance between viral control and immunopathology.

Although it was first isolated in 1947, significant research into Zika virus (ZIKV) only began relatively recently [7]. This is primarily due to the fact that it caused only a handful of isolated infections, inducing a mild febrile illness, from its initial isolation until the 21st century [8,9]. However, a series of recent outbreaks in Yap Island, Federated States of Micronesia (2007); French Polynesia (2013); South and Central America (with other outbreaks world-wide; 2015–2016); and India (2018) have demonstrated a novel epidemic capacity for ZIKV [10–14]. Even more striking were novel neurological symptoms associated with ZIKV infection, particularly following the French Polynesian and South and Central American outbreaks [8]. ZIKV has been identified as a potential trigger for Guillain-Barré syndrome (GBS), an autoimmune ascending paralysis that sometimes follows infection [8,15]. However, the most dramatic symptom now associated with ZIKV infection is fetal microcephaly, a neurodevelopmental defect that can cause lifelong complications for newborns [8]. These symptoms—and the outbreaks they were a part—represent a striking change in phenotype for a virus that caused only mild symptoms in its initial characterizations [16,17].

In response to these recent outbreaks and the novel neurological symptoms associated with infection, there has been significant progress in improving our understanding of T cell responses to ZIKV. Broad characterizations of T cell responses induced by ZIKV in humans and mice, including the epitopes of the virus to which they respond, have helped demonstrate protective roles for T cells. These studies have been complemented by cases in which T cell responses have pathogenic consequences for the host. Finally, given the similarities between ZIKV and Dengue virus (DENV), a number of studies have compared T cell responses directed against these two *flaviviruses* to determine whether they are cross-protective or pathogenic. In this review, we will summarize the current understanding of T cell responses during ZIKV infection and the models used to investigate these responses.

2. Profiling the T Cell Response to ZIKV Infection

2.1. T Cell Responses in Mice

A variety of mouse models have been used to interrogate T cell responses to ZIKV infection. Initially, most models used immunocompromised mice, which typically involved genetic deletion of the IFN-α/β receptor (IFNAR) either globally or in a subset of myeloid cells (LysMCre^{+}IFNAR$^{fl/fl}$ mice), or treating with an anti-IFNAR blocking antibody prior to infection [18–22]. The primary lesson from these models is the importance of type I IFN signaling in anti-ZIKV immunity. However, it is also important to consider the impact of IFN deficiency in the context of studying T cell responses to ZIKV. Type I IFNs play a crucial role in promoting the activation of both CD4 and CD8 T cells and are particularly important for enhancing CD8 T cell accumulation and antigen sensitivity [23–26]. Thus, immunocompetent mouse models represent a very useful tool for characterizing and understanding the CD4 and CD8 T cell responses to ZIKV infection.

Our group and others have demonstrated that, in immunocompetent mice, ZIKV establishes a self-limiting infection with transient mild weight loss as the only discernible symptom of infection [27,28]. However, infection induces a robust Th1 CD4 T cell response, which features expression of the transcription factor T-bet and production of effector cytokines IFN-γ, TNF-α, and interleukin (IL)-2 [27]. Furthermore, CD8 T cells upregulate expression of IFN-γ and TNF-α, produce the cytolytic molecule granzyme B, and present a highly activated phenotype following ZIKV infection [27,28]. Expansion of this antigen-experienced CD8 T cell population correlated with increased transcripts of type I IFNs [27]. No activation of CD4 or CD8 T cell responses was observed when mice were immunized with UV-inactivated virus, indicating that active infection with live ZIKV is required for the generation of CD4 and CD8 T cell-mediated immunity [27]. These models were used to identify an immunodominant CD8 T cell epitope in the ZIKV envelope protein, highlighting the specificity of the approaches used to quantify and characterize the T cell responses to ZIKV infection (Figure 1) [27,28]. An additional study using intracranial infection of immunocompetent mice described a functional role for T cells during ZIKV infection. When mice were infected intravenously, followed by intracranial infection four weeks later, they were protected from the high central nervous system (CNS) viral load and severe disease observed in mice that were only infected intracranially [29]. However, this protection was lost in T cell-deficient mice, demonstrating a key role for T cells in controlling intracranial ZIKV infection and pathology [29]. Together, these studies demonstrate that ZIKV actively infects immunocompetent mice, generating a robust and functional CD4 and CD8 T cell response to infection.

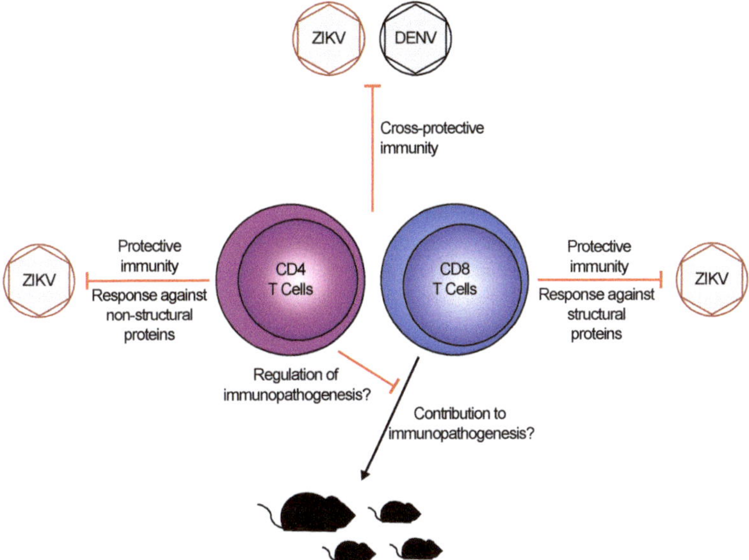

Figure 1. Zika virus (ZIKV) infection induces robust, cross-protective T cell immunity. In both humans and mice, ZIKV infection leads to the generation of Th1 CD4 T cell and effector CD8 T cell responses, which preferentially target epitopes in non-structural and structural proteins, respectively. Studies have shown that immunity to ZIKV is cross-protective against subsequent Dengue virus (DENV) challenge, and vice-versa. Although studies suggest CD8 T cells may contribute to immunopathogenesis in neonatal and adult mice, with CD4 T cells playing a potential regulatory role, this remains to be determined during human infection.

Similar findings have been observed in LysMCre$^+$IFNAR$^{fl/fl}$ mice, which lack IFNAR in mature macrophages and granulocytes, with a partial deletion in CD11c$^+$ splenic dendritic cells [18,30,31]. In this model, ZIKV infection caused an increased frequency of activated CD8 T cells in the spleen, which were shown to be positive for granzyme B [18]. This model was used to identify several CD8 T cell epitopes, including the aforementioned immunodominant epitope in the ZIKV envelope protein [18,27,28]. When CD8 T cells were depleted in this model, mice had higher viral burdens in the serum, CNS, and other tissues. This was reversed when CD8 T cell-depleted mice received a transfer of memory CD8 T cells [18]. Further studies in LysMCre$^+$IFNAR$^{fl/fl}$ mice have described a similar Th1 CD4 T cell response to what was observed in immunocompetent mice, as well as a T follicular helper (Tfh) cell response from 7 days post-infection (dpi) onward [19]. CD4 T cells were required for the generation of an immunoglobulin G (IgG) antibody response, but their depletion had no impact on the CD8 T cell response, and nor did it impact viral burden [19]. Thus, although IFNAR expression on certain myeloid cells is not strictly required for the generation of T cell responses, the impact its deletion has on T cell accumulation is unclear as only frequencies were reported in these studies. Similarly, it is unknown whether CD8 T cells in LysMCre$^+$IFNAR$^{fl/fl}$ mice are functionally impaired since analyses of CD8 T cell capacity to produce IFN-γ, killing capacity, or antigen sensitivity were either not undertaken or not compared to wild-type (WT) mice [18]. It also warrants further investigation into whether there was an impact on the generation of antigen-specific T cell responses since only the immunodominant CD8 T cell epitope in the ZIKV envelope protein is shared between this study and other studies [18,27,28].

Although IFNAR deficient mice are known to be susceptible to ZIKV infection [32–34], depleting CD4 T cells from 10–12 week old IFNAR knock-out (KO) mice caused higher viral loads, more severe paralysis, and reduced survival [20], and caused lethal infection in 3-4 week old IFNAR KO mice [21].

Each of these studies found that transferring memory CD4 T cells from ZIKV-immune mice, but not naïve mice, was protective against a subsequent lethal ZIKV challenge [20,21]. Lucas et al. found this protection to be dependent on IFN-γ signaling and B cells in the recipient mice, which suggests that CD4 T cells were important for promoting B cell and antibody responses against ZIKV (Figure 1) [21]. Similarly, ZIKV infection in CD8 T cell depleted, IFNAR-deficient mice was lethal, as was infection of IFNAR-blocking antibody-treated *Rag1*$^{-/-}$ mice (which lack both T and B cells) [21,35]. In the latter case, WT control mice treated with the IFNAR blocking antibody do not succumb to infection [35], demonstrating the importance of adaptive immunity in protection from lethal ZIKV infection. Finally, antibody-mediated blocking of IFNAR, followed by intravaginal infection, enables the virus to spread systemically despite increased frequency of tetramer-specific CD8 T cells and total CD4 T cells in the lower female reproductive tract [22]. Depleting both CD4 and CD8 T cells together, but not individually, led to a loss of viral control, suggesting that each subset is able to compensate for the loss of the other [22]. Thus, even in IFN-deficient mouse models, CD4 and CD8 T cells continue to play an important role in the immune response to ZIKV.

In all, these studies demonstrate the importance of robust CD4 and CD8 T cell responses during ZIKV infection. Broad characterizations of the T cell response in immunocompetent mice will serve as an important baseline, to which the T cell response to contemporary ZIKV isolates may be compared. Differences in T cell responses induced by epidemic strains of ZIKV could improve our understanding of how the virus has changed and whether this has had an impact on its pathogenesis. Although the studies in immunocompromised mice have highlighted the importance of T cell responses in these circumstances, human infection with ZIKV is rarely, if ever, fatal. Therefore, the findings from these models must be analyzed under the prism that such severe phenotypes are rarely observed during the course of natural infections in humans.

2.2. T Cell Responses in Humans

One focus of the limited number of studies characterizing human T cell responses to ZIKV has been to identify immunogenic epitopes and their locations within the ZIKV proteome. In a cohort of 45 American patients with confirmed ZIKV infection, highly polyfunctional CD4 and CD8 T cells responses were detected following stimulation with pools of 15mer peptides (overlapping by 11 peptides) from all ZIKV proteins [36]. They found that although 89% of patients' CD4 T cells responded to peptides from the capsid and envelope proteins, the most robust IFN-γ production was following stimulation with peptides from the non-structural (NS)1, NS3, and NS5 proteins [36]. Conversely, CD8 T cell responses against the NS3, NS4B, and NS5 proteins were detected in most patients, but the most robust IFN-γ responses were against the capsid and envelope proteins [36]. Similarly, a case report from Florida identified NS2-specific CD4 T cells and envelope-specific CD8 T cells in a returning traveler with ZIKV infection [37]. A case series featuring five returning American travelers with ZIKV infection identified very modest, but detectable CD4 and CD8 T cell responses (<1% cytokine-producing among total CD4 or CD8 T cells) against pooled peptides from the capsid, pre-membrane, envelope, and NS5 proteins, although no other proteins were tested [38]. Two additional case reports from the same group found consistent CD4 T cell responses against NS1, NS3, and NS5 proteins, and CD8 T cell responses against envelope, NS3, and NS5 proteins [39,40]. In all, a common theme of these characterizations is a tendency for CD8 T cells to respond to structural proteins (primarily capsid and envelope proteins), and for CD4 T cells to respond to NS proteins (mainly NS1, NS3, and NS5 proteins; Figure 1). It is interesting to note that these results are reflective of the results found in mouse studies, although it is unlikely that the epitope peptide sequences would be the same [18,19,27,28].

The other question often addressed by human studies relates to how T cell responses to ZIKV change over time. In the case report from Florida described above, T cell responses appeared 7 days post-onset of symptoms (POS), peaked 21 days POS, and memory T cell responses were detectable as late as 148 days POS [37]. Similarly, the T cell responses described in the 45-patient American cohort were tracked into memory time points as late as 10 to 12 months POS [36]. One study took an alternative approach, tracking cytokine responses and cellular dynamics in the blood over time [41]. In this cohort of 55 Singaporean patients, viremic patients in the acute phase of infection had reduced numbers of immune cells in their blood (including CD4 and CD8 T cells), and higher production of IFN-γ [41]. T cell cytokines IFN-γ and IL-12, and chemokine CCL5 (also known as Regulated upon Activation, Normal T cell Expressed, and Secreted, or RANTES) were maintained into the convalescent phase of infection (10–35 days POS) [41]. Interestingly, the authors found significantly more CD4 and CD8 T cells in the blood of non-viremic patients with moderate symptoms when compared to viremic patients with moderate symptoms, suggesting a possible involvement of T cell responses in the clearance of viremia [41]. Finally, in an Italian cohort, ZIKV infection activated both CD4 and CD8 T cells, but only CD4 T cells acquired an effector memory phenotype when compared to CD4 T cells from healthy controls [42]. This study also identified an increase in granzyme B-producing, double-negative T cells, which expressed the Vδ2 T cell receptor [42]. The authors highlight this observation because Vδ2 T cells have been implicated in recurrent abortions, although they have never been associated with ZIKV-induced fetal complications [42]. The overarching trend, however, is for ZIKV to induce long-lasting CD4 and CD8 T cell responses which span both structural and NS proteins (Figure 1).

3. T Cell Responses to DENV and ZIKV: Cross-Protective or Pathogenic?

Given the similarity between DENV and ZIKV and their shared regions of endemicity, obvious questions have emerged as to whether immunity to one virus can cross-protect against the other. However, immune responses against distinct DENV serotypes have also been suggested to worsen disease outcomes through distinct B- and T-cell dependent mechanisms. In the context of B cells, antibody-dependent enhancement (ADE) occurs when sub-neutralizing antibody responses generated during a primary infection recognize and bind to cross-reactive epitopes from the secondary, heterologous infection. When the virus particles and sub-neutralizing antibodies are subsequently internalized via Fc receptor-mediated endocytosis, the virus remains capable of replicating within the cell [43]. ADE has been associated with reduced antiviral responses and may enable DENV to infect cells that are normally non-permissive to infection [43]. With T cells, original antigen sin (OAS) occurs when a heterologous secondary infection activates memory T cells that recognize similar but distinct antigens that were present during the primary infection. The result is that these memory T cells mount an ineffective response against the secondary infection, which prevents a more effective T cell response from being generated [44]. As such, significant research has focused on the impact of prior DENV immunity on the immune response to ZIKV or the impact of ZIKV immunity on the immune response to DENV infection.

The majority of human studies of cross-reactivity have investigated the ability of DENV- or ZIKV-derived peptides to restimulate T cells from the heterologous infection. As a whole, these papers consistently identify cross-reactivity in both CD4 and CD8 T cell responses, targeting a variety of viral proteins (Figure 1) [45–49]. In particular, one study identified capsid and envelope protein-specific CD4 T cell responses following ZIKV infection [46]. Upon comparing the peptide sequences of the epitopes to previously identified epitopes in the capsid and envelope proteins from YFV, DENV, and tick-borne encephalitis virus (TBEV), they found that, while the epitopes were all located in similar regions of the proteins, surprisingly they did not share similar sequence identity. As such, this suggests that sequence identity was not the driving factor in the conservation of these epitopes [46]. Functionally, studies have also shown that prior DENV immunity has no impact on the ability of CD4 or CD8 T cells to produce IFN-γ or TNF-α [50]. However, patients co-infected with both DENV and ZIKV had a slight decrease in the frequency of IFN-γ^+ or TNF-α^+ CD4 T cells and similar frequencies of IFN-γ^+ or TNF-α^+ CD8 T cells compared to patients infected with DENV or ZIKV alone, although the implications of this finding remain unclear [50]. Another group found that while prior DENV exposure had no impact on the CD4 T cell response to ZIKV infection, patients in the acute phase of ZIKV infection had more IFN-γ^+ CD8 T cells after restimulation with ZIKV-derived peptides [47].

Further, the CD8 T cell response to ZIKV in DENV-naïve patients was more targeted to structural proteins, while DENV-immune patients' responses were shifted toward the NS proteins [47]. However, prior DENV immunity had no impact on either the transcriptional profile of the CD8 T cells, nor the capacity of CD4 or CD8 T cells to produce IFN-γ at later stages of infection [47,51]. Together, these studies demonstrate a high degree of cross-reactivity between T cell responses to DENV and ZIKV and, thus far, have given no indication that this may have a negative impact on disease outcomes.

Mouse models have been particularly useful for determining the impact of prior DENV or ZIKV immunity on infection outcomes and disease severity. For example, transgenic mice expressing human leukocyte antigen (HLA) class I were crossed with IFNAR KO mice to identify ZIKV-derived peptides that could contribute to human CD8 T cell responses [52]. Several of these peptides were found to restimulate CD8 T cells from DENV-infected mice, and immunization with the ZIKV/DENV cross-reactive peptides induced a CD8 T cell response that reduced ZIKV viral loads in the serum and brain [52]. Similarly, HLA class II transgenic mice were used to identify several potential CD4 T cell epitopes, many of which had homologous and cross-reactive sequences in DENV, WNV, and YFV [53]. Further evidence of a cross-protective role for CD8 T cells was described by transferring memory CD8 T cells isolated from DENV-immune IFNAR KO mice to naïve IFNAR KO mice, which provided protection from ZIKV-induced weight loss, morbidity, and mortality [54]. Finally, one study has analyzed the impact of prior *flavivirus* immunity during pregnancy [55]. They observed that DENV-immune pregnant females had lower ZIKV viral loads in the spleen and placenta and were completely protected from fetal resorption following ZIKV infection [55]. This protection was dependent on the presence of DENV-specific memory CD8 T cells, and their depletion resulted in higher viral loads in maternal tissues and fetal resorption [55]. There is clearly a high degree of cross-reactivity between the T cell responses against ZIKV and related *flaviviruses*, which may play an important role in host protection, particularly in the context of pregnancy. However, the ZIKV outbreak in South and Central America represented the introduction of ZIKV to an area where DENV and other *flaviviruses* already circulate. It is reasonable to assume that a high degree of prior *flavivirus* immunity exists in these populations, yet fetal microcephaly was one of the most striking symptoms associated with this outbreak [8]. It is also worth noting that the above study took place in IFNAR KO dams, raising the question of how the loss of a key part of the antiviral immune system impacts the observed phenotype. Further research is therefore needed to reconcile the protection from fetal pathology observed in DENV-immune dams with the increase in fetal microcephaly observed during the South American outbreak.

4. Role of T Cells in ZIKV Pathogenesis

Given the ability of CD8 T cells to directly kill infected cells, and the capacity of both CD4 and CD8 T cells to produce effector cytokines and cytolytic molecules, it is intriguing to ask whether the T cell response induced by ZIKV could contribute to pathogenesis. Immunopathogenic T cell responses have been described primarily in the context of influenza A virus infection, during which T cell infiltration and TNF-α production promote viral clearance while also contributing to lung pathology [56]. Similarly, Schmidt and colleagues observed exacerbated morbidity and mortality when mice with memory CD8 T cells (but no memory CD4 T cells or antibodies) against respiratory syncytial virus (RSV) were re-challenged with RSV [57]. CD8 T cell responses to TBEV have been linked to reduced survival in mouse studies, and increased cell death in infected human neuronal tissue [58,59]. In the context of ZIKV infection, a pair of studies took the approach of infecting WT neonatal mice (one day old), which caused chorioretinal lesions and neuronal degeneration [60,61]. This tissue-damaging phenotype correlated with CD8 T cell infiltration into the eye and CNS, respectively, suggesting a potential role for CD8 T cells in mediating the damage (Figure 1) [60,61]. These observations are reflective of several studies that have examined ZIKV infection in neonatal mice, which leads to central nervous system infection and pathology, and is often fatal [32,62–64]. Although this may provide some insight into pathogenesis in fetuses or infants, neonatal mice do not possess a fully mature immune system and are highly susceptible to infection with neurotropic viruses [65,66]. As such, using this model to analyze immune responses must be undertaken with these caveats in mind.

Using IFNAR KO mice, Jurado et al. have described a clearer role for CD8 T cells in causing brain damage and paralysis during ZIKV infection [67]. They found significant CD8 T cell infiltration into the brain following ZIKV infection, which correlated with paralysis and death of all mice by 9 dpi [67]. Upon depletion of CD8 T cells, they observed reduced paralysis and improved survival, although mice had higher viral loads in the brain, suggesting that while CD8 T cells are important for viral clearance, they can cause significant immunopathology in the brain (Figure 1) [67]. When both CD4 and CD8 T cells were depleted, mice had an intermediate phenotype with a significant decrease in survival, suggesting a regulatory role for CD4 T cells in the brain (Figure 1) [67]. This potential regulatory role is supported by depletion of CD4 T cells alone, which caused all mice to develop paralysis and succumb to infection [67]. Another group observed that CD4 T cell depletion had no impact on the CD8 T cell response, but caused significantly higher viral loads in the CNS, worsened disease, and decreased survival [20]. Together, these studies implicate CD8 T cells in contributing to ZIKV pathogenesis in immune-privileged sites, while they describe a potential regulatory role for CD4 T cells.

5. Conclusions

Since the beginning of the South American ZIKV outbreak, significant research has been conducted to further our understanding of T cell responses to ZIKV infection. In both humans and mouse models of infection, ZIKV induces robust T cell activation, which leads to the establishment of a memory T cell population, suggesting an important role for CD4 and CD8 T cells in the immune response to ZIKV. This is highlighted by depletion studies, in which loss of either CD4, CD8, or both T cell subsets together can result in worsened morbidity, mortality, or even fetal resorption. Identification of ZIKV epitopes, in particular broadly conserved epitopes between studies, and even among *flaviviruses*, could provide novel candidates for vaccine design. Given that the work done so far studying T cell cross-reactivity has demonstrated a protective role for these cells, it stands to reason that cross-reactive epitopes could be useful in vaccination against multiple, co-circulating *flaviviruses*. However, this remains to be formally tested, and the magnitude of the South and Central American ZIKV outbreak in a DENV endemic region suggests that prior DENV immunity may not provide complete protection. Finally, there may be a role for CD8 T cells in enhancing ZIKV pathogenesis, although thus far studies have been completed uniquely in extremely young or immunocompromised mice, raising questions as to whether CD8 T cells also play a role in ZIKV pathogenesis in healthy, immunocompetent adults. In the future, it will be of importance to continue to explore the impact of prior immunity to *flaviviruses* during pregnancy. Further research is also needed to understand whether ZIKV has improved its capacity to evade host immune responses, including T cell-mediated immunity, and whether this has contributed to the increased pathogenesis observed during recent outbreaks.

Author Contributions: Conceptualization, R.D.P. and M.J.R.; resources, M.J.R.; writing—original draft preparation, R.D.P.; writing—review and editing, R.D.P. and M.J.R.; visualization, R.D.P. and M.J.R.; supervision, M.J.R.; project administration, M.J.R.; funding acquisition, M.J.R.

Funding: The Richer Lab is funded by the Canadian Institutes of Health Research, grant number PJT-152903 and PJT-162212; the Natural Sciences and Engineering Research Council, grant number RGPIN-2016-04713; Fonds de Recherche du Québec—Santé, grant number 32807; and the Canada Foundation for Innovation. R.D.P is supported by the Frederick Banting and Charles Best Canada Graduate Scholarships—Doctoral Award from the Canadian Institutes of Health Research.

Acknowledgments: We would like to thank Stephanie Condotta, Marilena Gentile, and Stefanie Valbon for critical review of the manuscript.

Conflicts of Interest: The authors declare no conflict of interest.

References

1. Luckheeram, R.V.; Zhou, R.; Verma, A.D.; Xia, B. CD4[+] T cells: Differentiation and functions. *Clin. Dev. Immunol.* **2012**, *2012*, 925135. [CrossRef] [PubMed]
2. Smith, C.M.; Wilson, N.S.; Waithman, J.; Villadangos, J.A.; Carbone, F.R.; Heath, W.R.; Belz, G.T. Cognate CD4[+] T cell licensing of dendritic cells in CD8[+] T cell immunity. *Nat. Immunol* **2004**, *5*, 1143–1148. [CrossRef] [PubMed]
3. Valbon, S.F.; Condotta, S.A.; Richer, M.J. Regulation of effector and memory CD8[+] T cell function by inflammatory cytokines. *Cytokine* **2016**, *82*, 16–23. [CrossRef] [PubMed]
4. Slon Campos, J.L.; Mongkolsapaya, J.; Screaton, G.R. The immune response against flaviviruses. *Nat. Immunol.* **2018**, *19*, 1189–1198. [CrossRef] [PubMed]
5. Bassi, M.R.; Kongsgaard, M.; Steffensen, M.A.; Fenger, C.; Rasmussen, M.; Skjodt, K.; Finsen, B.; Stryhn, A.; Buus, S.; Christensen, J.P.; et al. CD8[+] T cells complement antibodies in protecting against yellow fever virus. *J. Immunol.* **2015**, *194*, 1141–1153. [CrossRef] [PubMed]
6. Wang, Y.; Lobigs, M.; Lee, E.; Mullbacher, A. CD8[+] T cells mediate recovery and immunopathology in West Nile virus encephalitis. *J. Virol.* **2003**, *77*, 13323–13334. [CrossRef] [PubMed]
7. Dick, G.W.; Kitchen, S.F.; Haddow, A.J. Zika virus. I. Isolations and serological specificity. *Trans. R. Soc. Trop. Med. Hyg.* **1952**, *46*, 509–520. [CrossRef]
8. Rajah, M.M.; Pardy, R.D.; Condotta, S.A.; Richer, M.J.; Sagan, S.M. Zika Virus: Emergence, Phylogenetics, Challenges, and Opportunities. *ACS Infect. Dis.* **2016**, *2*, 763–772. [CrossRef]
9. Kindhauser, M.K.; Allen, T.; Frank, V.; Santhana, R.S.; Dye, C. Zika: The origin and spread of a mosquito-borne virus. *Bull. World Health Organ.* **2016**, *94*, 675–686. [CrossRef]
10. Duffy, M.R.; Chen, T.H.; Hancock, W.T.; Powers, A.M.; Kool, J.L.; Lanciotti, R.S.; Pretrick, M.; Marfel, M.; Holzbauer, S.; Dubray, C.; et al. Zika virus outbreak on Yap Island, Federated States of Micronesia. *N. Engl. J. Med.* **2009**, *360*, 2536–2543. [CrossRef]
11. Oehler, E.; Watrin, L.; Larre, P.; Leparc-Goffart, I.; Lastere, S.; Valour, F.; Baudouin, L.; Mallet, H.; Musso, D.; Ghawche, F. Zika virus infection complicated by Guillain-Barre syndrome—Case report, French Polynesia, December 2013. *Eurosurveillance* **2014**, *19*, 20720. [CrossRef]
12. The History of Zika Virus. Available online: http://www.who.int/emergencies/zika-virus/history/en/ (accessed on 16 May 2019).
13. Swati, G. *Zika Spreads Rapidly in India, with 94 Cases Confirmed*; CNN: Atlanta, GA, USA, 2018.
14. Slater, J. *India Wrestles with First Significant Outbreak of Zika Virus*; The Washington Post: Washington, DC, USA, 2018.
15. Krauer, F.; Riesen, M.; Reveiz, L.; Oladapo, O.T.; Martinez-Vega, R.; Porgo, T.V.; Haefliger, A.; Broutet, N.J.; Low, N.; Group, W.H.O.Z.C.W. Zika Virus Infection as a Cause of Congenital Brain Abnormalities and Guillain-Barre Syndrome: Systematic Review. *PLoS Med.* **2017**, *14*, e1002203. [CrossRef]
16. Dick, G.W. Zika virus. II. Pathogenicity and physical properties. *Trans. R Soc. Trop. Med. Hyg.* **1952**, *46*, 521–534. [CrossRef]
17. Simpson, D.I. Zika Virus Infection in Man. *Trans. R. Soc. Trop. Med. Hyg.* **1964**, *58*, 335–338. [CrossRef]
18. Elong Ngono, A.; Vizcarra, E.A.; Tang, W.W.; Sheets, N.; Joo, Y.; Kim, K.; Gorman, M.J.; Diamond, M.S.; Shresta, S. Mapping and Role of the CD8[+] T Cell Response During Primary Zika Virus Infection in Mice. *Cell Host Microbe* **2017**, *21*, 35–46. [CrossRef]
19. Elong Ngono, A.; Young, M.P.; Bunz, M.; Xu, Z.; Hattakam, S.; Vizcarra, E.; Regla-Nava, J.A.; Tang, W.W.; Yamabhai, M.; Wen, J.; et al. CD4[+] T cells promote humoral immunity and viral control during Zika virus infection. *PLoS Pathog.* **2019**, *15*, e1007474. [CrossRef]
20. Hassert, M.; Wolf, K.J.; Schwetye, K.E.; DiPaolo, R.J.; Brien, J.D.; Pinto, A.K. CD4[+] T cells mediate protection against Zika associated severe disease in a mouse model of infection. *PLoS Pathog* **2018**, *14*, e1007237. [CrossRef]
21. Lucas, C.G.O.; Kitoko, J.Z.; Ferreira, F.M.; Suzart, V.G.; Papa, M.P.; Coelho, S.V.A.; Cavazzoni, C.B.; Paula-Neto, H.A.; Olsen, P.C.; Iwasaki, A.; et al. Critical role of CD4[+] T cells and IFNgamma signaling in antibody-mediated resistance to Zika virus infection. *Nat. Commun* **2018**, *9*, 3136. [CrossRef]

22. Scott, J.M.; Lebratti, T.J.; Richner, J.M.; Jiang, X.; Fernandez, E.; Zhao, H.; Fremont, D.H.; Diamond, M.S.; Shin, H. Cellular and Humoral Immunity Protect against Vaginal Zika Virus Infection in Mice. *J. Virol.* **2018**, *92*. [CrossRef]

23. Kolumam, G.A.; Thomas, S.; Thompson, L.J.; Sprent, J.; Murali-Krishna, K. Type I interferons act directly on CD8 T cells to allow clonal expansion and memory formation in response to viral infection. *J. Exp. Med.* **2005**, *202*, 637–650. [CrossRef]

24. Curtsinger, J.M.; Mescher, M.F. Inflammatory cytokines as a third signal for T cell activation. *Curr. Opin. Immunol.* **2010**, *22*, 333–340. [CrossRef]

25. Richer, M.J.; Nolz, J.C.; Harty, J.T. Pathogen-specific inflammatory milieux tune the antigen sensitivity of CD8$^+$ T cells by enhancing T cell receptor signaling. *Immunity* **2013**, *38*, 140–152. [CrossRef]

26. Longhi, M.P.; Trumpfheller, C.; Idoyaga, J.; Caskey, M.; Matos, I.; Kluger, C.; Salazar, A.M.; Colonna, M.; Steinman, R.M. Dendritic cells require a systemic type I interferon response to mature and induce CD4$^+$ Th1 immunity with poly IC as adjuvant. *J. Exp. Med.* **2009**, *206*, 1589–1602. [CrossRef]

27. Pardy, R.D.; Rajah, M.M.; Condotta, S.A.; Taylor, N.G.; Sagan, S.M.; Richer, M.J. Analysis of the T Cell Response to Zika Virus and Identification of a Novel CD8$^+$ T Cell Epitope in Immunocompetent Mice. *PLoS Pathog.* **2017**, *13*, e1006184. [CrossRef]

28. Huang, H.; Li, S.; Zhang, Y.; Han, X.; Jia, B.; Liu, H.; Liu, D.; Tan, S.; Wang, Q.; Bi, Y.; et al. CD8$^+$ T Cell Immune Response in Immunocompetent Mice during Zika Virus Infection. *J. Virol.* **2017**, *91*. [CrossRef]

29. Nazerai, L.; Scholler, A.S.; Rasmussen, P.O.S.; Buus, S.; Stryhn, A.; Christensen, J.P.; Thomsen, A.R. A New In Vivo Model to Study Protective Immunity to Zika Virus Infection in Mice With Intact Type I Interferon Signaling. *Front. Immunol.* **2018**, *9*, 593. [CrossRef]

30. Clausen, B.E.; Burkhardt, C.; Reith, W.; Renkawitz, R.; Forster, I. Conditional gene targeting in macrophages and granulocytes using LysMcre mice. *Transgenic Res.* **1999**, *8*, 265–277. [CrossRef]

31. Diamond, M.S.; Kinder, M.; Matsushita, H.; Mashayekhi, M.; Dunn, G.P.; Archambault, J.M.; Lee, H.; Arthur, C.D.; White, J.M.; Kalinke, U.; et al. Type I interferon is selectively required by dendritic cells for immune rejection of tumors. *J. Exp. Med.* **2011**, *208*, 1989–2003. [CrossRef]

32. Lazear, H.M.; Govero, J.; Smith, A.M.; Platt, D.J.; Fernandez, E.; Miner, J.J.; Diamond, M.S. A Mouse Model of Zika Virus Pathogenesis. *Cell Host Microbe* **2016**. [CrossRef]

33. Rossi, S.L.; Tesh, R.B.; Azar, S.R.; Muruato, A.E.; Hanley, K.A.; Auguste, A.J.; Langsjoen, R.M.; Paessler, S.; Vasilakis, N.; Weaver, S.C. Characterization of a Novel Murine Model to Study Zika Virus. *Am. J. Trop Med. Hyg.* **2016**, *94*, 1362–1369. [CrossRef]

34. Dowall, S.D.; Graham, V.A.; Rayner, E.; Atkinson, B.; Hall, G.; Watson, R.J.; Bosworth, A.; Bonney, L.C.; Kitchen, S.; Hewson, R. A Susceptible Mouse Model for Zika Virus Infection. *PLoS Negl. Trop. Dis.* **2016**, *10*, e0004658. [CrossRef]

35. Winkler, C.W.; Myers, L.M.; Woods, T.A.; Messer, R.J.; Carmody, A.B.; McNally, K.L.; Scott, D.P.; Hasenkrug, K.J.; Best, S.M.; Peterson, K.E. Adaptive Immune Responses to Zika Virus Are Important for Controlling Virus Infection and Preventing Infection in Brain and Testes. *J. Immunol.* **2017**, *198*, 3526–3535. [CrossRef]

36. El Sahly, H.M.; Gorchakov, R.; Lai, L.; Natrajan, M.S.; Patel, S.M.; Atmar, R.L.; Keitel, W.A.; Hoft, D.F.; Barrett, J.; Bailey, J.; et al. Clinical, Virologic, and Immunologic Characteristics of Zika Virus Infection in a Cohort of US Patients: Prolonged RNA Detection in Whole Blood. *Open Forum Infect. Dis.* **2019**, *6*, ofy352. [CrossRef]

37. Ricciardi, M.J.; Magnani, D.M.; Grifoni, A.; Kwon, Y.C.; Gutman, M.J.; Grubaugh, N.D.; Gangavarapu, K.; Sharkey, M.; Silveira, C.G.T.; Bailey, V.K.; et al. Ontogeny of the B- and T-cell response in a primary Zika virus infection of a dengue-naive individual during the 2016 outbreak in Miami, FL. *PLoS Negl. Trop. Dis.* **2017**, *11*, e0006000. [CrossRef]

38. Lai, L.; Rouphael, N.; Xu, Y.; Natrajan, M.S.; Beck, A.; Hart, M.; Feldhammer, M.; Feldpausch, A.; Hill, C.; Wu, H.; et al. Innate, T-, and B-Cell Responses in Acute Human Zika Patients. *Clin. Infect. Dis.* **2018**, *66*, 1–10. [CrossRef]

39. Edupuganti, S.; Natrajan, M.S.; Rouphael, N.; Lai, L.; Xu, Y.; Feldhammer, M.; Hill, C.; Patel, S.M.; Johnson, S.J.; Bower, M.; et al. Biphasic Zika Illness With Rash and Joint Pain. *Open Forum Infect. Dis.* **2017**, *4*, ofx133. [CrossRef]

40. Waggoner, J.J.; Rouphael, N.; Xu, Y.; Natrajan, M.; Lai, L.; Patel, S.M.; Levit, R.D.; Edupuganti, S.; Mulligan, M.J. Pericarditis Associated With Acute Zika Virus Infection in a Returning Traveler. *Open Forum Infect. Dis.* **2017**, *4*, ofx103. [CrossRef]

41. Lum, F.M.; Lye, D.C.B.; Tan, J.J.L.; Lee, B.; Chia, P.Y.; Chua, T.K.; Amrun, S.N.; Kam, Y.W.; Yee, W.X.; Ling, W.P.; et al. Longitudinal Study of Cellular and Systemic Cytokine Signatures to Define the Dynamics of a Balanced Immune Environment During Disease Manifestation in Zika Virus-Infected Patients. *J. Infect. Dis.* **2018**, *218*, 814–824. [CrossRef]

42. Cimini, E.; Castilletti, C.; Sacchi, A.; Casetti, R.; Bordoni, V.; Romanelli, A.; Turchi, F.; Martini, F.; Tumino, N.; Nicastri, E.; et al. Human Zika infection induces a reduction of IFN-gamma producing CD4 T-cells and a parallel expansion of effector Vdelta2 T-cells. *Sci. Rep.* **2017**, *7*, 6313. [CrossRef]

43. Halstead, S.B. Pathogenic exploitation of Fc activity. In *Antibody Fc: Linking Adaptive and Innate Immunity*; Ackerman, M.E., Ed.; Academic Press: Cambridge, MA, USA, 2014; pp. 333–350.

44. Klenerman, P.; Zinkernagel, R.M. Original antigenic sin impairs cytotoxic T lymphocyte responses to viruses bearing variant epitopes. *Nature* **1998**, *394*, 482–485. [CrossRef]

45. Lim, M.Q.; Kumaran, E.A.P.; Tan, H.C.; Lye, D.C.; Leo, Y.S.; Ooi, E.E.; MacAry, P.A.; Bertoletti, A.; Rivino, L. Cross-Reactivity and Anti-viral Function of Dengue Capsid and NS3-Specific Memory T Cells Toward Zika Virus. *Front. Immunol.* **2018**, *9*, 2225. [CrossRef]

46. Koblischke, M.; Stiasny, K.; Aberle, S.W.; Malafa, S.; Tschouchnikas, G.; Schwaiger, J.; Kundi, M.; Heinz, F.X.; Aberle, J.H. Structural Influence on the Dominance of Virus-Specific CD4 T Cell Epitopes in Zika Virus Infection. *Front. Immunol.* **2018**, *9*, 1196. [CrossRef]

47. Grifoni, A.; Pham, J.; Sidney, J.; O'Rourke, P.H.; Paul, S.; Peters, B.; Martini, S.R.; de Silva, A.D.; Ricciardi, M.J.; Magnani, D.M.; et al. Prior Dengue Virus Exposure Shapes T Cell Immunity to Zika Virus in Humans. *J. Virol.* **2017**, *91*. [CrossRef]

48. Paquin-Proulx, D.; Leal, F.E.; Terrassani Silveira, C.G.; Maestri, A.; Brockmeyer, C.; Kitchen, S.M.; Cabido, V.D.; Kallas, E.G.; Nixon, D.F. T-cell Responses in Individuals Infected with Zika Virus and in Those Vaccinated Against Dengue Virus. *Pathog. Immun.* **2017**, *2*, 274–292. [CrossRef]

49. Herrera, B.B.; Tsai, W.Y.; Brites, C.; Luz, E.; Pedroso, C.; Drexler, J.F.; Wang, W.K.; Kanki, P.J. T Cell Responses to Nonstructural Protein 3 Distinguish Infections by Dengue and Zika Viruses. *MBio* **2018**, *9*. [CrossRef]

50. Badolato-Correa, J.; Sanchez-Arcila, J.C.; Alves de Souza, T.M.; Santos Barbosa, L.; Conrado Guerra Nunes, P.; da Rocha Queiroz Lima, M.; Gandini, M.; Bispo de Filippis, A.M.; Venancio da Cunha, R.; Leal de Azeredo, E.; et al. Human T cell responses to Dengue and Zika virus infection compared to Dengue/Zika coinfection. *Immun. Inflamm. Dis.* **2018**, *6*, 194–206. [CrossRef]

51. Grifoni, A.; Costa-Ramos, P.; Pham, J.; Tian, Y.; Rosales, S.L.; Seumois, G.; Sidney, J.; de Silva, A.D.; Premkumar, L.; Collins, M.H.; et al. Cutting Edge: Transcriptional Profiling Reveals Multifunctional and Cytotoxic Antiviral Responses of Zika Virus-Specific CD8$^+$ T Cells. *J. Immunol.* **2018**, *201*, 3487–3491. [CrossRef]

52. Wen, J.; Tang, W.W.; Sheets, N.; Ellison, J.; Sette, A.; Kim, K.; Shresta, S. Identification of Zika virus epitopes reveals immunodominant and protective roles for dengue virus cross-reactive CD8$^+$ T cells. *Nat. Microbiol* **2017**, *2*, 17036. [CrossRef]

53. Reynolds, C.J.; Suleyman, O.M.; Ortega-Prieto, A.M.; Skelton, J.K.; Bonnesoeur, P.; Blohm, A.; Carregaro, V.; Silva, J.S.; James, E.A.; Maillere, B.; et al. T cell immunity to Zika virus targets immunodominant epitopes that show cross-reactivity with other Flaviviruses. *Sci. Rep.* **2018**, *8*, 672. [CrossRef]

54. Wen, J.; Elong Ngono, A.; Regla-Nava, J.A.; Kim, K.; Gorman, M.J.; Diamond, M.S.; Shresta, S. Dengue virus-reactive CD8$^+$ T cells mediate cross-protection against subsequent Zika virus challenge. *Nat. Commun.* **2017**, *8*, 1459. [CrossRef]

55. Regla-Nava, J.A.; Elong Ngono, A.; Viramontes, K.M.; Huynh, A.T.; Wang, Y.T.; Nguyen, A.T.; Salgado, R.; Mamidi, A.; Kim, K.; Diamond, M.S.; et al. Cross-reactive Dengue virus-specific CD8$^+$ T cells protect against Zika virus during pregnancy. *Nat. Commun.* **2018**, *9*, 3042. [CrossRef]

56. Peiris, J.S.; Hui, K.P.; Yen, H.L. Host response to influenza virus: Protection versus immunopathology. *Curr. Opin. Immunol.* **2010**, *22*, 475–481. [CrossRef]

57. Schmidt, M.E.; Knudson, C.J.; Hartwig, S.M.; Pewe, L.L.; Meyerholz, D.K.; Langlois, R.A.; Harty, J.T.; Varga, S.M. Memory CD8 T cells mediate severe immunopathology following respiratory syncytial virus infection. *PLoS Pathog.* **2018**, *14*, e1006810. [CrossRef]

58. Gelpi, E.; Preusser, M.; Laggner, U.; Garzuly, F.; Holzmann, H.; Heinz, F.X.; Budka, H. Inflammatory response in human tick-borne encephalitis: Analysis of postmortem brain tissue. *J. Neurovirol.* **2006**, *12*, 322–327. [CrossRef]

59. Ruzek, D.; Salat, J.; Palus, M.; Gritsun, T.S.; Gould, E.A.; Dykova, I.; Skallova, A.; Jelinek, J.; Kopecky, J.; Grubhoffer, L. CD8$^+$ T-cells mediate immunopathology in tick-borne encephalitis. *Virology* **2009**, *384*, 1–6. [CrossRef]

60. Manangeeswaran, M.; Ireland, D.D.; Verthelyi, D. Zika (PRVABC59) Infection Is Associated with T cell Infiltration and Neurodegeneration in CNS of Immunocompetent Neonatal C57Bl/6 Mice. *PLoS Pathog* **2016**, *12*, e1006004. [CrossRef]

61. Manangeeswaran, M.; Kielczewski, J.L.; Sen, H.N.; Xu, B.C.; Ireland, D.D.C.; McWilliams, I.L.; Chan, C.C.; Caspi, R.R.; Verthelyi, D. ZIKA virus infection causes persistent chorioretinal lesions. *Emerg. Microbes. Infect.* **2018**, *7*, 96. [CrossRef]

62. Li, S.; Armstrong, N.; Zhao, H.; Hou, W.; Liu, J.; Chen, C.; Wan, J.; Wang, W.; Zhong, C.; Liu, C.; et al. Zika Virus Fatally Infects Wild Type Neonatal Mice and Replicates in Central Nervous System. *Viruses* **2018**, *10*, 49. [CrossRef]

63. Miner, J.J.; Sene, A.; Richner, J.M.; Smith, A.M.; Santeford, A.; Ban, N.; Weger-Lucarelli, J.; Manzella, F.; Ruckert, C.; Govero, J.; et al. Zika Virus Infection in Mice Causes Panuveitis with Shedding of Virus in Tears. *Cell Rep.* **2016**, *16*, 3208–3218. [CrossRef]

64. Wu, Y.H.; Tseng, C.K.; Lin, C.K.; Wei, C.K.; Lee, J.C.; Young, K.C. ICR suckling mouse model of Zika virus infection for disease modeling and drug validation. *PLoS Negl. Trop. Dis.* **2018**, *12*, e0006848. [CrossRef]

65. Couderc, T.; Chretien, F.; Schilte, C.; Disson, O.; Brigitte, M.; Guivel-Benhassine, F.; Touret, Y.; Barau, G.; Cayet, N.; Schuffenecker, I.; et al. A mouse model for Chikungunya: Young age and inefficient type-I interferon signaling are risk factors for severe disease. *PLoS Pathog.* **2008**, *4*, e29. [CrossRef]

66. Pedras-Vasconcelos, J.A.; Puig, M.; Sauder, C.; Wolbert, C.; Ovanesov, M.; Goucher, D.; Verthelyi, D. Immunotherapy with CpG oligonucleotides and antibodies to TNF-alpha rescues neonatal mice from lethal arenavirus-induced meningoencephalitis. *J. Immunol.* **2008**, *180*, 8231–8240. [CrossRef]

67. Jurado, K.A.; Yockey, L.J.; Wong, P.W.; Lee, S.; Huttner, A.J.; Iwasaki, A. Antiviral CD8 T cells induce Zika-virus-associated paralysis in mice. *Nat. Microbiol.* **2018**, *3*, 141–147. [CrossRef]

MDPI

St. Alban-Anlage 66

4052 Basel

Switzerland

Tel. +41 61 683 77 34

Fax +41 61 302 89 18

www.mdpi.com

Cells Editorial Office

E-mail: cells@mdpi.com

www.mdpi.com/journal/cells